CZECHOSLOVAK ACADEMY OF SCIENCES

SEISMICITY

OF THE

EUROPEAN

AREA

Part 2

CZECHOSLOVAK ACADEMY OF SCIENCES

Scientific Editor

Prof. Dr. Alois Zátopek

Member

of Czechoslovak Academy of Sciences

Reviewer

Dr. Jiří Vaněk, DrSc.

SEISMICITY

of the

European

Area

Part 2

VÍT

KÁRNÍK

1971

SPRINGER-SCIENCE+BUSINESS MEDIA, B.V.

SBN 90 277 0179 2

Additional material to this book can be downloaded from http://extras.springer.com

ISBN 978-94-010-3080-9   ISBN 978-94-010-3078-6 (eBook)
DOI 10.1007/978-94-010-3078-6

Published in co-edition with ACADEMIA, Publishing House of the
Czechoslovak Academy of Sciences, Prague

Preface

The second part of the "Seismicity of the European Area" contains further results of the survey of European seismicity, performed during the period 1959 – 1967. It was intended to publish the second part with the shortest possible delay. The first part appeared in 1968 when the manuscript of the second part was handed over to the publisher. Unfortunately, due to unexpected technical difficulties the publication had to be postponed until the end of 1970. Thus the manuscript reflects the state of knowledge corresponding to the year 1967 when the work on the manuscript was concluded. It was only possible to make minor changes, or to insert additional, very brief information into the catalogue

Prague, July 20, 1970   Vít Kárník

CONTENTS

1.   Introduction

This volume on the "Seismicity of Europe" pursues farther the investigations initiated in 1954 by the *European Seismological Commission*. The first volume contains the earthquake catalogue for the period 1901–1955 and also some of the statistical characteristics and discussions on the seismic activity which were based directly on the unified seismological information given in [1]. During the study of the seismological data for the period 1901–1955 it was felt that additional information covering a longer period of time should be used.

The main purpose for the investigation of the seismicity of Europe is to find some regularities in seismic activity and its upper threshold. For most regions the period of fifty five years is probably insufficient for the manifestation of earthquake generating forces.

The 19th century still yields reliable macroseismic material, despite the fact that its accuracy decreases for the earlier events and varies from region to region. In general we may trust this information because in many European countries the interest in natural sciences was roused at the end of the the 18th century and systematic earthquake observations had been organised. The seismic activity of a region is in fact determined by the stronger shocks and it may be assumed that shocks of $I_0 \geq$ $\geq$ VII have not escaped notice.

Thus the first part of this volume contains a summary *earthquake catalogue* 1801–1900, $I_0 \geq$ VII, as a source of basic earthquake parameters. Its form is somewhat different from that used for the period 1901–1955. For obvious reasons the lower intensity threshold has been elevated compared with that used in the catalogue for 1901–1955. The boundaries of the investigated area as well as the seismic regions have not been altered [1]. Serious difficulties were encountered in unifying the information concerning the maximum intensity of individual shocks and often epicentres were difficult to locate. The classification of old historical records involves many non-seismological problems and belongs partly to the cultural and administrative history of each country. Information from the years prior to 1801 is much more inaccurate and unhomogeneous, particularly for the more active regions of the Balkan and of the Eastern Mediterranean.

Our compilation of earthquake parameters is being extended also to the years prior to 1801 and following 1955 but the final version could not be realized in short time to allow for publication in this volume.

Homogeneous seismological information, compiled in the first volume, was used for the construction of *seismic maps* of a different type. Epicentres, maximum intensities and seismic energy are mapped to the scale 1 : 7 500 000. These maps give a general survey of seismic activity within the investigated area and can serve also for practical purposes. Their preparation was laborious and delayed the publication of this volume. They are appended in a special envelope. No detailed maps for individual earthquake zones were constructed because it was understood that the project should yield unified earthquake parameters for an overall comparative study of seismicity of the European area. Regional or local studies can than be made on the basis of a standard catalogue and maps by specialists from the corresponding countries. Finally, the information was used for an additional *statistical treatment* of data, and for the discussion of the magnitude-frequency relation because of its primary importance. In this way the analysis of the $N(M)$-function made in the first volume was continued and the investigation was aimed primarily at the problem of regional and time variation of the parameter b in the magnitude-frequency relation, at the limits of applicability of this simple relation and at the problem of smallest and largest shocks. The trends in activity revealed by a comparison of the strain-release curves were studied in the example of the North Anatolian fault.

The volume ends with two brief chapters on the focal mechanism of earthquakes and on the occurrence of tsunami in the European area. This review which completes the basic information on earthquakes in the catalogues was compiled using the studies and reports of the most competent experts, Dr. A.R. Ritsema and Dr. N.N. Ambraseys. It must be noted that both volumes represent a certain stage in the treatment of seimological information in the European area. Such research is never-ending and goes ahead step by step revealing newer and newer problems so that neither the author nor the reader can be fully satisfied with the results reached at a certain stage. The large number of new monographs dealing with European seismicity which appeared after the deadlines of both manuscripts give evidence of the growing interest in this field and about new and important achievements.

Acknowledgements

The compilation of catalogues and the handling of large sets of data require the cooperation of many specialists. I wish to express my sincere thanks to all those who contributed either with their catalogues or who assisted in selecting and classifying the parameters, in the calculations, the drawings, etc. My appreciation is due particularly to Dr. N.N. Ambraseys for valuable remarks on the catalogue, to Professors A. G. Galanopoulos, P. Caloi, M. de Panfilis, M. Peronaci, N. Öçal, K. Ergin, D. Trajić and M. Trapp for very efficient assistance in searching for additional information during my visits to Rome, Istanbul, Belgrade and Vienna.

This study trip was aimed at the improvement of this basic catalogue and maps and was supported by UNESCO.

The discussions within the Sub-Commission for the Seismicity of the European area were always encouraging and helpful for the continuous progress of the project; I am grateful to the old and new members of the Sub-Commission (Båth, Galanopoulos, Hiller, Medvedev, Munuera, de Panfilis, Ritsema, Rothé, Savarensky and Sponheuer) for the very useful exchange of ideas. The theory of largest values was applied in collaboration with Dr. Z. Hübnerová who performed also the corresponding calculations. The tables and graphs were made by Mr. B. Závorka and Mr. J. Nykles, the manuscript was typed by Mrs. S. Černíková. Mrs. O. Zenklová assisted very actively in the evaluation of historical reports.

Finally, I am very grateful to the Czechoslovak Academy of Sciences for good working conditions which enabled me to fulfil the task set by the European Seismological Commission to the Publishing House "Academia" for the publication of both volumes and also to "Kartografia" in Prague for the print of the seismic maps.

1.2.   SOURCE LITERATURE

The compilation of earthquake parameters for the years prior to 1901 must be based almost entirely on non-instrumental, macroseismic observation, although the end of the 19th century is already marked by the first attempts to record earth vibrations by physically defined devices. The quality and distribution of the first seismoscopes and seismographs, however, was not sufficient to allow for the instrumental location of epicentres and for objective statistics. On the other hand, many outstanding pioneers of modern seismology started with the preparation and publication of earthquake catalogues, containing a detailed compilation of data for historical shocks, or an exhaustive description of recent shocks.

This period stimulated also intensified attempts to establish an intensity scale as a tool for the classification of earthquakes [3]. Thus the catalogues from the 19th century differ in the mode of classification of earthquake effects. Understandably it was felt by the investigators that some rating of the earthquake force must be introduced before any study could be started. The oldest catalogues of shocks from the 19th century only listed the observations and arranged them according to date, place and type of effects; later works introduced a simple classification of weak, medium and strong earthquakes. In 1858 Mallet already used three intensity classes I, II, III with defined meanings, the same system being applied by Milne in 1911. Rossi and later also Forel already divided the shocks into ten classes. The combined Rossi-Forel scale was more widely used than any other of the old scales. This scale, or its modified version, was used in Italy, Switzerland and Spain and this fact must be taken into consideration when accepting the classifications made in the 19th century. Later catalogues compiled in the 20th century used mostly the twelve-degree scale of Mercalli – Cancani – Sieberg.

Before the compilation of the European catalogue 1801 – 1900 was started, the member countries of the E. S. C. had been asked to compile tabular lists of shocks within their countries, the intensities of which were $I_0 \geqq$ VII. The response was not as favourable as during the previous campaign concerning shocks of the 20th century, however, the representatives of the following countries prepared and forwarded the lists: Algeria, Czechoslovakia, France, Finland, Hungary, Norway, Poland, Portugal, Rumania, Spain, Sweden, Turkey, Yugoslavia.

For Albania, Austria, Belgium, Holland, Germany, Denmark, Greece, Italy, Israel, Morocco and the USSR, uniformly compiled catalogues had been published before. Only some of them, however, contain epicentre coordinates and the information about the time used (local, universal) is given very rarely. It is difficult to locate an epicentre when only macroseismic observation is available and an estimate of its probable position is very valuable, particularly for a person without a detailed knowledge of the geography of the respective country.

All the publications or manuscripts used for the compilation of the catalogue 1801 – 1900 are listed in para 2.42. A few catalogues appeared during the preparation of the European list 1801 – 1900 or reached the author after the compilation work had been finished in 1967, e. g. the catalogue of Ergin et al. for Turkey (EGU)*.

However, the national catalogues were not sufficient and world or regional catalogues were found to be a very useful source of information for regions not covered by national catalogues of for confrontation with other source literature. The most important of them are those by R. Mallet and J. W. Mallet (MMC) and by A. Perrey (PAA, etc.). These authors did not classify the intensity of the shocks but gave a description of each event. However, the catalogues cover only the period till 1872. The remaining period until 1885 is partly summarized in the catalogue by C. V. Fuchs (FC), which is inconsistent with the two previous ones as far as the details are concerned. There is no world or European catalogue for the time span 1886 – 1900 and we must rely entirely on national sources. It follows from the list of references that most of the information was obtained from "second hand" sources, not from the original ones. This fact implies the danger of possible distortions brought about by faulty transcription or by changes made by the authors. The second hand source means that the original records in chronicles, newspapers, letters, yearbooks, etc., were not studied but that summarizing catalogues, frequently quoting the original texts, were compiled. Most of the information in the catalogue 1801 – 1900 is taken from these basic catalogues.

* For abbreviations see 2.42.

2.　Earthquake catalogue 1801—1900

Data on earthquakes prior to 1901 are less plentiful than for the 20th century. Therefore the tabular form has been abandoned and the catalogue 1801 – 1900 has a simplified form compared with that of 1901 – 1955 [1]. The basic information is arranged in the following sequence: Date, time of origin, epicentre coordinates, number of the region, maximum macroseismic intensity I_0, remarks on the depth of focus, on the size of the shaken area, on the availability of an isoseismal map and additional information on epicentre location, intensity, tsunami, etc. This is followed by a brief description of earthquake damage with a list of localities shaken by maximum and minimum intensities. The sources of more detailed information are quoted as well as the publications used for the compilation. Unfortunately, the requirement of homogeneity in time, space and intensity cannot be fulfilled completely. This condition is important for statistical treatment and comparative studies. Because of the absence of seismological stations in the 19th century we must expect that most of the oceanic earthquakes either escaped attention or were assigned a lower intensity. In many countries a systematic collection of macroseismic observations was non-existent and again some medium-intensity shocks might be missing in regional catalogues. Thus, we have chosen a higher threshold value for the focal intensity, i. e. $I_0 = $ VII, however, it is very probable that our catalogue is not even complete for this class. This can be seen, e. g. along the North Anatolian fault, the most active region of the investigated area. No shocks of $I_0 \geqq$ VII (except one in 1829) were reported for the first forty years of the 19th century although in the following years they are numerous, at least 55 events of $I_0 \geqq$ VII.

The quality of information was very variable, from detailed monographs giving an exhaustive description of an event to earthquakes described only with one or two words (e. g. damaging, disastrous, etc.). For some events the original text is quoted in full, because it is impossible to use the scanty information for the determination of basic parameters.

We cannot avoid some gaps in the lower intensity classes and in oceanic areas and readers of this catalogue must take these facts into account.

2.1.   EARTHQUAKE PARAMETERS

2.11.   DATE, TIME OF ORIGIN

The exact date is lacking for some events. In a few countries (Bulgaria, Russia, Turkey) the old Julian calendar was still in force in the 19th century and the difference of 12 days had to be removed for the data of some publications. There were several evident mistakes in dating the original information, some earthquakes with an identical description differed slightly in dates, i.e. errors of the type of VI instead of IV, etc. The time of origin is given where possible. If it was indicated in Greenwich time, the capitals UT are added. Frequently it was difficult to find out the truth since the data had been transcribed by different investigators from one publication to another and the original source was not available. Thus it must be assumed that the remaining times of origin correspond to the local time, however, errors cannot be excluded.

2.12.   EPICENTRE COORDINATES

The epicentres were located as the centres of the most heavily shaken areas or simply the locality of highest intensity was taken as an approximate epicentre. It is known that such practices, particularly the latter, are not correct, the instrumental epicentre being usually shifted to one side of the pleistoseismal area. However, there was no other possibility of indicating the possible location of the focus. The coordinates determined in the second way are put in parentheses. On the other hand, when the distribution of stations is unfavourable or their number small, the instrumental epicentres are wrong and shifted 200 km or 300 km or even more, from the true position. Such erroneous ISS epicentres are not rare early in the 20th century.

Some regional catalogues (e. g. that of Greece) give epicentre coordinates which were accepted. For some events, however, it is impossible to locate the probable epicentre, the question mark then remains for epicentre coordinates. If more epicentre determinations are available, they are added at the end under the heading of additional information. The epicentre coordinates are followed by the number of the corresponding region. The delineation of 39 focal regions has been maintained and is the same as in the first volume [1, Fig. 1].

For some events in Greece or Rumania a large surface of the shaken area indicates a focus deeper than normal and the authors of national catalogue classified these shocks as intermediate-depth shocks. Such information is inserted after the coordinates and the notation $h = i$ is used.

2.13. EPICENTRAL INTENSITY

There is often much divergence in epicentral intensities published by different authors for the same earthquake. For some shocks a detailed description is available and the epicentral intensity can be determined uniformly according to an appropriate scale. We applied the *macroseismic scale* in the version presented by Medvedev, Sponheuer and Kárník [2] and recommended by the E. S. C. This version is identical with the MCS or MM scales, respectively, only the text has been somewhat rearranged and supplemented. The slight differences cannot influence the classification of shocks in both catalogues, i. e. 1901 – 1955 and 1801 – 1900, respectively. The discrepancies originate mainly by the way in which different investigators apply the principles of a scale. We estimate the intensity according to the extent of damage or destruction, the number of houses affected or people killed, taking into account the prevailing type of masonry. Observations concerning geological effects are rare.

It is necesary to mention some difficulties arising in the assessment of the intensity. We should consider the type and state of structures that were damaged or destroyed, the role of the ground conditions and particularly the natural exaggerations involved in the reports of early observers and chroniclers. However, it is not easy to make a realistic determination of intensity using a brief report and to estimate the degree of exaggeration, caused by personal impressions or by a free translation. The reported effects usually represent the behaviour of the worst structures and of the worst ground. Most of the populated area are just on alluvial ground in the valleys or on the shores. In the local houses are poorly built and weakened by age, floods or previous shocks. It was difficult to assign intensity on the basis of very brief reports which do not allow one to estimate the intensity with and reasonable certainty. Which intensity belongs e. g. to such a report as "a wall collapsed at ..." not accompanied by reports from other localities? Such examples may be numerous and only bear evidence of the accuracy of some information. Nevertheless, such events are included in the catalogue whenever it is felt that they fall into the category of $I_0 \geqq$ VII. Sometimes not even an approximate estimation is possible and the question mark follows after the symbol I_0. The lack of a complete description of an event might lead to some striking differences in intensities assessed by different authors, e. g. in Spain, (MU) and (BRS), or in Turkey, (ONB) and (EGU).

Some authors evidently tend to overestimate the observed damage, e. g. Montandon (MF) and Réthly (RA). The majority of investigators make quite a realistic estimation of I_0, differing within about one half or one degree and their classification has been accepted. The value of I_0 accepted or determined using the description of macroseismic effects is given first and is followed by values published by other investigators. A simple review shows that our estimations of I_0 correspond to the lowest or medium values.

The highest epicentral intensity $I_0 =$ XII cannot be proved during the investigated interval and even the intensity $I_0 =$ XI is rare and never do all effects give clear evidence of such assessment. Some substantial changes of the landscape may be attributed to slumps of soft ground, etc. The joint treatment of shocks of the 19th and 20th centuries is possible also by the *conversion of I_0 to M* for shocks of the 19th century. This unification cannot, however, be used for a more detailed investigation. The simple conversion of I_0 to M, without taking the depth of focus into account, yields only very approximate mean value; the few extreme magnitude values which are of the greatest importance for the strain or energy calculations are lost during this simplified procedure. When comparing the magnitude-intensity graphs plotted for data of 1901−1955 [1] we can observe the dispersion reaching ± 1 magnitude unit. We must expect that among the earthquakes observed in 1801 to 1900, e. g. in Turkey, of intensity $I_0 = X$ and XI, i. e. with $M = 7-7\frac{1}{2}$ on the average, one or two reached $M = 8 \pm$, as was the case in the 20th century. The macroseismic scale is not convenient for statistics and for energy calculations where such difference plays an important role.

Since our estimation of I_0 in 1801−1900 is lower than the average of the values published by other investigators, a correction of $+0.5$ should be added to M when using the conversion formulae from the first volume [1]; they are valid for I_0 taken over from national catalogues.

2.14. DESCRIPTION OF MACROSEISMIC EFFECTS, FORESHOCKS, AFTERSHOCKS

The basic information of the heading is followed by a brief summary of macroseismic effects, when abundant data are available, or by a simple quotation of the text, when only a brief note exists. The macroseismic information is reviewed in such a way that the highest and smallest effects are described and the corresponding localities are listed. This summary allows the reader to check the epicentre location, the epicentral intensity and the extension of the shaken area. More detailed information can be found in the sources quoted; it is not necessary to repeat all the data already published. Whenever possible, the origin and effects of tsunami waves are mentioned.

The description of effects on men, buildings, landscape, information on tsunami occurrence etc. are followed by lists of foreshocks and aftershocks of $I_0 \geqq VII$ whenever such information is available.

Serious difficulties arose in connection with the exact English transcription of the names of localities, rivers, mountains and countries. The problem of local names is also connected with the cultural and political history of each country. Different administrations changed some names, some villages probably do not exist any more, some names exist in several versions (local language, official language), etc. Since the catalogue is written in English, the English transcription of names was

accepted according to the Times Atlas of the World 1955—59, 1967. Later on the Institute of Cartography in Prague was asked to check the names again. The specialists of the Institute searched in different manuals, atlases, and individual maps, but only about one half of the total number of names has been identified. Finally, seismologists in the corresponding countries were consulted and the number of unidentified names somewhat decreased. There remained, however, many names about which we are not certain whether they are correct. Their wording might be substantially changed during several transcriptions so that they cannot be identified. These names are listed in the appendix to the catalogue in paragraph 2.41.

Some examples illustrate the difficulties involved. For example the largest Ionian Islands Kérkira and Kefallinía are frequently named Corfu (Corfu, Korfu) and Cephalonia, respectively, the town Edremit (Turkey) is called also Adramanti or Adramiti; the other name of Istanbul, i. e. Constantinople, is well known, but smaller localities in Eastern Anatolia are now quoted in their official Turkish version, however in older publications the Armenian names were also used. In some countries there are two official languages, e. g. in Belgium the same town is called Lutych or Liège, etc. The most confusing are changes in names made for political reasons. Also different administrations use different names in regions with a mixed population or some administration converts the name in the local language into the official language.

2.2. STATISTICAL TABLES

It is of interest to compile similar statistical tables as has been done for the 1901 to 1955 period. Since the classification of shocks from the 19th century is done by I_0

Table 1

Europe and Mediterranean area 1801—1900

$N(I_0)$, $h = n$

I_0 \ Region	2	3	6	8	9	10a	10b	11	12	13	14	15
VII	2	3	1	4	1	4	4	10	6		1	6
VII—VIII					2							2
VIII	1	1	1		2	1	1	1	1			6
VIII—IX												
IX									1	1		3
IX—X												3
X												1
X—XI												
XI												

I_0 \ Region	16	17	18a	18b	19	20	21a	21b	21c	21d	22a	22b
VII	13	4	18	15	57	22	7	12	3	3	16	39
VII−VIII	3	2	5	2	17	9	3	3		2	2	3
VIII	7		6	2	18	7	1		1	2	2	9
VIII−IX	4	1	4	1	6	3	1	2			1	2
IX	2		1		4	6		1			1	2
IX−X					2	4						
X	1		1	1	1	2						
X−XI					2							
XI	1											

I_0 \ Region	23	24	25	26a	26b	26c	27	28	29	30	31	32
VII		4	7	15	1	28	5	1	1			12
VII−VIII	1		7	12	1	4	2	1				2
VIII		1	7	11		15	1		4	1	1	
VIII−IX			14	5		5	2		1			3
IX		1	6	4	1	7	6			1		3
IX−X		1	9	6		2	1		1	1		1
X		1	3	5		2	1			2		1
X−XI				3		3						
XI				1								

I_0 \ Region	33	34	35	37	38
VII	8	3	2	1	
VII−VIII	3				
VIII	5	1	1		
VIII−IX					
IX	5		2	1	
IX−X	3	1			1
X	1		2	1	
X−XI					
XI					

$h = i$

I_0 \ Region	21d	26a	26b	26c
VII	3		2	
VII−VIII	1		2	
VIII	1	1		1
VIII−IX	2	1	1	
IX	1		1	
IX−X				
X		1		1
X−XI		1	1	

Note: 74 earthquakes could not be classified because of unsufficient description, they may be of $I_0 \geqq$ VII.

the first table gives the numbers of shocks arranged according to I_0 for each region. When using the conversion intensity-magnitude table given in the first volume the comments in para 2·13 must be taken into account. The reliability of the $N(I_0)$-table is lower, the material has all the shortcomings already mentioned in the previous paragraphs and can be used only for an approximate comparison. The tabulated values can be plotted in the intensity-frequency diagrams. The range of intensities is relatively small, mostly between VII and X, i. e. one and a half magnitude unit only. As already mentioned earlier, the observations are not homogeneous for several regions in the Balkan area or in the Eastern Mediterranean area.

Table 1 shows relatively low seismicity in the regions Nos 27, 29, 30, 31, 33, 34 comparing with the situation in the 20th century. This discrepancy may be caused by incomplete information which does not allow to decide about the intensity of the strongest shocks.

2.3.   SUPPLEMENTARY CATALOGUES FOR THE YEARS PRIOR TO 1801 AND AFTER 1955

Even the 1801−1955 period is not completely illustrative for the seismic activity in the investigated area. There are regions where the last outburst of seismic energy occurred in the 18th century or earlier or where two periods between two destructive shocks are longer than 150−200 years. Such earthquake zones are e. g. the one near Komárno in S. Slovakia, the region of Basel in Switzerland, the region of Skopje in Macedonia, etc. Hence, additional catalogues would be desirable. The difficulties encountered in compiling historical reports increase, however, with the decreasing date. We enter into times without any organized collection of data on earthquake occurrence and some information can be found only in chronicles together with other natural disasters, floods, storms, volcanic explosions, pests, famine, etc. The reports lose their objectivity and are more influenced by the temper and world outlook of the author. Thus, it becomes more and more difficult for an investigator to obtain a realistic account from the text. We can only judge that in the region under consideration there was an earthquake, probably damaging or destructive, i. e. we have some evidence of the seismic activity, the location and intensity remain very uncertain. We should be able to quote only the information published in a few catalogues and we could hardly add anything new. Therefore the summarizing catalogue in this volume does not go back to the years prior to 1801, although such a review has been compiled and serves as additional information in the archives of the Geophysical Institute, Czechosl. Acad. Sci., Prague.

The parameters for European earthquakes from the years following 1955 can be collected relatively simply. They are upblished in the ISS (since 1964 I. S. C. Bulletin), in the Bulletin du BCIS, in the P. E. D. cards of the USCGS or in the preliminary bulletins of the Institute of the Physics of the Earth in Moscow. Their accuracy is satisfactory except for the depth of focus. There is some inhomogeneity in magnitudes which are not given for all shocks and are not determined by the same method. The discrepancies originate by the application of different calibrating functions, different instruments and wave types. We have compiled a catalogue for 1956–1967 according to the same principles as for 1901 to 1955 and determined the magnitudes as the means of all values available. For many shocks with $M > 4$ ($I_0 \geqq$ VI), which were included into the supplementary catalogue, only the magnitudes for Prague and since 1959 also those for Průhonice are available. They were determined by the standard procedure already applied for the previous years [1] with a station correction $\delta M = +0.2$. The extended catalogue was used in some statistical calculations, mainly in the study of the activity along the North Anatolian fault system (see para 4.4). As far as epicentre coordinates are concerned, those published by the BCIS have been preferred. BCIS is the centre for the routine localization of European shocks and the Director of BCIS, Professor J. P. Rothé, prepared a continuation of the work by Gutenberg and Richter "Seismicity of the Earth" for the years 1953–1965 including European shocks, too [221].

Europe and Mediterranean area

2.4. CATALOGUE OF EARTHQUAKES

1 8 0 1 — 1 9 0 0

$I_0 \geqq VII$

Note: The parentheses after some localities contain other versions of the local names or the district.

The symbols and the accuracy of information are treated in 2.1.

1800

Febr. 26, 21^h, (38·7°N, 9·2°W), R.12, $I_0 = $ VII?; $I_0 = $ VII (BRS), not mentioned in (FH). Buildings damaged in Lisboa (MMC, DCE); slight in Lisboa (MU); Febr. 26 to Febr. 27 seismic shocks in Lisboa, houses shaken (GS).

Oct. 7, $07^h 43^m$, 46·6°N, 15·6°E, R.22b, $I_0 = $ VII, $r = 105$ km, $h = 8$ km, $M = 4·6$ (RV), not in (MMC).

Dec. 29, $10^h 15^m$ UT, 41·7°N, 12·8°E, R.19, $I_0 = $ VII (MR), $h = $ sup?
Several chimneys and stones were thrown down from the church at Velletri, all houses damaged, felt strongly at Marino and Castel Gandolfo (BI), not in (PAI).

1801

July (?)   According to (MMC) "houses were thrown down at Eskilstuna (Central Sweden), mountains overthrown and great damage done", $I_0 = $ IX—X (MF), $I_0 = $ VIII (MJD), not mentioned in (BMF), doubtful, storm?

Sept. 7, 06^h UT, (56·4°N, 4·0°W), R.10a, $I_0 \leqq $ VII?, $I_0 = $ IX (MF), $I_0 = $ VII (DCB). Several vibratory shocks in Edinburgh, epicentre at Comrie? (MMC).

Oct. 8, $07^h 53^m$ UT, $44\frac{1}{4}$°N, $11\frac{1}{4}$°E, R.19, $I_0 \leqq $ VII?
A few chimneys thrown down at Bologna, strong at Cesena, slight at Vicenza and Padova (MMC, PAI, BI, MJD).

Beginning of Dec., "A violent earthquake at Ljubljana (Laibach), in Carniola and at Eger (Hungary) where part of the fortification fell" (MMC), very doubtful, not mentioned in (RV, CD, RA).

1802

?   (34°N, $36\frac{1}{4}$°E), R.37, $I_0 \leqq $ VII?
At Bikaa, Baalbek and in several localities in Lebanon moderate damage (SAM, PK), not in (KAD, SNP).

Jan. 4, $07-08^h$, (45·3°N, 14·4°E), R.22a, $I_0 \geqq $ VII, $I_{max} = $ VII (TD), epicentre in the sea?
A very violent shock at Trieste, violent at Fiume and Bakar, slight at Ljubljana (Laibach) "in Carinthia several hills disappeared and new elevations were formed" (MMC), tsunami at Fiume and Bakar (ANS), not mentioned by (RV).

Jan. 17—18, (20^h), 38·1°N, 0·7°W, R.16, $I_0 = $ VIII? (BRS), $I_0 = $ VII (MJD), $I_0 = $ V (MU), $I_0 = $ IX—X (MF).
A few houses were destroyed at Torrevieja (MMC, CS).

May 12, $09^h 30^m$ UT, $45\frac{1}{2}$°N, 10°E, R.18b, $I_0 = $ VIII—IX, $I_0 = $ IX (MR, CAM), $I_0 = $ X—XI (MF), isos. map in (BI).
A disastrous shock in N. Italy, at Soncino five houses collapsed, several houses seriously injured or partly destroyed, at Callignano and Brescia several houses collapsed, at Orzinovi 400 (80%) houses seriously damaged, considerable damage also at Crema, Ticengo, Mantova, etc., felt at Parma, Pavia, Bologna, Torino, etc. (MMC, PAI, BI).

Sept. 14, $24^h 00^m$, Nov. 8, $23^h 30^m$, 48·6°N, 7·8°E, R.8, $I_0 = $ VII?
A sequence of shocks Sept. 11 — Nov. 8; many chimneys damaged in Strasbourg (RF, RSF, SW), some walls cracked (MMC).

Oct. 26, $10^h 55^m$ UT, 45·8°N, 26·6°E, $h = i$, R.21d, $I_0 = $ VIII (ES, MSM), $I_0 = $ IX, $M = 7·1$ (PRA, FA, RA, MF), 46·2°N, 26·0°E (ES), isoseis. map in (MSM).

1802

          Many buildings largely damaged in Bucureşti, sulphur water flowed from cracks in the ground, houses damaged and walls of churches fissured at Craiova, Iaşi, Vidin, Ruse, Chernovtsy, Varna, Braşov, Sibiu, felt at Istanbul. Widely felt in the Balkans and in Ukraina (MMC, PRA, FA, WS), allegedly also in Moscow, Peterburg (Leningrad), Tula, Kaluga, Orel (PAD, MJD).

Nov. 7, 11^h45^m, 36·5°N, 2·9°E, R.15, $I_0 = ?$, $I_0 = $ VII (MU), $I_0 = $ VIII (MJD), $I_0 = $ IX (CAM). A violent shock at Alger and Blida (Algeria), felt on ships near the coast.

1803

Febr. 2, $(43\frac{1}{2}°N, 5\frac{1}{2}°E)$, R.18a, $I_0 = $ VII? (RF).
          Chimneys were thrown down in Marseille (MMC, MJD).

Oct. 29,   (41·6°N, 44·5°E), R.32, $I_0 = $ VII? (BZ), $M = $ 5·2 (AV).
          Walls fissured at Tbilisi (BZ, MO), Oct. 17 (PAD).

1804

Jan. 13. 17^h06^m, 36·8°N, 3·2°W, R.16, $I_0 = $ VIII, $I_0 = $ VII (GS, BRS), 37·1°N, 4·2°W, $I_0 = $ IX (MU).
          A sequence of shocks in the region of Granada and Malaga, the most violent ones caused cracks in the walls of churches and the collapse of some houses at Motril (GS, MF, MMC), several walls fissured at Melilla (RG).

June 8, 03^h, $38\frac{1}{4}°N$, $21\frac{3}{4}°E$, $h = n$, R.26a, $I_0 = $ IX, three shocks (GG, GGT, CAM), $I_0 = $ VIII−IX (GGK, GG), $I_0 = $ X−XI (MF), $I_0 = $ VIII (MJD).
          Almost complete destruction of Pátrai, casualties, many houses overthrown, violent in (Zante) Zákinthos and in Morea (GG, MMC), destructive tsunami (ANS).

Aug. 22−28, 36·9°N, 2·4°W, R.16, $I_0 = $ VIII−IX (GS), $I_0 = $ VII (BRS), $I_0 = $ VIII (MU, MJD), $I_0 = $ IX (CAM), $I_0 = $ X−XI (MF).
          Roquetas was for the most part ruined by the main shock on Aug. 25, 07^h, salinas in the Bay of Roquetas submerged, houses collapsed at Canjayar (four victims), Turon and Dalias, felt at Madrid and Cartagena (GS, MMC).

Oct. 11, 17^h (41·7°N, 44·8°E), R.32, $I_0 \leq $ VII?, several shocks.
          An old wall collapsed and an old house was damaged at Tbilisi (BZ, PAD, MO, MJD).

Oct. 18, 20^h UT, $43\frac{1}{2}°N$, 11°E, R.19, $I_0 = $ VII? (MR).
          At S. Gemignano local houses were damaged, panic, felt at Colle and Poggibonsi (BJ, MMC, MJD).

1805

July 3,   at sunrise, 36°N, 24°E, R.26b, $I_0 = $ VII, $h = i$, $r = $ 810 km (GG, MMC, SAM), $I_0 = $ VII (MJD), $I_0 = $ IX (MF).
          Crete, "the towns of Khaniá (Chania) and Réthimnon suffered most damage" (MMC), felt in Sicily and at Napoli (BI).

July 26, 20^h57^m UT, 41·2°N, 14·7°E, R.19, $I_0 = $ X−XI, $I_0 = $ XI−XII (MR), $I_0 = $ XI (MF), isos. map in (BI).
          A catastrophic shock in Central Italy, most violent in the province of Molise, about 5600, victims. Baranello, Busso, Cameli, Cantalupo, Carpinone, Casalciprani,

Frosolone, Guardieregia, Macchiagodena, Mirabello, S. Giuliano, S. Polo, Spineto, Vinchiaturo were completely destroyed, in many other villages houses were partly ruined, for details see (BI), large deep cracks, at many places long fissures in the ground (up to 2 miles) and vertical mouvements (up to 2 m), a new lake was formed by a rockslide near S. Giorgio la Molara, changes in springs, etc. (BI, MMC, PAI, MR).

Oct. 13. 41°N, $14\frac{1}{2}$°E, R.19, $I_0 = $ VII?
Napoli and the neighboring country, at S. Maria di Capua Vetere and Nola several buildings were damaged, one ruined (BI, MMC, PAI, MJD); on the 15th Vesuvius was in eruption (MMC, PAI).

Nov. 16—17, 38°N, 24°E, R.26a, $I_0 = $ VII, $h = n$ (GG, GGK), not mentioned by (MMC, PAP).
A heavy shock in Attica, serious damage in Athínai (Athens), blocks of stone thrown down from the western part of Parthénion.

1806

Febr. 12, 02^h30^m UT, 44·8°N, 10·8°E, R.19, $I_0 = $ VII.
At Novellara and Correggio various houses were damaged and chimneys collapsed, Gualtieri, Guastalla, Fabrico, Reggiolo, Campanola and Carpi suffered also by damage, felt at Milano, Mantova, Verona, Genova, Vicenza (BI, MMC, PAI, MJD).

April 9 and 10, (38.1°N, 15·6°E), R.20, $I_0 = $ VII?, $I_0 = $ IX (MF).
At Reggio and in other places in Calabria rather violent shocks injured some houses (MMC, PAI, BI, MJD).

Aug. 26—30, $41\frac{3}{4}$°N, $12\frac{3}{4}$°E, R.19, $I_0 = $ VIII, $I_0 = $ IX (CAM), $I_0 = $ IX—X (MF).
Strong shocks caused slight damage in Roma, more severe at Velletri and Frascati, the monastery and the church at Nemi were much fissured, the church at Ariccia collapsed, felt as for as Napoli (BI, MJD, MMC, MR), epic. appeared to be the mount La Fajola (?) (Monti Albani).

Sept. 22, 19^h45^m UT, 47·8°N, 18·1°E, R.21a, $I_0 = $ VII—VIII (RA, MMC, KC, KR), $P = 31\ 000$ km^2 (RA), $I_0 = $ VII (CD).
Cracks in walls, chimneys fell down in Komárno, some houses collapsed (?) at Komárno, Budapest VI, felt at Bratislava, Sopron, Vesprem, Székesfehérvár IV, Tolna, Högyész (RA).

Oct. 27, Nov. 1—17, 37·3°N, 3·6°W, R.16, $I_0 = $ VIII—IX, $I_0 = $ X (MF), $I_0 = $ VIII (MU), $I_0 = $ IX (BRS), Nov. 1—17 (MMC).
Violent shocks in the region of Granada where a village was ruined, several houses were injured in Granada; epicentre in the Sierra Elvira (BRS, MMC, GS, MU).

1807

March 30, 11^h13^m, $45\frac{3}{4}$°N, 3°E, R.11, $I_0 = $ VII? (MJD), not mentioned in (RF), intensity questionable.
Strong shock in the northern part of Puy de Dôme, some old ruins were thrown down (MMC, PAF).

Nov. 10, $40\frac{1}{4}$°N, $15\frac{3}{4}$°E, R.19, $I_0 = $ VII—VIII, $I_0 = $ IX—X (MF), $I_0 = $ VII (MJD), not mentioned in (MMC, PAI).

1807

At Tremutola, Viggiano, Montemurro, Sarconi, Moliterno and Saponara considerable damage, some walls collapsed, the monastery at Saponara was rended uninhabitable (BI).

Nov. 18, $(36.7°N, 3°E)$, R.15, $I_0 = $ VIII? (MJD), not mentioned in (GA), $I_0 = $ IX (CAM). Exact epicentre unknown, at Algiers violent shocks, buildings were thrown down (RK, MMC).

1808

April 2, 16^h43^m UT, $44.8°N$, $7.3°E$, R.18a, $I_0 = $ VIII—IX, $I_0 = $ IX (MR), $I_0 = $ X—XI (MF), $r = 200$ km.
A sequence of shocks in the valley of Pellice and Chisone, centre seems to be at Pinerolo (Pignerol), Piemonte. The main shock was ruineous at Luserna, S. Giovanni, Bricherasio, Bibbiena, Torre, Pellice, felt as for as the Rhône valley, Milano, Montbrison and Bern. Severe aftershocks on April 2, 20^h30^m (VIII?), April 15, (VII?), April 16, (VII—VIII?), April 20, (VII?), May 17, (VII?) (BI, PAI, MMC, MR).

Oct. 26, 21^h UT, $43\frac{1}{2}°N$, $10\frac{1}{4}°E$), R.19, $I_0 = $ VII?, $I_0 = $ VII (MR, MJD).
Several shocks at Livorno, chimneys were thrown down; epic. in the sea? (PAI, MMC, BI).

1809

May 3—5, $39\frac{1}{2}°N$, $20\frac{1}{2}°E$, R.25, $I_0 = $ VIII, $h = n$ (GG), $I_0 = $ IX—X (MF), disastrous (MCA), $I_0 = $ VII (MJD).
Damaged houses in Kérkira (Corfu), destructive effects at Ioánnina (Jannina) (GG, MMC, PAP, MCA).

Aug. 25, 13^h, $43\frac{1}{4}°N$, $13\frac{1}{2}°E$, R.19, $I_0 = $ VII—VIII, not mentioned in (MR), $I_0 = $ VIII (MJD).
Violent shock damaged churches and houses at Macerata, many chimneys thrown down, seven or eight houses ruined (PAI, MMC, BI).

Nov. 17, 21^h40^m, $49.0°N$, $21.2°E$, R.21a (KC), $I_0 = $ VII?, $I_0 = $ VII—VIII (RA).
Damaging shock in the valley of Toryša, cracks in walls, two chimneys fell in Prešov, some damage at Solivar and Velký Šariš (KR, KC, RA).

1810

Between 1810—1815, $(35°N, 24\frac{1}{2}°E$, R.26b, $I_0 = $ VIII?) a strong shock destroyed allegedly the monastery Asomatos (GG).

Jan. 14, 17^h09^m, $47.4°N$, $18.2°E$, R.21b, $I_0 = $ IX, $I_0 = $ VIII (MJD), $I_0 = $ IX—X (RVP), $I_0 = $ X (MF), $P = 50\ 000$ km^2, $h = 5—8$ km (RA, KR, CD), isos. map in (RAB).
A destructive earthquake near Mór, at Mór many houses destroyed, all buildings uninhabitable, long fissures in soft ground, heavy damage (VIII—IX) at the neighbouring villages Bodajk, Csákberény, Isztimér, Fehérvácsurgó, Ondód, Csurgó, Guth, Balinka, VII—VIII at Metsér, Sikátor, Csernye, Veleg, etc., 3 persons killed, felt as far as Bratislava (RA, MMC, CD, RVP, RAB).
Aftershock on Jan. 21, $13—14^h$, $I_0 = $ VII—VIII, $I_0 = $ VIII—IX (RAB).
Damage to buildings (RA).
(RAB) gives other damaging aftershocks with $I_0 = $ VIII—IX (?) on May 14, 15, 27, Dec. 21, 18^h11^m, Sept. 6, they are not mentioned in (CD).

1810

Febr. 16, $35\frac{1}{2}°$N, 25°E, R.26b, $I_0 = $ IX, $h = i$, $r = 110$ km (GG), $I_0 = $ XI–XII (MF), $I_0 = $ X (CAM).

The town Iráklion was ruined and 2000 persons perished (MMC, SAM, PAP), many houses collapsed ($\frac{1}{3}$), 200 victims, large shaken area (GGK, GG, PAP, MMC, ANC, MJD), felt strongly in Malta, at Napoli and in N. Africa (PAP), slightly in Cyprus (ANC).

March 20 or 25, 28·2°N, 16·6°W, R.13, I_0?, volcanic?, $h = $ sup, $I_0 = $ IX (MJD).

Island of Teneriffe, a very violent earthquake, many people perished beneath the ruins of the houses.

July 18,   47·6°N, 14·5°E, R.18b, $I_0 = $ VII?, *local* shock, not mentioned in (MMC). Admont (Steiermark, Austria) (TT).

1811

July 15,   44·5°N, 10·8°E, R.19, $I_0 = $ VII?, isos. map in (BI), $I_0 = $ VII (MJD).

At Sassuolo some chimneys thrown down, in the mountains also some walls and chimneys collapsed, some damage at Montebaranzone, strong at Reggio nell'Emilia, Parma, Livorno and Genova (BI, MR, MMC).

1812

March 22, 02^h20^m UT, ($44\frac{3}{4}°$N, $12\frac{1}{2}°$E) R.19, $I_0 = $ VIII?, $I_0 = $ IX (MF), not mentioned in (MR).

In Roma considerable damage done (PAI, MMC), walls fissured in some churches and houses (BI, MJD).

March 27, 23^h UT, 43·7°N, 5·7°E, R.18, $I_0 = $ VIII, $r = 90$ km, $I_0 = $ IX (RO, RF, CAM), $I_0 = $ IX–X (MF).

Heavy damage at Beaumonte, one church partly destroyed, ten houses collapsed, 150 houses suffered by cracks, slight damage at Avignon, damage to buildings at Pierrevert, Sainte-Tulle, Manosque, felt at Marseille, Aix, Antibes, Digne (PAF, RO, MJD).

May 2, 11^h, 47·2°N, 1·5°W (RF), R.11, $I_0 = $ VII?

Damage at Nantes and its environs, some chimneys and plaster thrown down (MMC, PAF, MJD).

Sept. 11, 12^h UT, $43\frac{1}{2}°$N, $11\frac{1}{4}°$E, R.19, $I_0 = $ VIII, $I_0 = $ VII (MJD), isos. map in (BI).

Some houses destroyed at Poppiano, S. Quirico, Guicciardini, Innocenti, Rodolfi, Cicia Porci, serious damage in other localities SW of Firenze (BI, MMC, PAI, MR).

Oct. 25, 07^h UT, $46\frac{1}{2}°$N, 13°E, R.18b, $I_0 = $ VIII, $I_0 = $ VII (MR), not mentioned by (TT, RV), isos. map in (BI).

A disastrous shock in Friuli-Carnia, some houses demolished at Cavasso, Sequals and Fanna, serious damage at Belluno, S. Casciano, Caneva, Sardene, Pordenone, etc. (BI), felt in Bavaria, Tirol and Lombardy (MR, MMC, MJD).

1813

Febr. 1–2,   in the night, epicentre?, $I_0 = $?

Three strong shocks were felt in Bucureşti, walls cracked (MMO), no damage reported by (FA).

1913

Aug. 17, 00^h UT, 46·7°N, 15·8°E, R.22, I_0 = VII, r = 135 km, h = 10 km (RV).
Strong shock in Sloven'ja, Carinthia and Styria (MMC).

Sept. 21, 07^h45^m UT, $44\frac{1}{4}$°N, 12°E, R.19, I_0 = VII−VIII, I_0 = VIII (MR), I_0 = VII (MJD).
A strong shock at Imola, buildings were much damaged, epic. near Faenza (all chimneys thrown down, all houses fissured) (MMC, PAI, BI).

Dec. ?, I_0 = ?, R.26a, I_0 = VIII (MJD), I_0 = IX (GG, CAM), I_0 = X−XI (MF).
Starachoritza (Sorachovitzas? near Ioánnina) almost entirely overthrown by an earthquake accompanied by a storm (MMC, PAP); destruction in the region of Ioánnina (MCA); damage caused probably by a storm (GGK).

Dec. 26, 14^h30^m, 45·3°N, 18·4°E, R.22b, I_0 = VII?, h = sup?, not mentioned by (MMC), local. Fall of chimneys at Dzhakovo, panic (TD).

1814

Nov. 6, 05^h45^m, epicentre? R.18, I_0 = ?, I_0 = VIII (MJD), I_0 = IX (MF, CAM), not mentioned in (RF), questionable, storm?
Two severe shocks at Lyon and along the whole line from Mâcon to Vienne, some houses were thrown down; a heavy rain (MMC, PAF)!

1815

Date? $38\frac{3}{4}$°N, $20\frac{1}{2}$°E (GG), R.26a, I_0 = VIII, h = n, not mentioned in (PAP, MMO), I_0 = VII−VIII (GGK).
Levkás, Ionian Islands, casualties, many houses collapsed (GG, GGL).

Dec., ?,   35°N, $25\frac{3}{4}$°E, R.26b, I_0 = IX, h = n; not mentioned in (PAP).
S. and E. Crete shaken, a greater part of Ierápetra destroyed, solid houses in Itéa also destroyed as well as the monastery Acrotiriani (GG, SAM), serious damage at Toplu (GGK).

1816

Date?   ($41\frac{1}{4}$°N, $19\frac{1}{2}$°E), R.25, I_0 = VII−VIII?, I_0 = VIII (MD).
An extremely strong shock at Durrës and Shijak (MD, MCA, GGK).

Febr. 2, 00^h40^m, (35°N, 10°W), R.12, I_0 = VII, I_0 = IX (MU), (38·7°N, 9 2°W), I_{max} = = III (FA), I_0 = VII (BRS).
A strong shock shattered Lisboa and was felt on ships 300 leagues to the west, also felt on Madeira and in Holland; in Lisboa the people quitted their houses (V) (MMC), some houses were destroyed, tsunami (GS), epicentre probably in the sea.

Aug. 13, 22^h45^m, Inverness (57·5°N, 4·2°W), R.10a, I_0 = VIII (?) (DCB), I_0 = VII (MJD), no description available.

1817

March 11, 21^h10^m, 45·9°N, 6·8°E, R.18a, I_0 = VII−VIII, r = 180 km, I_0 = VIII (RF, RO).
Haute Savoie, several violent shocks damaged churches at Houches and Saint-Gervais, fissured walls at Grand-Bornand, avalanches (RF), furniture was thrown

down, arches were broken (MMC), felt at Chamonix, Lausanne, Bern, Neuchâtel, Genève (RO).

March 18, 10^h45^m, 42·2°N, 2·1°W, (BRS), R.17, $I_0 =$ VIII−IX, $I_0 =$ VIII, 42·4°N, 2·4°W (MU, GS, MJD), $I_0 =$ IX (CAM), $I_0 =$ X−XI (MF).
Violent earthquake in NE Spain, at Prejano (Logrono) most houses became un-inhabitable, some houses were destroyed at Arnedillo, Arnedo, churches damaged at Ausejo and Calahorra; felt from Santander to Tarragona and between Toledo and the mountains of Cuenca (MMC, GS).

Aug. 23, 08^h UT, $38\frac{1}{4}$°N, $22\frac{1}{4}$°E, R.26a, $I_0 =$ IX−X, $h = n$, $I_0 =$ IX (GG), $I_0 =$ IX−XI (GGK), $I_0 =$ X −XI (MF).
A complete destruction of Aíyion (Aigion), subsidence of terrain (GG), the town of Vostitsa was destroyed, strongly felt at Pátrai, weakly at Kórinthos (Korint), tsunami swept the downtown section of Aíyion and the Cape Alíki (ANS), many violent shocks during the following 8 days (GG, MMC, PAP, GGT).

1818

Febr. 20, $(18^h)10^m$ UT?, 37·6°N, 15·0°E (PET), R.20, $I_0 =$ IX, $I_0 =$ X (CAM), $I_0 =$ X−XI (MF), $I_0 =$ IX−X (MR), isos. map in (BI).
A disastrous shock at Catania and in Calabria; at Catania great masses of stone were thrown from the tops of buildings, at many other places Mascalucia, Nicolosia, Trecastagni, Viagrande, Aci Catena, S. Gregorio, S. Giacomo, S. Lucia buildings were thrown down and 72 persons were killed, some statues moved a little (MMC, PAI).

Febr. 23, 18^h UT, 43·9°N, 8·1°E (RF), R.18a, $I_0 =$ VII−VIII, $I_0 =$ VIII (MJD), $I_0 =$ = IX−X (MF), $I_0 =$ IX, $r = 200$ km (RO, CAM), isos. map in (BI).
Severe shock in Liguria. Some houses damaged at Diano-Castello, Alassio, S. Remo and Vence, serious damage occurred only exceptionally (BI), houses destroyed at Vence (MMC), epicentre in the sea? (MR).

Febr. 24−25, night, $37\frac{3}{4}$°N, 14°E, $h =$ sup, R.20, $I_0 =$ VII?, not mentioned in (PAI, MR, BI).
A shock of very limited shaken area in the Madonies (Sicily), cracks in buildings at Geraci, here and there considerable damage (MMC), local.

Febr. 28, 08^h30^m UT, $(37\frac{1}{4}$°N, $14\frac{1}{2}$°E), R.20, $I_0 =$ VII?, volcanic?, epic. very uncertain according to (MR).
Two shocks which did great damage at Catania (MMC, MJD), "an earthquake at Catania" (PAI), epic. at Mineo-Caltagirone where houses were cracked and churches seriously damaged (BI).

April 25?, (42·8°N, 23·3°E, R.23), in Sofia many houses and churches (djamia) collapsed, panic, cold and warm water springs ceased (WS), $I_0 =$ VIII (GGB), epicentre?

Sept. 8, 09^h37^m UT, $37\frac{3}{4}$°N, 14°E, $h =$ sup?, R.20, $I_0 =$ VII−VIII?, $I_0 =$ VIII (MR, MJD), $I_0 =$ IX (CAM), $I_0 =$ X(MF), isos. map in (BI).
Houses damaged at Petralia and Polizzi, some damage in localities around the mountains of Madonie (MR, MMC, BI).

Dec. 9, 18^h UT, $(44\frac{3}{4}$°N, 10·4°E), R.19, $I_0 =$ VII, $I_0 =$ IX−X (MF).
Many chimneys fellt at Parma, Modena, Genova, Vicenza (PAI, MMC, BI, MJD).

1819

Jan. 8, 22^h30^m UT, ($43\frac{3}{4}°$N, 8°E), R.18a, $I_0 =$ VII?
Houses at Porto Maurizio and San Remo were damaged, felt at Alassio and Genova (PAI, MMC, BI).

Febr. 24, 23^h24^m UT, 37·9°N, 14°E, R.20, $I_0 =$ VII—VIII, $I_0 =$ VIII (MR), $h =$ sup?, isos. map in (BI).
Much damage at Castelbuono, Geraci, Collesano, S. Mauro, Lascari, strong at Palermo (BI, PAI, MMC), some houses collapsed (PAI).

Febr. 28,   ($41\frac{3}{4}°$N, $44\frac{3}{4}°$E), R.32, $I_0 =$ VII?
Many shocks at Tbilisi (Georgia), some old houses were thrown down (MMC, BZ).

March, ?,   35,4°N, 0·1°E, R.15, (GA, RK), $I_0 =$ IX, $I_0 =$ X (GA), 35·7°N, 0·6°W, $I_0 =$ IX (MU).
Disastrous shock in Algeria, almost all houses at Mascara thrown down and a great number of the inhabitants perished, at Oran only some houses slightly fissured (GA, RK, MMC).

May 26, 17^h? UT, ($42\frac{1}{4}°$N, $11\frac{3}{4}°$E), R.19, $I_0 =$ VII—VIII?, $I_0 =$ VII (MJD), $I_0 =$ VIII (MR), $I_0 =$ IX—X (MF).
Corneto-Montalto-Civitavecchia (MR), churches heavily damaged at Corneto (BI) and a great number of people perished (?), felt along the Mediterranean coast (MMC, PAI).

Aug. 31, 15^h—16^h, 65°N, 17°E, R.3, $I_0 =$ VII, $r = 500$ km, $M = 6·1$ (BMF), map of the shaken area in (KF).
Strong earthquake in Sweden and Norway, shaken area from Karungi to Stockholm and from Kolahalfön to Trondheim (KRS). Particularly felt at Salten and Helgeland; at Heurnoes chimneys were thrown down (MMC, KRS, BMF, KF, MJD, PAD, AV).

Dec. 20, 10^h, 37·7°N, 1·7°W (BRS), R.16, $I_0 =$ VII?, $I_0 =$ V (MU), not mentioned in (MMC). Violent shock at Lorca, felt at Murcia and Totana (GS).

1820

Date?   A strong earthquake caused considerable damage at Mineo (Catania) (BI).

Date?   The town Kilis (36·8°N, 37·1°E) was slightly damaged and the inhabitans lived 40 days in tents (PL), $I_0 =$ VII? It might correspond to 1822, Aug. 13.

March 17, $38\frac{3}{4}°$N, $20\frac{1}{2}°$E (GG), R.26a, $I_0 =$ IX?, $I_0 =$ VIII (MJD), $I_0 =$ VIII—IX (GGK), $h = n$, tsunami.
Subsidence in the centre of the town Levkás where churches and most stone-built houses collapsed; many aftershocks (MMC, GGK), (SAM, MCA) give Febr. 21 or 20.

March 17, $38\frac{1}{4}°$N, $26\frac{1}{4}°$E (GG), R.26c, $I_0 =$ VII?, $h = n$; questionable.
Khíos, much damage during a violent shock accompanied by a storm (SAM, MMC, GH).

July 17, 06^h30^m UT, 47·3°N, 11·7°E, $h =$ sup?, R.18b, $I_0 =$ VII (SW, TT), *local*.
At Schwaz all houses damaged in different extent, even vaults and walls were thrown down, rockslides, felt in the lower valley of the Inn, Innsbruck (SW).

Dec. 29, 05^h, $37\frac{3}{4}°$N, $21\frac{1}{4}°$E (GG), R.26a, $I_0 =$ IX, $I_0 =$ VIII (MJD), $I_0 =$ X (MF), $r = 125$ leagues (MMC).

The town of Zákinthos (Zante) partly destroyed (PAP), many houses (79) at Zákinthos thrown down or injured, 8 men killed (MMC), felt in the Ionian Islands
and in Morea; in Elis and Arcadia changes in water ground level (GG, GGR),
felt in Malta (MCA).

1821

Jan. 6, 18^h45^m, $37\frac{3}{4}°N$, $21\frac{1}{4}°E$, R.26a, $I_0 = IX-X$, $h = n$; $I_0 = X$ (GG), $I_0 = VIII-X$
(GGK).
Much damage was done in the villages around the town of Zákinthos, the town of
Sala in Morea was almost entirely destroyed by these shocks and those of De-
cember, numbers of people perished beneath the ruins (MC, MMC, PAP, MCA),
destructive tsunami at Pátrai (GGK, ANS).

April 8, May 9, violent shocks destroyed several houses and the fortification at Melilla, I_0?,
(35·4°N, 3·0°W, R.15) (RG, GS).

Aug. 2, 02^hUT, (38·9°N, 16·6°E), R.20, $I_0 = VII-VIII$? Sept. 12, $I_0 = IX-X$ (MF),
$I_0 = VII$ (MJD).
Some houses destroyed at Catanzaro, slightly felt at Napoli (BI).

Nov. 17, 13^h45^mUT, (45·8°N, 26·6°E), R.21d, $I_0 = VII$?, $I_0 = VI$ R.F. (AI). Large area indicates a deeper focus, i. e. epicentre with $h = i$ in the E. Carpathians, not mentioned in (PRA). Identical information is given by (PAD, FA, MJD) for Sept. 17.
Strong shocks in Moldavia, Valachia, S. Russia, some damage in Kiev, Lvov, Iaşi
and Bucureşti, fissures in walls (AI, FH, MMC, MJD, EU), (FA) assumes the
epicentre near Crimea, no data, however, in (PV).

Nov. 22, 01^h15^m UT, (42·1°N, 15·5°E), R.19, $I_0 = VII-VIII$, $I_0 = VIII$ (MR), $I_0 = IX-X$
(MF).
A strong shock near Napoli, particularly at Porto Cannone, Termoli (42·0°N,
15·0°E) and Tremiti Islands, where buildings were fissured, some of them destroyed (PAI, MMC, BI).

1822

Jan. 10,   38·1°N, 0·1°W, R.16, $I_0 = VII$, not mentioned by (GS, MMC, MU), doubtful?
250 shocks at Torrevieja, Alicante (BRS).

Febr. 18, 16^h15^m UT, 47·8°N, 18·2°E, R.21a, $I_0 = VIII$, $I_0 = VII$ (CD), $I_0 = IX$ (RA),
a damaging foreshock on Febr. 6 (RA).
At Komárno the chimneys fell and many buildings cracked, the church, the castle,
and the monastery were damaged, changes in the level ground water, some buildings
became uninhabitable, at Iža many houses collapsed and the top of the tower of
the church fell down, felt also in Budapest (RA, KC, KR, MMC).

Febr. 19, 08^h15^m, 45·8°N, 5·8°E, R.18a, $I_0 = VIII$, $r = 150$ km, $I_0 = VIII-IX$ (RO),
$I_0 = VII$ (MJD), $I_0 = IX$ (MF).
Violent earthquake in SE France and in Switzerland, epicentre near Chindrieux,
many chimneys fell and walls cracked at Chambéry, Yenne, Chindrieux, Rumilly,
Annecy, Seyssel (two houses collapsed). Felt at Belley, Dijon, Clermont, Lyon,
Bourg, Genève, Lausanne, Bern, Zürich and Aix (RF, MMC, RO, RAF, BI).

April 5, 6, 10, 37·8°N, 14·8°E, R.20, $I_0 = VII$?, a very local shock, $h = $ sup; volcanic?
At Nicosia (Sicily) several shocks damaged houses seriously (MR, MMC, BI, MJD).

1822

July 6, 06^h45^m, $38 \cdot 7°N$, $9 \cdot 2°W$, R.12, $I_0 = ?$, $I_{0 \cdot} = IV(MU)$, $I_0 = VII$ (BRS), not mentioned in (FA).
A violent shock at Lisboa (MMC, GS, DCE).

July 29, 01^h, $36 \cdot 8°N$, $3 \cdot 2°W$ (BRS), R.16, $I_0 = VII$, $37 \cdot 2°N$, $3 \cdot 6°W$, $I_0 = V$ (MU).
At Granada many buildings were injured, felt at Almeria (MMC, MJD).

Aug. 13, 20^h, $36°N$, $36°E$ (PK), $I_0 = XI$ (SAM, SNP), $I_0 = X-XI$ (ONB), isos. map in (SAM), p. 35.
Destructive earthquake, particularly at Antakya (Antiochia) and Halab (Aleppo), two thirds of the town destroyed, thousands of inhabitants perished, less damage at Djesr and Ladhiqiya (Latakia), tsunami at Beirut, Jerusalem, Iskenderun and in Cyprus (ANC, MMC, KAD, SNP, PAP, ONB, CAM).

Sept. 5, $36°N$, $36°E$, R.35, $I_0 = X?$, $I_0 = X-XI$ (ONB).
Another disastrous shock at Halab (Aleppo) which destroyed what had resisted the former one, more than 20 000 persons are said to have lost their lives (MMC, CAM).

Oct. 8, $38 \cdot 1°N$, $0 \cdot 8°W$, R.16 (BRS), $I_0 = V$ (MU), $I_0 = VII$ (BRS).
Orihuela, several shocks (MMC), intensity comparable with the largest shock in 1802 (GS).

1823

Jan. 5, 02^h, $47 \cdot 9°N$, $23 \cdot 9°E$, R.21d, $I_0 = VII$, $h = n$, $I_0 = IX$ (PRA), $I_0 = VIII$ (RA,KR).
Rather severe shock in Rumania, strongly felt at Baia Marie, Sighet and in the Maramureş region, fissures in the walls (RA, FA).

Jan. 10, $38 \cdot 1°N$, $0 \cdot 8°W$, $I_0 = ?$, $I_0 = IV$ (MU), $I_0 = IX-X$ (MF), not mentioned in (BRS), $I_0 = IX$ (CAM), $I_0 = VIII$ (MJD).
Violent shocks in the region of Murcia, Cartagena, Alicante and Torrevieja, felt also at Valencia (GS), several houses fell (MMC).

March 5, 16^h37^m UT, epic. in the Tyrrhenian Sea, R.20, $I_0 = IX?$, $I_0 = X$ (MF, MR, CAM), isos. map in (BI).
Disastrous shock at Palermo and at some other localities in Sicily; tsunami; great damage at Palermo, Roccapalombo, Pozzilo, S. Agata, Isnello, Castelbuono, etc., and particularly at Naso ($38 \cdot 2°N$, $14 \cdot 8°E$), mud volcanoes at Terrapilata, changes in the warm springs of Termini (MMC, PAI, MR); a detailed description in (BI).

June 11, 11^h45^m, $47 \cdot 8°N$, $21 \cdot 2°E$, R.21b, $I_0 = VI-VII?$, $I_0 = VIII$ (RA), not in (CD, MMC).
Walls fissured at Füzesgyarmat, Szeghalom and Gyarmat (RA).

June 19, $39\frac{3}{4}°N$, $20\frac{3}{4}°E$, R.26a, $I_0 = VIII?$, $I_0 = VII$ (MJD), $I_0 = VIII-IX$ (GGK), $I_0 = IX$ (GG, MD, CAM), $I_0 = XI$ (MF), June 21 (MCA).
Suli (Thesprótikon) was completely destroyed (MCA); remainder of the fortifications destroyed (MMC); felt strongly at Ioánnina (Janina) and Gzrig (MD); epicentre near Shkodër (Scutari) (GGK).

Aug. 7, 05^h15^m, $42 \cdot 6°N$, $18 \cdot 1°E$, R.22a, $I_0 = VII-VIII$, $r_{max} = 165$ km, $I_0 = IX$ (TD), epicentre at the coast?
An explosion type earthquake at Dubrovnik and Gruž, churches cracked, heavy damage to walls up to the distance of 15 miles, felt at Split (MMC, PAP).

1823

Two foreshocks with $I_0 =$ VII on Aug. 7; a damaging aftershock on Aug. 20 (MMC, PAP), strong tsunami (ANS), not in (TD).

Aug. 25,   at Kuşadasi (Scala Nova) (Anatolia, R.26c) houses were thrown down by violent shocks (MMC, PAP), $I_0 =$ IX (CAM), west off Izmir (Smyrna), $I_0 =$ VIII (MJD), doubtful, not in (ONB, PL, GG).

Sept. 2, 19^h, 39·4°N, 0·4°W, R.16, $I_0 =$ VII? (BRS), $I_0 =$ III (MU).
A very severe shock at Valencia (MMC, GS).

1824

Febr. 21, 20^h, Island of Levkás (Ionian Islands, R.26a), many houses were injured (PAP, MMC), $I_0 =$?, not mentioned in (GG).

Dec. 10,   39·5°N, 16·6°E, R.20, $I_0 =$ VIII (MR, MJD), $I_0 =$ IX (CAM), $I_0 =$ X (MF), local, $h =$ sup?
Some shocks at Corigliano and Longobucco, several houses were thrown down, three persons perished (PAI, MMC, BI).

1825

Jan. 19, $11-12^h$, $38\frac{3}{4}$°N, $20\frac{3}{4}$°E, R.26a, $h = n$, $I_0 =$ X—XI, $I_0 =$ XI (GG, MF), $I_0 =$ X (CAM).
All houses in the town Levkás collapsed, 34 victims (GG, GGM), the town Levkás almost totally destroyed, many inhabitants perished, at Préveza some houses were thrown down, the earth opened (PAP, MMC), also the villages in the mountains of the island Levkás suffered much, felt from Kérkira (Korfu) to Zákinthos (Zante). Strong tsunami (ANS, GGM).

March 2, 07^h, 36·4°N, 2·8°E, R.15, $I_0 =$ X (CAM), $I_0 =$ X—XI (GA), $I_0 =$ IX (MU).
Disastrous earthquake in Algeria, the town of Blida was completely destroyed, several thousands of inhabitants perished, a village was buried between two-hills, at Alger V (RK) or great damage reported by (MMC), felt throughout the mountains of the "Petit" Atlas (RK, GA, MMC, GS), 11 very violent aftershocks till March 6.

1826

Jan. 26,   39°N, $20\frac{1}{2}$°E, R.26a, $h = n$, $I_0 =$ VI—VII (GGK), $I_0 =$ VII (GG, MJD), $I_0 =$ IX (MF).
The town of Préveza was much injured by a heavy shock (PAP, MMC, MCA, SAE).

Febr. 1,   40·6°N, 15·7°E, R.19, $I_0 =$ VIII—IX, $I_0 =$ VIII (MJD), $I_0 =$ IX (MR, CAM), $I_0 =$ IX—X (MF), isos. map in (BI), $h =$ sup?
The church and many houses were partly destroyed at Tito; Potenza also much suffered, serious damage at Brienza, Balvano, Montemurro, Tramutola, Sala, Atena, Sasso and Pietrafessa, felt at Napoli and Avellino (PAI, MMC, BI).

April 21—June 17, 36·8°N, 3·2°W, R.16, $I_0 =$ VII (MJD), $I_0 =$ VIII (BRS).
A series of shocks at Granada, several buildings were more or less injured, panic (GS, MMC).

Oct. 1, 08^h45^m, 47·4°N, 19·4°E, R.21b, $I_0 =$ VII (RA), $I_0 <$ VI (CD).
Szentpéter (Hungay), felt also in Budapest (RA), not mentioned in (CD), a violent shock felt in Obuda (Ofen), Pest, Pilis, Monor and Gyömrö (MMC).

1826

Oct. 26, 18^h, a violent shock damaged houses at Manduria (40·4°N, 17·6°E), Terra d'Otranto, Italy, (R.19), felt in the province of Bari (BI), $I_0 =$?, $I_0 =$ VII (MJD).

1827

Date?   Many strong earthquakes in Tbilisi and almost in all regions of Georgia, at Pambak, Shoragial, Borchaloia, a disastrous shock in Armenia in the region Darachichaga and the lake of Sevan where many churches were injured (BZ), (40·6°N 44·7°E), R.32.

Febr. 1,   48·5°N, 0·5°E, R.11, $I_0 =$ VII (RF), Jan. 2 (MMC, PAF).
Normandy, chimneys thrown down and panes of glass broken at Mortagne, damage at Aleçon VI (RF, MMC, PAF).

April 2,   epicentre? R.18, $I_0 =$ VI?, $I_0 =$ VIII (MFS).
The vault of the church at Ardez (46·8°N, 10·2°E) and some walls at Fetan fissured during a shock felt also at Bevers in the Upper Engadine (MFS).

Apr. 17, 13^h30^m UT, 42·6°N, 18·1°E, R.22a, $I_0 =$ VII (MJD), $I_0 =$ VIII, $r_{max} = 285$ km (TD).
Felt slightly at Venezia (MMC). Observed strongly at Makarska, Dubrovnik, Opuzen Ston, Obrovac, Zadar (TD).

June 7,   38·4°N, 0·8°W, R.16, $I_0 =$ VII? (BRS), $I_0 =$ V (MU).
Elche-Crevillente, a very strong shock (GS); not mentioned in (MMC).

June or July, 40·7°N, 36·6°E, R.29, $I_0 =$ VII?, $I_0 =$ IX—X (MJD), could be also 1826?, $I_0 =$ X (CAM, ONB), $I_0 =$ VII (EGU), two shocks?
Large parts of the towns Tokat (Sivas) and Erbaa were destroyed, the damage extended also to the surrounding country (MMC, PAP), damage at Tokat, Erbaa and vicinity, epicentre probably on the line Turhal—Tokat—Almus (PL).

Oct. 20,   noon, (40·6°N, 44·7°E), R.32—34?, $I_0 =$ VII (BZ), $M = 5·2$ (AV).
A severe shock at Tbilisi, walls of many houses cracked, as well as at Stavropol and Jerevan (Erevan) (BZ, GC, MMC, MO, PAD, MJD).

Dec. 12—13, (38·0°N, 0·7°W), R.12, $I_0 =$ VI—VII?, $I =$ II (MU), $I_0 =$ VII (MJD).
At Lisboa the walls of the houses cracked and church bells were made to toll (GS, MMC).

1828

Febr. —,   Fliotschild, Landeyjar (Iceland), $I_0 =$?, $I_0 =$ IX—X (MF), Febr. 6—13 (AV), not mentioned in (TR, MMC), Febr. 19, $I_0 =$ VIII (MJD), Febr. 6—13 (AV).

Febr. 2, 14^h15^m, 40·7°N, 13·9°E, $h =$ sup, volcanic? R.19, $I_0 =$ IX (CAM), $I_0 =$ VIII (MJD), $I_0 =$ IX—X (MR), $I_0 =$ X (MF).
One of the most violent earthquakes in the island of Ischia, very limited shaken area, not felt on the adjacent coast or islands, epicentre between Fango and Casamenello, in Casamicciola a part of houses collapsed, 20 men killed, no damage in Serrafontana, Forio and Testaccio, but Lacco suffered remarkably (PAI, MMC, BI).

Febr. 8, 13^h10^m UT, 48·4°N, 9·3°E, R.7, $I_0 =$ VI—VII (SW), $r = 60$ km (SG).
A severe shock in the Swabian Alb, Kohlstetten, Engstingen, Hollingen, Holzelfingen most heavily shaken, damaged houses, fall of chimneys (SW, MMC, PAB).

1828

Febr. 14, $40 \cdot 7°$N, $13 \cdot 9°$E, $h =$ sup, R.19, $I_0 =$ VIII? (MMC).
Another violent shock at Casamicciola where some houses collapsed (MMC, PAI), aftershock (BI).

Febr. 23, $08^h(30^m)$, $(50 \cdot 8°$N, $5 \cdot 0°$E), R.9, $I_0 =$ VIII, $I_0 =$ VII$-$VIII (SW).
An intensive earthquake in Belgium, the N. France, and the basins of the Meuse, the Rhine and the Moselle, at Tirlemont many chimneys thrown down, walls cracked; chimneys fell also at Maastricht, Andenne, Liège, Tongeren and Perwez, some damage ($I =$ VI) at Ath, Aubel, Glabbeek and Namur, etc. (LAB, TL, PAB, PAF, SW, MMC), the limits were Longuyon, Dunkerque, Avesnes, Brugge (Bruges), Middelburg, Dordrecht, the Rhine valley (MMC).

March 7,   a disastrous shock at S. Onofrio, Monteleone, $I_0 =$ IX (MF), not mentioned in (MMC, PAI, MR, BI), very doubtful.

March 12,   Calabria, two small houses thrown down at Palmi (Sicily), R.20, and some others fissured, $I_0 =$ VI? (BI, MMC, PAI, MJD).

April 11 or 14, a strong shock (VI) at Urbino (Italy), a small cupola was allegedly destroyed at Serpieri, $I_0 =$? (BI), $I_0 =$ VII (MJD).

May 18,   Marsala ($37 \cdot 8°$N, $12 \cdot 4°$E), R.20, $I_0 =$?, $I_0 =$ VIII (MR), $I_0 =$ VII (MJD); not mentioned in (MMC, PAI, BI).

June 15, 05^h, some houses injured by two local shocks at Izmir (Smyrna), R.26c (MMC, PAP, PL), $I_0 =$?

Aug. 9, $40 \cdot 7°$N, $48 \cdot 5°$E, R.32, $I_0 =$ X (BZ), $I_0 =$ IX$-$X (MJD).
Four damaging shocks in the E. Caucasus, epicentre near Shemakha where 233 houses were destroyed, 197 damaged, houses ruined also at Baskhani (59 houses), Zeiva, Iegordzhevan, Mugdy, Megdi-Malaya, Kala-Zeiva, Sarsura, Surakhani, Nuret, Zirgova, Kyuchetan, Sardanus, Ingar, Lulut, Nugdy, Pir-Abdul-Kosul, Shishdanak, Tirzhil, Zaidakhani, Zergiran, Bilistan, Zarnava, Belyam, Keyvandy, Salyany heavy damage, $2\frac{1}{2}$ km long fissure, the village of Muganly buried bellow a landslip, half of the village Chagan swallowed up by the earth, changes in springs (BZ, MMC, MO, PAD).

Sept. 15, 05^h, 06^h, 17^h, $38 \cdot 0°$N, $0 \cdot 7°$W, R.16, $I_0 =$ VIII (GS, MJD), $I_0 =$ VI (MU), $I_0 =$ IX (CAM).
Violent shocks which threw down many houses at Torrevieja, Guardamar and La Mata, damage at Lorca and Orihuela, epicentre between Torrevieja and Guardamar (GS, MMC).

Oct. 9, 02^h20^m UT, $44\frac{3}{4}°$N, $9\frac{1}{4}°$E, R.18a, $I_0 =$ VIII? (MJD), $I_0 =$ IX (MR), $I_0 =$ IX$-$X (MF), isos. map. in (BI).
Some houses were destroyed at Gamminella, Montesegale, Trebbiano Nizza, Cecina, Villa di Monto (BI), epicentre near Voghera, many buildings were injured by cracks, the damage done by fall of others was considerable (MMC); no considerable damage (PAI); tsunami in the port of Genoa, felt at Toulon, Firenze, Livorno and Marseille (MMC, PAI), epic. Gamminella (Godiaco) (MR).

1829

Jan. 21, $21-22^h$, Hekla region, R.2, $I_0 =$ VIII? (MJD), volcanic?
A severe shock felt throughout the south of Iceland, great damage near Hekla,

some of the peasants cabins were completely ruined and others much injured (MMC, AV).

March 21, 18^h, 38·2°N, 0·9°W, h = (7) km, R.16, I_0 = IX (MU, SAE), I_0 = IX—X (BRS), I_0 = XI (MF), isos. map in (SAE).

Disastrous earthquake in the province of Murcia, epic. in the valley of Segura, several thousands of houses destroyed, thousands of people perished in the regions of Murcia, Guardamar, La Mata and Torrevieja, felt also in Madrid; sand and mud volcanoes, changes in the ground water level and in the course of the river Segura (GS, MMC).

April 13 or 23, 16^h, ($40\frac{3}{4}$°N, $24\frac{1}{2}$°E), R.24?, I_0 = IX—X?, I_0 = XI (MF).

A violent shock with the epicentre near the island of Thásos or in the Rodopi mountains, felt in Macedonia and W. Turkey as far as Edirne (Adrinople), many houses thrown down at Kavalla, Pravi and Xanthi; a minaret and several houses collapsed at Edirne (PAP, MMC), S. Watzoff reports also about collapsed chimneys, roofs and parts of minarets in S. Bulgaria and about fissures in the ground at the coast (WS). These reports might correspond to May 5? Not mentioned by (GG, ONB).

April 11—May 2, 38·2°N, 9·9°W, R.12, I_0 = VII (BRS), I_0 = VIII (MU).

Aftershocks in the region of Murcia, that of April 18 (09^h45^m) is said to be almost as violent as the main shock of March 21 (GS), felt at Orihuela, Almoradi, Torrevieja, Elche, Cartagena, Salinas and Guardamar (GS, MMC).

May 5, afternoon, $41\frac{1}{4}$°N, $24\frac{1}{2}$°E, $h = n$, R.24, I_0 = X, I_0 = IX (GG), I_0 = IX—X, P = 430 000 km^2 (GGK).

Violent shocks at Thessaloníki (Saloniki), houses, mosques and part of the town walls were thrown down, felt over the coast of Macedonia and Thrace, also at Istanbul and Bucureşti; the little town of Drama was totally destroyed and many of surrounding villages as well as the towns Kavalla and Sérrai suffered much (MMC, PAP, WS), large changes of the terrain (GGK), Edirne slightly damaged, felt at Istanbul (PL), I = VII—VIII in the Struma valley (GGB).

May 21—22, night, (41·7°N, 12·7°E), R.19, I_0 = VII (MR, MJD), I_0 = IX (MF), h = sup? A violent shock at Albano, Genzano, Riccia, etc. and particularly at Castel Gandolfo, one house collapsed (MMC, PAI), houses were fissured (BI).

May 23, 41°N, 29°E, R.27, I_0 = V (EGU), I_0 = VII (MJD).

Two shocks at Istanbul, on the Asiatic coast buildings were injured (MMC, PAP), not mentioned in (ONB, GG), tsunami at Istanbul (ANS), buildings only shaken in Istanbul (PL).

July 1, 03^h37^m UT, 47.6°N, 22·3°E, R.21b, h = (6) km, I_0 = VII—VIII, I_0 = VII (CD), I_0 = IX (PRA), P = 79 700 km^2, I_0 = IX—X (RA). Several shocks between 02^h and 16^h40^m (MMC).

A severe earthquake in the valley of the Ierul(?) (Bihor) (FA), felt from Eger (Hungary) to the Carpathian Mts. At Khust, Bestereczen (Bistriţa), Săuceni (Szekelyhíd), Tokaj and Pişcolt chimneys collapsed and walls fissured, at Hegyalja and Sătmar (Szatmár) some houses were destroyed, epicentre near Dengeleg (small fissure in the ground), Andrid (Endréd) and Pişcolt (RA, MMC). Main shock on July 4, 08^h30^m, foreshocks on July 1, 03^h, strongly felt, July 2, 04^h37^m, I_0 = = VII? (FA).

1829

Aug. 3—4, houses damaged at Nagy Károly, Endréd, Dengeleg, Iriny and Vezendiu (Porte-lek) (RA, MMC), R.21b, $I_0 =$ VII, $I_0 =$ VIII (MJD), epic.?

Sept. 6, $15^h 25^m$ UT, (45°N, 10°E), R.19, $I_0 =$ VII (PAI, MJD), $h =$ sup?
A violent earthquake at Cremona (Italy), cracks in the vault of a church, several buildings injured (MMC, PAI), 150 chimneys fell down, smaller intensity in the surroundings (BI).

Nov. 26, $01^h 40^m$ UT, 45·8°N, 26·6°E, $h = i$, R.21d, $I_0 =$ VIII—IX (PRA), $I_0 =$ IX R.F.(AI), $I_0 =$ IX—X (RA, MJD), $I_0 =$ X—XI (MF), $M = 6·9$ (PRA), 45·0°N, 26·5°E, $I_0 =$ VIII (ES), isos. map in (MSM).
Largely extended earthquake in the Carpathians, region of Vrancea (Vrincioaia); felt in Romania and Ukraina, 15 houses ruined at Bucureşti (115 with cracks in walls), cracked walls at Sibiu, in many villages fall of plaster and considerable damage to chimneys and ovens (FA), damage reported from Kishinev, Kherson, Tiraspol, Dabossary and Reni, felt at Kiev, Yekaterinoslav, Odessa and Nikolaev (MMC, MO, EU).

1830

Febr. 8, 10^h, (46°N, 15$\frac{1}{2}$°E), R.22b, $I_0 =$ VII (MMC), doubtful.
At Zagreb (Agram) cracks in walls of many houses, panes of glass broken; not mentioned in (RV, TD), (RV) gives only a slight shock at Ljubljana ($I_0 =$ III, IV) on March, 8, 05^h.

March 9, 10^h (local time?)), 43·5°N, 47·5°E (BZ), R.32, $I_0 =$ VIII—IX, $I_0 =$ VII—VIII (GC, PV), $I_0 =$ IX—X (MJD).
A damaging shock with epicentre in N. Caucasus, most violently felt at Andreev-skaya (collapse of more than 100 houses, several persons killed), Tarki (200 houses ruined and 200 damaged, two persons killed) and at Fort Vnezapnaya (all barracks seriously damaged); felt at Georgievsk and surrounding villages, Kicherskoe, Kizlyar-Shandrukovskaya (rockslide), Mozdok, Krasnodar (Ekaterinodar), Go-riachie Vody, Astrakhan, Tbilisi (BZ, PV, MO, PAD), according to (MMC, GC, PV) $H = 13^h 10^m$, at Andreevskaya a church fell and 400 inhabitants perished beneath the mud roofs of their houses.
Aftershock: June 25, $I_0 = $?, $I_0 =$ VIII (MJD).

July 1, 20^h, 48$\frac{1}{4}$°N, 23$\frac{1}{4}$°E, R.21a, $I_0 =$ VII? (MJD), $I_0 =$ VIII (RA), 05^h, 21^h (FA).
An earthquake in the Maramureş region (Ukraina), walls of many houses fissured, felt at Khust, Sighet, Rona and in the mines of Sugatag and Slatina (RA, FA, MMC, EU). A. Réthly classifies the shock as an impact earthquake (Einsturzbeben) (RA).

Nov. 22, 9^h? 44·9°N, 37·3°E, R.32, $I_0 =$ VII (PV, MJD).
At Anapa many houses damaged, fall of some chimneys, some damage in the forts Bugag, Temryuk, more slightly felt at Fanagoria and Taman (GC, PV, MO), not mentioned in (MMC, BZ).

1831

Jan. 2, 14^h UT, (40°N, 15$\frac{3}{4}$°E)? R.19—20, $I_0 =$ VIII (MJD), $I_0 =$ IX (CAM), $I_0 =$ X (MF).
At Lagonegro (region of Basilicata, Napoli) houses and churches injured by a

shock, ten houses and a neighbouring church(?) fell (MMC, BI, PAI), some buildings collapsed also at Lauria Inferiore and Superiore (BI). On the same day a strong shock damaged houses at Cajeta (Calabria) (PAI), and at Aieta (Paola) (BI).

Jan. 28, 17^h30^m UT, $38 \cdot 2°$N, $15 \cdot 2°$E, R.20, $I_0 = $ VII? (MR), questionable.

At Milazzo some small houses were destroyed (BI), Milazzo was ruined by lasting shocks (MMC, PAI).

April 3, $37\frac{3}{4}°$N, $27°$E, R.26c, $h = n$, $I_0 = $ VII? (GG), $I_0 = $ V−VII (GGK).

A sequence of strong shocks in the Sámos island caused a rockfall which killed 7 persons (MMC, GGK, PKP).

May 26, 10^h30^m UT, $(43\frac{1}{2}°$N, $7\frac{3}{4}°$E), R.18a, $I_0 = $ VIII−IX, $I_0 = $ VIII (MJD), $I_0 = $ IX (MR, RO), $I_0 = $ X (MF), $43 \cdot 8°$N, $7 \cdot 8°$E, $r = 150$ km (RO).

At Castellaro the church and 52 houses ruined, 49 uninhabitable, 5 victims; houses thrown down also at Bussana, much damage at Taggia, felt at Marseille, Genova, Torino and S. Remo, Pompeiana, tsunami (BI, RO, MMC, PAI), not in (RF).

June 28, 17^h, $37 \cdot 0°$N, $12 \cdot 6°$E, R.20?, $I_0 = $?, $I_0 = $ VIII (MJD), $I_0 = $ IX−X (MF); a volcanic eruption?

Severe shocks in Sicily, especially in Sciacca (Isola San Guilia), followed by the upheaval of a new island between Sciacca and Pantelleria (MMC, PAI, BI).

Sept. 11, 19^h30^m, $44\frac{3}{4}°$N, $10\frac{1}{2}°$E, R.19, $I_0 = $ VIII (MR), $I_0 = $ VII (MJD), $I_0 = $ IX−X (MF), isos. map in (BI).

A violent shock at Reggio, 200 chimneys fell, a palace seriously damaged; at Parma 140 chimneys collapsed, roofs and walls damaged, panic, and at Sorbolo several persons buried beneath the ruins, felt at Venezia, Modena (strong) and Castelnuovo ne'Monti (BI, MMC, PAI, PA).

1832

Jan. 13, $14-15^h$, $43 \cdot 0°$N, $12 \cdot 7°$E, R.19, $I_0 = $ VIII−IX, $I_0 = $ VIII (MJD), $I_0 = $ XI (MF).

At Bastia the fortifications collapsed, houses were destroyed, the church seriously injured, Bevagna and Cannara almost totally destroyed, serious damage also at Foligno (13 victims), Perugia, Assisi (all buildings damaged), Spello (many houses uninhabitable), Montefalco, Castellacio (9 victims), etc., felt at Roma, Firenze and Parma (BI, PA, MMC).

Jan. 29,   at Trevi (near Foligno), R.19, a disastrous aftershock ruined 48 houses (BI, MMC, PAI), $I_0 = $ VIII?

Febr. 7 − Apr. 7?, $(45 \cdot 4°$N, $24 \cdot 2°$E), R.21d, $I_0 = $?, $I_0 = $ VIII (PRA), $I_0 = $ IX−X (MF), questionable.

Eleven shocks in the Carpathians Mts., no damage reported by (FA), not mentioned by (RA, MMC), only fissures in the ground and no information on damage given by (MF).

March 8, 19^h about, $39 \cdot 0°$N, $16 \cdot 8°$E (PET), R.20, $I_0 = $ IX?, $I_0 = $ X (MR), $I_0 = $ X−XI (MF), isos. map in (BI).

A disastrous shock in Calabria, Cutro ruined, great damage in Soveria, Papanice, Marcedusa, S. Severino, Cotronei, Isola de Capo Rizzuto, Policastro, Catanzaro, Scandale, Castello, Ciro, S. Mauro; Roccabernarda, Rocca di Neto partly ruined (BI, MMC, PAI).

New ruines in Catanzaro on the beginning of April.

1832

March 13, 03^h20^m UT, $44\frac{3}{4}°$N, $10\frac{1}{2}°$E, R.19, $I_0 =$ VIII (MR, MJD), $I_0 =$ IX (CAM), $I_0 =$
= XI (MF).
At Reggio all chimneys thrown down, the barracks uninhabitable, churches seriously injured or partly collapsed as well as some houses, some destruction at Cam-
pegino, Bagnolo, S. Bernardino, many chimneys fell down and all houses were
seriously damaged at Parma, houses damaged at Poviglio, Traversetolo, Ciano,
Corregio, etc., felt at Genoa, Venezia, Udine, Milano, Torino (BI, MMC, PAI).

Nov. 24, 10^h30^m, violent tremors near Etna ($37,8°$N, $15·0°$E), R.20, $h =$ sup, $I_0 =$ VIII
(MJD), $I_0 =$ IX—X (MP). The bell tower at S. Giovanni seriously damaged
(later collapsed), 5 houses destroyed at Nicolosi, 3 victims (BI, MMC). Probably
identical with the report on the shock of Dec. 24: several houses thrown down at
Nicolosi and Belpasso, casualties (PAI).

1833

Jan. 11, 00^h50^m, $46·0°$N, $14·6°$E, R.22b, $I_0 =$ VI?, $I_0 =$ V (RV), $I_0 =$ VIII (TD).
Two violent shocks at Ljubljana (MMC).

Jan. 19,   ($40\frac{1}{2}°$N, $19\frac{1}{2}°$E), R.25, $I_0 =$ IX? $I_0 =$ X (MD), $I_0 =$ IX—X (MF, GGK).
Disastrous shock in the bay of Vlorë (Valona), high tsunami inundated the island
of Sazan (Saseno) (ANS), serious destructions at Vlorë and Kanino (MCA),
according to (MD) $I =$ X: Vlorë, Sazan, Kanino, Narta, $I =$ IX: Smokthine,
Velçan, Karbunarë, Tepelenë, Peshtani, Maricaj, Vasiari, Klisyra, Turani, Dukaj,
etc.; felt also in Italy at Lecce, Monteparano, Bari, Potenza, Foggia (MCA, BI).

1834

Febr.,   three shocks of different intensity along the NE coast of the Black Sea, felt strongly
at Bugat and Apapa ($44.9°$N, $37·4°$E, R.32) where old houses fell down (BZ, MO,
GC), $I_0 =$?, $I_0 =$ V—VI (GC), $I_0 =$ VII (MJD), (MMC) gives March 9.

Febr. 14, $44\frac{1}{2}°$N, $10°$E, R.19, $I_0 =$ VIII—IX, $I_0 =$ VIII (MJD), $I_0 =$ IX—X (MR), $I_0 =$ X
(MF), isos. map in (BI).
At Pontremoli all the biuldings were seriously injured, most chimneys collapsed,
and in some villages (Caprio, Zeri, Guinadi, Bratto, etc.) five or six miles to NW,
belfries, churches and ill-built houses fell, 60 (?) persons perished, epic. Monte
Molinatico; felt at Parma, Milano, Mantova, Torino (MMC, PAI, BI).

May 23, 06^h, $31°$N, $35\frac{1}{2}°$E, R.37, $I_0 =$ IX, $I_0 =$ X (SNP), $I_0 =$ VII (MJD), isos. map in (SNP).
At Jerusalem several churches, the city wall, many houses and cisterns seriously
damaged, a minaret collapsed, at Bethlehem much damage to monasteries, many
people killed, at Deir Mar Saba a tower cracked, large blocks of asphalt floated on
the Dead Sea (KAD, MMC). The epicentre was in the district of Lisan, the pleistoseismal area extended from south of the Dead Sea to Kerak, Jenin, Amman and the
Judean hills; causalties, houses ruined in Nablus, damaged houses at Karak,
serious damage at Gaza, some damage at Jerusalem nad Bethlehem, some faulting
in the Lebanon and in Baqa (Baka) (SNP).

June 5, $38°$N, $21°$E, R.26a, $I_0 =$ VII—VIII (GGK).
Strongest aftershock (main shock?) on June 18 (MCA), $I_0 =$ VIII (MJD), some
houses thrown down in the Kefallinia on June 18 (MMC, PAP), not mentioned by
(GG), (GGK) gives July 5.

1834

Sept. 31,   37°N, 1·2°W, R.16, I_0 = VII (MU, GS), I_0 = VIII (BRS).
Murcia—La Alberca; not mentioned by (MMC).

Oct. 4,   19^h UT. Walls fissured and many chimneys fell down at Bologna (R.19), I_0 = = VII? (MJD), felt slightly at Parma, Venezia and Ferrara (BI).

Oct. 15,   06—07^h UT, 47·6°N, 22·3°E, R.21b, I_0 = VIII—IX, I_0 = VII—VIII (PRA, .CD), I_0 = IX—X (RA, MF), P = 80 000 km^2, isos. map (RA), Oct. 17 (FA).
Epicentre near Dengeleg—Gálospetri, sand and mud ejected near Andrid (Érendréd), Gálospetri (church cracked, the tower the chimneys and houses collapsed), at Dengeleg, Andrid (Érendréd) and at Carei (Károly) many chimneys thrown down, houses and churches seriously damaged, some of them became uninhabitable, the monastery at Kaplony and the church at Szaniszló destroyed, etc., for details see (RA, FA), strongly felt up to the Carpathians (RA, FA, PAD, MO).

1835

Febr. 6,   18^{h}50^m UT, 44°N, 11$\frac{1}{2}$°E, R.19, I_0 = VII? (MR, MJD).
Damage at Vicchio, walls fisured and chimneys thrown down at Borgo S. Lorenzo (BI, MMC).

March 23,   08^{h}30^m UT (44·3°N, 7·6°E), R.18a, I_0 = VII (MJD), I_0 = VII—VIII (MR), local?
A shock at Boves near Cuneo (Piedmont) threw down a great number of chimneys (BI, MMC), not mentioned by (PAI), (MMC) gives May 23.

Aug. 23,   17^h, 38·5°N, 35·5°E, R.30, I_0 = IX—X? (ONB, MJD), I_0 = X (CAM), I_0 = VII (EGU).
At Kayseri more than 200 houses fell, 150 persons perished, in all villages within 30 miles around this place almost all houses were totally destroyed, many persons perished, two thirds of Welkeri ruined; Kumetri "swallowed by the earth" and a lake originated on its place. About twenty other villages suffered seriously by the earthquake (PAP, MMC). Epicentre on the line Ecemiş-Kayseri, twenty villages damaged (PL).

Oct. 12,   midnight, 39·3°N, 16·3°E (PET), h = sup.?, R.20, I_0 = IX, I_0 = X(MR, CAM), I_0 = XI (MF), isos. map in (BI).
At S. Pietro in Guarano, Zumpano, Lappano most houses were destroyed, 27 inhabitants perished, at Cosenza the buildings were seriously fissured (PAI, MMC, BI).

Oct. 28,   04^{h}35^m, (43°N, 0·2°E), R.17, I_0 = VII? (RF).
Largely extended shock, ceilings cracked at Bagnères (RF), some walls cracked (MMC), felt at Saint Bernard de Comminges, in the valley of Gavarnie and in the villages of the Pyrenees (GS).

Oct. 29,   (47$\frac{1}{4}$°N, 9$\frac{1}{2}$°E), R.18b, I_0 = VII?, I_0 = VIII (MFS).
Walls cracked at Appenzell, felt at Rorschach, Frauenfeld, Shaffhouse, Winterthur and Zurzach (MFS).

1836

April 24,   39·6°N, 16·6°E (PET), R.20, I_0 = X(MR, CAM), I_0 = XI (MF).
At Rossano 25% houses thrown down, Crosia completely destroyed, Caloveto, Cropalati, Scala and Paludi were partly destroyed, tsunami, long and deep fissures

in the earth, subsidence of the terrain, 589 victims, felt also at Napoli (BI, MMC, PAI).

June 12, 02^h35^m UT, (46°N, $11\frac{1}{2}$°E), R.19, $I_0 =$ VIII—IX, $I_0 =$ VIII (MJD), $I_0 =$ IX (MR, CAM), $I_0 =$ IX—X (MF), isos. map in (BI).

Severe shocks on June 11—18, main shock on June 12 felt heavily in the district of Treviso and Ascoli, at Liedolo, Fonte, S. Ilario houses were thrown down and churches seriously damaged, felt at Venice (V), Ferrara, Mantova, Brescia (PAI, MMC, MR, BI).

July 20,   about noon, (46°N, $17\frac{1}{2}$°E), R.19, $I_0 =$ VII—VIII?

At Bassano many houses were newly injured, felt allegedly at Brixen IV, Innsbruck, München, Verona, Parma, Modena, Ferrara (PAI, MMC, BI), aftershock on June 12.

Nov. 11,   at night, 46·2°N, 16·2°E, R.22b, $I_0 =$ VII, $I_0 =$ IX (TD), local?

At Zajezda fissures in the walls, chimneys thrown down.

Nov. 12, 24^h, aftershock, $I_0 =$ VII? at Zajezda (TD).

Nov. 18, 04^h, 05^h, 46·2°N, 15·9°E, R.22b, Two shocks with $I_0 =$ VII (TD), $I_0 =$ VIII (TD). Chimneys fell down at Krapina. Nov. 13—18, Violent shocks in Croatia (MMC).

Nov. 20, 07^h UT, 40·1°N, 15·8°E, $h =$ sup?, R.20, $I_0 =$ VIII—IX, $I_0 =$ VIII (MJD), $I_0 =$ IX (MR, CAM), $I_0 =$ IX—X (MF), isos. map in (BI).

A violent shock felt at Napoli, at Lagonegro, several houses were ruined, all others fissured, cracks in the ground caused landslipe, neighbouring localities Montesano, Nemoli, Rivero, Trecchina, etc., suffered much, 12 victims (BI, MR, MMC, PAI)

Jan. 1, 03^h, 33°N, $35\frac{1}{2}$°E, R.37, $I_0 =$ X (PK, CAM), $I_0 =$ XI (SAM, KAD, SNP), $I_0 =$ = IX +, isos. map (ANS, PK).

Destructive earthquake with the epicentre near Safad where all houses on steep slopes fell, fissures in the ground, 5000 victims, destructive effects (IX—X) at El Jish, Er Reina, Ein Zeitun, Tiberias (city walls overthrown, the lakeswept the shores, 700 victims), Sejera IX?, Sidon VIII—IX, Sur VIII—IX, Hunin VIII?, Qaditta and Lubya severe, Nazareth VI—VII, Nablusstrong, Kafr, Kanna V, Tsippori (Saffuriya), Jerusalem, Bethlehem, Hebron moderate, many casualties (KAD). Devastation from Beirut (Beyrouth) to Safad, deep fissures in solid rocks, new hot springs (PAP, MMC). Epicentre probably in the Tiberias depression; Beirut not particularly heavy, Nazareth great cracks in houses, Jericho slight, Tiberias city ruined, Esh Sham (Damascus) city affected, Hauran and Gaulan considerable (SNP). Strong tsunami along Syrian — Israeli coasts of Tiberias (ANS).

Jan. 24?   epicentre? R.18a, $I_0 =$ VII?, $I_0 =$ VIII (MFS).

Many chimneys fell and walls fissured at Brig, this shock corresponds probably to reports about tremors felt at Meiringen, Frutingen (MFS).

March 14, 15^h40^m UT, 47·5°N, 15·5°E, R.18b, $I_0 =$ VII (TT, RA, MJD, KC).

At Mürzzuchlag houses were fissured, some rooms were uninhabitable, walls fissured also at Reichenau and Schottwien, rockslides. Region Mürzzuschlag-Semmering, March 15 (MMC), felt in Austria, Czechoslovakia (Praha, Olomouc, Brno, Bratislava, etc.), Hungary (Sopron) (KC, RA).

1837

March 20, 08^h UT, $37\frac{1}{2}°$N, $23\frac{1}{2}°$E, R.26a, $h = n$, $I_0 =$ VII−VIII, $I_0 =$ VII (GG), $I_0 =$ = VI−VII (GGK), $I_0 =$ VIII (MJD), $I_0 =$ IX (CAM), $I_0 =$ IX−X (MF), $r =$ = 210 km (GG).
In the island of Ídhra (Ydra) some houses were thrown down and others were injured, damage in the islands of Páros, Syros, Spetsai and Thíra (Santorin), felt also in the interior of Greece at Kalámai (Kalámata) and Messíni (PAP, MMC, GGM), in Athínai blocks of marble fell down (GGK).

April−May, during a sequence of strong shocks in the region of Montecassino, R.19, some houses were slightly damaged (BI), $h =$ sup? $I_0 =$ VII? (MJD).

April 11, 17^h UT about, $(44\frac{1}{4}°$N, $10\frac{1}{2}°$E), R.18b, $I_0 =$ VIII−IX, $I_0 =$ VIII (MJD), $I_0 =$ IX (MR, CAM), $I_0 =$ IX−X (MF), isos. map in (BI).
Houses thrown down at Argigliano, at Uglian Caldo 90% houses ruined, 5 victims, serious damage at Montefiore, Regnano, Minucciano, felt at Modena, Parma, Genova and Firenze (BI, MMC, PAI, MR).

Aug. 15, $38°$N, $22°$E, R.26a, $h = n$, $I_0 =$?, $I =$ VII (GG), $I_0 =$ VI−VII (GGK). Not mentioned by (MMC, PAP).
A roof collapsed at Pírgos (Pyrgos), rockslides in the mountains, fissures in the ground (GGK).

Sept. 22, 12^h, $45\cdot8°$N, $16\cdot0°$E, R.22b, $I_0 =$ VII? $I_0 =$ VIII (TD), local?
All walls cracked in Zagreb (MMC), chimneys thrown down.

Oct. 4, $42\cdot1°$N, $19\cdot3°$E, R.25, $I_0 =$?, $I_0 =$ VIII (MD), $I_0 =$ VII (MA).
The castle of Shkodër damaged by an earthquake (MCA); not mentioned by (PAP, MMC).

Oct. 6, many houses thrown down at Agram (Zagreb) (MMC), a doubtful report, not mentioned by (TD, PAP).

Oct. 31, $01−02^h$? $38\cdot1°$N, $6\cdot7°$W, R.16, $I_0 =$ VII (BRS), $I_0 =$ V (MU).
Violent at Murcia, at Torrevieja solid buildings were violently shaken (GS, MMC).

1838

Date?, Jaffa, (Tel-Aviv), great destruction, probably identical with Jan. 1, 1837 (KAD, PK).

Jan. 23, 18^h45^m UT, $45\cdot8°$N, $26\cdot6°$E, R.21d, $h = i$, $I_0 =$ IX, $M = 7\cdot1$ (PRA), $I_0 =$ X (RA), $I_0 =$ XI (MF), $r \geqq 190$ km, $M = 7\cdot3$ (KR), $45\cdot3°$N, $26\cdot5°$E, $I_0 =$ VIII (ES). isos. map in (MSM).
A disastrous earthquake with the epicentre in the Vrancea (Vrincioaia) region, very large shaken area; in the E. Carpathians, Valachia and S. Moldavia destruction of many churches nad houses. In Craiova the walls fissured, the ceilings and chimneys collapsed, in the district of Vîlcea 39 churches were demolished or seriously injured, in the district of Romanați 53 churches, the district of Mehedinți 4 churches, the district of Olt 17 churches. Heavy damage in the districts of Dimbovița, Prahova, Ilfov, Săcueni, Ialomița, Buzău, Rîmnicul Sărat, Putna, in the last two districts the earth was fissured, in some villages ground water rised up and flooded the houses; in București (Bucarest) all houses were injured in different extent, 8 victims, largest damage in the most massive houses; Brașov-vibrations, Prejmer − a tower demolished, Vărghiș and Sighișoara − churches seriously demaged, felt throughout the Transsylvania and Banat with less or

more damage, damage also at Odessa, Perekop etc., the river Bacas was blocked up by a rockfall. The border line of the shaken area included Budapest, Lvov (Lwow), Tarnopol, Novomoskovsk, Sevastopol, Odessa, Istanbul (FA, MMC, RA, PAD, MO, EU, MJD), in the southern part of Crimea several houses were damaged (PV).

June 11, destruction in the region between Skagafjördhur and Eyafjördhur, particularly in Olafsfjördhur and Siglufjördhur, large rock falls at the coast line, $I_0 = IX$? (MJD), $I_0 = X-XI$ (MF), R.2.

June 23, 10^h18^m, (43·9°N, 12·9°E), R.19, $I_0 = VII$? $I_0 = VII$ (MJD), $I_0 = VII-VIII$ (MR), local?
Walls fissured at Pesaro, slight at Venezia (Venice), panic, felt strongly along the coast to Fano and Sinigaglia (BI, PAI, MMC).

Aug. 10, $20-21^h$, 45·3°N, 14·5°E, R.22a, $I_0 = VII$, $r = 15$ km (TD), $I_0 = IX-X$ (MF).
At Bakar (Bukkari) the great tower of the church fell, Rjeka (Fiume) V. At Kralevica, Kostrena some chimneys fell, demaged houses, strong in Krk, felt at Trieste (MMC, PAI); foreshock at 02^h, $I_0 = VII$ (TD), not in (BI).

Aug. 26, (46·3°N, 16·3°E), R.22b, $I_0 = VII$? $r = 60$ km (TD), $I_0 = VIII$? (RA).
Earthquake in the frontier region of Hungary and Yugoslavia, "much damage at Racz-Kanisza and Varazdin where cracks were formed in many houses and some of them thrown down" (MMC), houses damaged at Varazdin (RA).

1839

March 26, the main shock of a sequence lasting from Dec. 1838 to March 1840 at S. Giovanni di Moriana (Savoia, R.18a) and surrounding localities fissured walls and threw down chimneys, isos. map (BI), $I_0 = VII$?

April 14, 14^h, a shock was strongly felt at Constantine and Alger where previously damaged houses collapsed, epicentre position unknown (RK, GS, MMC).

July 11, 12, 13 and 16, 47·4°N, 19·7°E, R.21b, $I_0 = VII$ (CD), $I_0 = VIII$ (RA).
Main shock on July 13, 15h, felt.

Oct. 23, 22^h UT, (56·4°N, 4·0°W), R.10a, $I_0 = VII$? (MJD), $I_0 = VIII$ (DCB), Comrie, Pertshire (MJD).

1840

Jan. 5, an earthquake in the Pyrenees, chimneys thrown down (MMC), $I_0 = VII$ (MJD), not mentioned by (RF, GS).

April 26--30, 49·4°N, 20·4°E, R.21, $I_0 = VII$? (RA, KC).
A damaging shock in E. Slovakia, at Spišská Stará Ves and surroundings (?) chimneys fell, bells sounded, walls and ground near the river cracked, felt in the Carpathians (RA).

June 20—July 28, 39·5°N, 43°E, R.33, $I_0 = IX-X$ (ONT), 39,8°N, 44·4°E (BZ, AV), $I_0 = VIII$ (MJD), 40·1°N, 43·4°E, $I_0 = VII$ (EGU, TM).
A destructive earthquake in Armenia near Agri Dagi (Mount Ararat) on July 2, 14^h?; in the districts of Nakhichevan 3000 houses heavily damaged, 33 people perished, the village of Ahura (Akhuri, Akori, Arguri) and the monastery of St. Yakov were buried (1000 victims) by a landslide, Kara-Asanlu injured, at Shusha

no damage, some houses damaged or seriously injured at Nakhichevan and at Yerevan (Erevan), at Tatevo an old Armenian monastery injured, felt at Tbilisi, Lenkoran, Alexandropol; at Maku (Iran) and Caldiran (Bayazid, Turkey) many houses totally ruined, long landslips from the slopes of Agri Dagi (Ararat) destroyed villages, changes in streams and ground water level (BZ); numerous fissures along the rivers Araxes and Arpatchai (Arpa, Arpaçay), sand craters (MMC), Nakhichevan totally destroyed, two districts in Armenia ruined (PAP), Kagizman, Igdir seriously demaged, epicentre probably on the line Kagizman—Tuzluca or Digor—Ani (PL). Foreshocks on June 20—28, $I_0 =$ VIII—IX (ONT).

July 6, 39°45′N, 44°25′E, new landslides on Agri Dagi (Ararat), devastation of the country 20 km around (BZ).

July 15, another violent shock, terrible damage by landslips, 3000 houses were thrown down, many victims (MO, MMC), the effects correspond to the description of (BZ) for July 27.

July 26, 39°45′N, 44°25′E, a strong earthquake in the region of Agri Dagi (Ararat) felt at Tbilisi (BZ).

Aug. 14—18, new landslides in the region of Agri Dagi (Ararat) (MO, PAD).

Aug. 27, 12^h05^m UT, 46·2°N, 14·7°E, R.18b, $I_0 =$ VII—VIII, $h = 8$ km, $M = 4·9$, $r =$ $= 140$ km (RV), $I_0 =$ VII (TD, MJD), $I_0 =$ IX (MF).
Much damage in Styria (MMC), G. Grad VII, Kamnik VII, at Ljubljana (VII), several houses vere fissured, chimneys and tiles were thrown down, severe also at Görtschach, Lack, Lusttal, Stein and St. Oswald.

Oct. 30,   38°N, 21°E, $h = n$, R.26a, $I_0 =$ IX—X (MJD), $I_0 =$ IX (GG), $I_0 =$ X (MP, CAM), $I_0 =$ VIII—X (GGK).
"This shock was the most destructive of ever felt in Zákinthos (Zante)", one village was ruined, the buildings with foundations on limestone escaped well, the island of the Trente-Nova (?) sank into the sea (MMC, PAP), felt in Ipiros (Épeiros) (MCA), alltogether 1271 houses collapsed (35 at Zante), 12 persons perished, most destruction at Skulikadon and Agios Demetrios (St. Dimitrios) (GGK).

Nov. 29,   aftershock of July 2?, R. 33, $I_0 =$ VII—VIII? $I_0 =$ VIII (MJD), Dec. 7 (BZ).
A strong earthquake in the district of Sharur(?), 3 houses injured, several people perished; at Nakhichevan cracks in walls of many houses (BZ, MMC, MJD),

1841

Febr. 21,   41·7°N, 15·6°E, R.19, $I_0 =$ VII—VIII, $I_0 =$ VIII (MR, MJD), $I_0 =$ X (MF), $h =$ sup? At Foggia several houses fissured, at S. Marco in Lamis 2 houses collapsed, 200 houses were fissured (BI), no damage reported in (MMC, PAI).

Febr. 26,   island of Zákinthos (Zante) (37·7°N, 20·8°E), R.26a, a few houses thrown down, some others injured (MMC), an earthquake in Zante (PAP), not mentioned by (GG), $I_0 =$?

April 3, 02^h16^m UT, 57·0°N, 8·0°E, R.3, $I_0 =$ VII (LI), $I_0 =$ VIII, $r = 250$ km, $M = 5·9$ (BMF), $I_0 =$ VIII (MJD), $I_0 =$ VIII—IX (MF).
In Jutland houses were violently shaken and chimneys thrown down (MMC).

May 18, 22^h, 39·4°N, 45·2°E, R.34, $I_0 =$ VII? $I_0 =$ IX (MJD).
At Kevrag 6 houses injured, at Nakhichevan walls fissured (MO, PAD).

1841

June 8—9, 42·2°N, 13·9°E, R.19, $I_0 = $ VII? $I_0 = $ VII (MJD), $h = $ sup? isos. map in (BI).
Some houses fissured at Torre de Passeri (Abruzzo Ult.), felt at Alanno, Pescara, Chieti (BI).

June 10, (42°N, 14·1°E), R.19, $I_0 = $ VII—VIII, $I_0 = $ VIII (MR, MJD), $I_0 = $ IX—X (MF), isos. map in (BI).
Various houses damaged at Taranto, Torricella Peligna, Palena, 2 houses collapsed (BI, MMC, PAI, MR).

July 1, "The town of Bayazid (Çaldiran), (40·3°N, 45·1°E), R.32, in Georgia was swallowed up in consequence of an earthquake" (MMC), a very doubtful newspaper report (PAD), not mentioned by (BZ, MO).

July 13, $12^h 30^m$, 47·8°N, 16·2°E, R.18b, $I_0 = $ VII (TT, KC, MJD).
At Wiener Neustadt many buildings and walls were injured (MMC, E. Suess).

July 15, $16-17^h$, At Holbaek (55·7°N, 11·7°E), R.3, several walls were thrown down (MMC, AV), very doubtful, not mentioned by (LI, BMF).

July 30, $14^h 30^m$, (56·4°N, 4·0°W), R.10a, $I_0 = $ VII? (MJD), $I_0 = $ VIII (DCB).
Comrie (Perthshire), chimney-tops broken (MMC).

Oct. 24, $12^h 10^m$ UT, 47·8°N, 18·1°E, R.21a, $I_0 = $ VII—VIII (RA, KC), $I_0 = $ VII (CD), $I_0 = $ VIII (MJD), $I_0 = $ IX—X (MF), $r_{max} = $ 75 km.
A damaging shock at Komárno, walls cracked, many chimneys fell, plaster of most houses fell off, felt at Budapest (RA, KC), wooden houses thrown down (MMC).

Oct. 24, $13^h 08^m$ UT, a severe shock at Köln (Germany), R.9, threw down chimneys and cracked walls (SW, RU, MMC), local?, $I_0 = $ VII?

Dec. 2, $18^h 53^m$, (45·8°N, 5·8°E)? R.18a, $I_0 = $ VII (RF, MJD), $I_0 = $ VII—VIII, $r = $ 120 km (RO).
Chimneys thrown down at Annecy, Rumilly and Chambéry, ceilings cracked at Rumilly (RF), felt at Grenoble, Lyon, Belley, Genève, Seyssel, Nantua, Chalons sur Saône, Mâcon and Saint-Gervais (RO, PAF).

Dec. 25, 7^h?, (44·9°N, 37·4°E), R.32, $I_0 = $ VII? (MJD), $I_0 = $ VI—VII (GG, PV).
At Anapa, Nikolayevskaya and neighbouring places old houses injured, cracks in walls, fall of some chimneys (GC, PV, MO, MMC).

1842

Jan. 2, 40·5°N, 50·0°E (BZ), R.32, $I_0 = $ IX (MJD), $I_0 = $ VIII; $M = $ 5·2 (AV).
A severe earthquake in the Apsheron peninsula, at Mashtagi all the 700 houses ruined, near Surakhani a fissure in the ground, felt at Nardoran, Kurdakhani VI?, Zabrat, Baku (strong) (BZ).

April 18, $08^h 30^m$? (GG), $18^h 17^m$ (MMC), $36\frac{1}{2}$°N, $22\frac{1}{4}$°E, R.26b, $h = i$, $I_0 = $ VIII—X (GGK), $I_0 = $ IX (GG), $I_0 = $ IX—X (MF).
A damaging shock in Greece, houses and churches injured at Kalámai and Androusa, in the province of Mani some inhabitants were crushed beneath the ruins, at Pátrai little damage, felt at Sparti and Athínai (PAO, MMC).

Aug. 31, $18^h 30^m$ UT, 46·5°N, 17·0°E, R.21b, $I_0 = $ VII—VIII, $I_0 = $ VII (CD), $I_0 = $ VIII (RA), local?

At Nagykanizsa e⸱en in the most solid houses (monastery) walls and vaults heavily cracked, many chimneys collapsed (RA).

Sept. 9, several houses injured, felt within the area of 6—8 leagues (MMC, MJD). July 31, many houses damaged, two chimneys and a wall collapsed (Erdbebenwarte, p. 42, V. Jg. 1905—6).

Nov. 18, $37\frac{1}{2}°$N, 15°E, R.20, h = sup?, I_0 = VII—VIII? I_0 = VIII—IX (MR), I_0 = IX—X (MF).

At Belpasso some houses thrown down, strongly felt at Nicolosi and Pedara (PAI, MMC, BI).

Dec. 4, 03^h about. Some houses injured at Alger (36,7°N, 3°E)) (MMC, GS), not mentioned by (GA, RK), I_0 = VII (MJD).

1843

Sept. 5, (41·2°N, 20·1°E), R.25, I_0 = VIII—IX? (GGK), I_0 = IX (MF, MD, MCA).

Damaging shock at Elbasan (Albania), a part of the town ruined (MD, MCA), an earthquake in Albania (PAP).

Sept. 14, 16^h, 17^h, epicentre in the sea? (42·6°N, 18·1°E), R.22a, I_0 = ?, I_0 = VIII (TD).

Violent shocks at Zegna (?) (Croatia), Ombla, in the island od Šipan (Giuppana) and in the neighbouring localities, Split, Opuzen, Slano, Rjeka, Zadar and Kotor. Estimated damage to houses was 2000 forints, felt in Hercegovina, in the island of Korčula, at Dubrovnik (PAP); tsunami at Dubrovnik and Graz (Gradac?) (ANS). Aftershocks with I_0 = VII at 16^h30^m, Oct. 2, 20^h, Oct. 6, 05^h, Dec. 25, 05^h UT (TD), strong aftershock on Sept. 15, 01^h, 13^h (PAP).

Oct. 18, $36\frac{1}{4}°$N, $27\frac{1}{2}°$E, R.26c, $h = n$, I_0 = IX? (GG), I_0 = VII—IX (GGK), 36°N, 28°E, I_0 = X—XI (ONB).

Earthquake in Ródhos (Rhodos), most violent in the island of Khálki, houses thrown down, rockslide, casualties (SAM, SAE, PAP).

Oct. 26, 11^h30^m, 40°N, 41·5°E, R.33, I_0 = VII—VIII, I_0 = VIII (MJD), I_0 = IX (ONB), I_0 = VIII, 39·9°N, 41·3°E (ANV).

A violent shock at Erzurum, chimneys thrown down, panic, 4 or 5 victims (PAP), not mentioned by (PL).

Oct. 27, at Barberino, Vernio, (Toscana, Italy), some houses vere seriously fissured (BI), I_0 = VII? I_0 = VII—VIII (MR), no data in (PAI).

Dec. 22, 15^h53^m, epicentre?, R.10b, I_0 = ?, I_0 = VII R. F., P = 45 000 km^2 (ME).

Affected places: Birkham, St. Malo, Cherbourg, Herm. Jethou (ME).

1844

Febr. 27, $10^h?30^m$, 42·6°N, 18·1°E, R.22a, I_0 = VII (TD).

Aftershock of Sept. 14, 1843; no damage reported in (PAP).

March 10, at Forli, Italy, a shock destroyed many chimneys and fissured some walls (BI), h = sup?, I_0 = VII? (MJD).

March 22, $10^h?13^m$, 43·4°N, 16·7°E, R.22a, I_0 = ?, I_0 = VII (TD), I_0 = VI (PAP).

(PAP) gives March 21, 09^h15^m, a severe shock at Zadar (Zara) which made fissures in the ceilings of some houses; on March 23 light tsunami at Dubrovnik (ANS).

1844

May 12?   41°N, 35°E ca, R.29, $I_0 = $ VIII—IX? (ONB), $I_0 = $ VIII (MJD), $I_0 = $ IX (CAM), $I_0 = $ VII (EGU).
Ankara, Osmancik (CAM), a newspaper reports on violent shocks between Ankara and Osmancik, many houses thrown down, 200 victims (PAP), epic. near Osmancik (PL).

July 17,   some damage at Palestrina (BI) and Genezzano, R.19, $I_0 = $ VII? (MJD), $I_0 = $ IX—X (MF), epic.?, $I_0 = $? (MR).

Dec. 11, $05—06^h$, 43·1°N, 05°W (RF), 42·7°N, 0°long, R.17, $I_0 = $?, $I_0 = $ VII (BRS), 42·7°N, 1·5°E, $I_0 = $ IV (MU).
A severe shock at Gabas and Louvie-Juzon in the Pyrenees (GS).

1845

May 24,   in the night, 41·6°N, 43·5°E (AV), R.32, $I_0 = $ VII (BZ), $I_0 = $ VIII (MJD).
A strong earthquake in the district of Akhaltsikhe, many houses fissured, at Bolshaya Kondura one house collapsed, 4 victims (BZ, MO).

June 23,   (38·6°N 27·5°E), R·26 c, $I_0 = $ VIII — IX(GGK).
Heavy destruction at Manisa, some damage and rockfalls at Izmir (Smyrna) and surroundings (GGK).

Oct. 11,   39·1°N, 26·2°E, R.26c, $h = n$, $I_0 = $ X?, $I_0 = $ IX—XI (GGK), $I_0 = $ XI, $r = $ 320 km (GG), $I_0 = $ IX (CAM), $I_0 = $ VIII (MJD), $I_0 = $ X (SAM).
Rockslides (GG, SAM), (PAP, CAM, MJD) give Oct. 15, 04^h45^m; Oct. 9, $I_0 = $ VI (EGU).
The village Lisvori (Liskoli) almost completely ruined, at Visari several houses and the church half ruined, at Plomárion (Ploumari) 8 houses collapsed, 60 houses injured, at Mitilíni several houses injured, at Ayiássos the church and some houses fissured, at Acras nine houses completely collapsed, large rockslide near Priscia (?) and on the Olimbos (Olympus), violent at Khíos, Foglieri (?), Karaburun and also at Istanbul and Izmir (Smyrna) (PAP).

Dec. 21, 20^h40^m, 46·1°N, 14·5°E, $h = 7$ km, R.18, $I_0 = $ VII—VIII, $r = 125$ km, $M = 4·8$ (RV), $I_0 = $ VII (MJD).
At Ljubljana all houses damaged, 50 chimneys fell down (RV), felt strongly at Rijeka (Fiume), felt at Trieste, Venezia and Klagenfurt (BI, PAP).

1846

Jan. 11,   some houses were damaged at Nakhichevan (39·2°N, 45·4°E), R.32 (MO), $I_0 = $?, $I_0 = $ VII (MJD).

March 28, 15^h, 36°N, 25°E, $h = i$, R.26b, $I_0 = $ VII—VIII (GGK, MJD), $I_0 = $ VII, $r = $ 1100 km (GG), $I_0 = $ IX—X (MF).
Considerable damage in Kríti (Crete), 100 houses at Iráklion (Heraklion) heavily injured, at Khaniá 20 houses injured, severe in Malta and Gozo, felt in S. Sporades, in Ródhos (Rhodos), in Syria, at Iskandariya (Alexandria) and Al Quahira (Cairo), in the direction to the west over Greece, Zákinthos (Zante), Sicily, Lecce and Napoli (SAM, BI, SF).

June 10,   37°N, 22°E, $h = n$, R.26a, $I_0 = $ X, $r = 470$ km (GG), $I_0 = $ XI (MF), $I_0 = $ X (CAM), isos. map in (GGK).

Epic. area Messíni—Mikromani—Mesopotamia (Aslanaga); fissures, sand craters, casualties, a very strong aftershock 10 hours later (GG). (PAP) gives June 11, 04^h, the most violent shock, almost all houses ruined at Nisi, Mikromani (fissures in the ground, sand craters), Asprochoma, Kalamos, Mesopotamia, Balion and Garizogli, partly demolished were Vastas, Gliata, Kourtkousi, Phoutzala, Pharmezi, Delimeri, Veis-Aga, Kalami, Katsikon (Katzikovi), Hospitakia, Kartepoli, Kalámai, Peperiksa, Anaziri, Aristodimion, Kalamaria, Alepochori and Mavromati; at Kalámai (Calamatta) all houses injured, fissures in the ground, sand ejections, at Athínai (?) 28 or 30 victims, in total 2500 houses ruined, more details in (GGM, SF), highest intensities observed on soft, alluvial ground.

June 21, 38·5°N, 27°E, R.26c, $I_0 =$ VIII (GG, MJD), $I_0 =$ VII—VIII (GGK), $I_0 =$ VIII to IX (ONB), $I_0 =$ IX (CAM).
A heavy shock in Sámos and in the neighbouring peninsula of Samsun-Dağ (Mikale), a rockslide near Kerkis (GG, SAM), (GG, GGK) give June 13.

July 29, 20^{h}40^m, 50·2°N, 5·4°E, R.9, $I_0 =$ VII, $P =$ 212 000 km^2 (SW, MJD).
A damaging shock in the Rhine valley, highest intensities of VII (walls fissured, fall of chimneys, damaged roofs, fall of plaster) observed in the districts of St. Goar (Badenhard, Biebernheim, Kaub, Norath, St. Goarshausen, Werlau) and Simmern, at Ketting, Ehrenbreitstein, Mainz and Kostheim, (SW)) gives an isoseismal map and a detailed list of localities shaken with intensities VI, V an IV, respectively (SW, TL, RU).

Aug. 9, (40·4°N, 16·4°E), R.20, $I_0 =$ VII? (MJD), $I_0 =$ VI (MR), $I_0 =$ IX—X (MF).
At Craco, Campomaggiore and Potenza houses fissured (BI, MR, PAI).

Aug. 14, 16^h? 43·4°N, 10·6°E, R.19, $I_0 =$ IX—X (MJD), $I_0 =$ X(MR, CAM), $I_0 =$ XI (MF), isos. map in (BI).
Disastrous shock in Toscana, at Orciano (98% houses ruined), Luciana, Monte Scudaio, Lorenzana, Castelnuovo di Mis. and Riparbella many houses destroyed, the remainig ones were seriously damaged, 56 victims, felt at Siena, Firenze, Milano, Genova, further details in (BI, PAI, MF).

Aug. 17, (46·8°N, 6·6°E), R.18a, $I_0 =$ VII? (MJD), $I_0 =$ VIII (MFS).
Two chimneys fell and walls fissured at Yverdon, at Orbe the bells sounded, felt throughout the canton of Vaud (MFS).

1847

April 7, 18^{h}30^m UT, 50·4°N, 11·0°E), R.6? $I_0 =$ VII? (MJD).
A damaging shock in the region of Thuringia (chimneys fell, walls were fissured), reports from Eisfeld, Coburg, Grabfeld, Gera, felt also in N. Bavaria at Altdorf near Nürnberg IV—V, at Adelheim, Weiherhaus, Feucht IV, Burgthann and Pühlheim III (SW, PAD).

Aug. 7, (29½°N, 30½°E), R.38, $I_0 =$ IX—X, $I_0 =$ VIII (MJD), $I_0 =$ IX (CAM), an isoseismal map in (SAM).
A disastrous earthquake in the region of El Faiyûm (Egypt), at Faiyûm 3000 houses and 42 mosques destroyed, in the province of Central Egypt about 1000 houses and 27 mosques ruined, heavy damage in El Quahira (Cairo), the European-style buildings, however, withstood; at Barkûks the upper part of the mosque collapsed, at Gamalieh three walls collapsed, at Gamamieh 16 houses and 2 mosques partially destroyed,

1847

at Esbekîje one house and one wall collapsed, 40 houses partially ruined, at Kalifa 2 houses and a mill destroyed, houses and mosques were injured also at Abdîne, Bâb-el-Charie, Kism et-Kaissua, Fostat and El Iskandariya (Alexandria) (SAM).

Oct. 15, 05^h, $46 \cdot 2°N$, $21 \cdot 3°E$, R.21b, $I_0 = $ VIII? $I_0 = $ VII—VIII (RA), $I_0 = $ IX (PRA), local?

An earthquake felt at Arad and surroundings (FA), chimneys thrown down, walls cracked in larger houses (RA).

1848

Jan. 11, $(37\frac{1}{4}°N, 15\frac{1}{4}°E)$, R.20, $I_0 = $ VIII—IX?, $I_0 = $ VIII (MJD), $I_{max} = $ VII—VIII? (MR), $I_0 = $ XI (MF).

Epicentre in the sea in front of Augusta, Sicily (MR); the town of Augusta almost completely ruined, only 27 houses stood up; serious damage at Siracusa, Noto and Catania, strong at Messina (MF, BI, PAD).

Febr. 11, $35 \cdot 2°N$, $3 \cdot 0°W$, R.16? $I_0 = $?, $I_0 = $ VIII (MU).

A "very destructive" earthquake felt in Morocco (Maroc), great damage done at Melilla, accompanied by a violent uragan (GS, RG), during March and April numerous shocks felt at Melilla where they caused some damage (RG).

April 23? According to a letter from the year 1890 there was a strong earthquake in Sofia, R.23, where some buildings fell (WS, MF), $I_0 = $ VIII (GGB), $I_0 = $ IX—X (FM), epic.?

Sept. 24, $(40 \cdot 7°N, 48 \cdot 6°E)$, R.32, $I_0 = $ VII (MJD).

At Shemakha, E. Caucasus, in some local houses the walls fissured, the plaster fell, the vault of a mosque was injured (BZ, MO).

Oct. 2, a foreshock to the following earthquake, $I_0 = $?, slight oscillations (GS), $I_0 = $ II (MU), $I_0 = $ VIII (BRS).

Oct. 3, 11^h30^m, 15^h30^m, 18^h, $40 \cdot 3°N$, $1 \cdot 3°W$, R.17, $I_0 = $ VII—VIII, $I_0 = $ VIII (BRS, GS, MU), $42 \cdot 0°N$, $1 \cdot 1°W$ (BRS), $I_0 = $ VI (MU), $I_0 = $ IX—X (MF).

A violent shock in the region of Teruel, at Tramacastilla and Monterde de Albarracin some tall buildings, especially the church, suffered much, walls cracked in many parts, at Noguera the tower of the church heavily injured, some houses damaged, strongly felt at Orihuela del Tremedal, felt at Torres de Albarracin (GS).

1849

Nov. 28, 18^h UT, R.19, $I_0 = $ VII (MR, MJD), Febr. 28 (MR).

At Borgotaro almost all houses were fissured and chimneys thrown down, felt at Pontremoli, Pisa (BI, MR).

1850

Date? $39 \cdot 9°N$, $41 \cdot 3°E$, R.33, $I_0 = $ VII, region of Erzurum (EGU).

Date? $38 \cdot 4°N$, $27 \cdot 4°E$, R.26c, $I_0 = $ VIII, Bati Anadolu in SW Anatolia (EGU).

Jan. 1, $37\frac{1}{2}°N$, $15°E$, R.20, $I_0 = $ VIII (MJD), local, $I_0 = $ VIII—IX (MR), $I_0 = $ IX (CAM), $I_0 = $ IX—X (MF), $h = $ sup?

At Belpasso and Biancavilla (Etna) some houses were ruined, felt at Catania (BI).

1850

Febr. 9, night, 36·3°N, 4·8°E, R.15, I_0 = VIII?, I_0 = IX (GA), I_0 = VII (MJD), local?
Houses destroyed at Zamora-el-Guenzet, some damage at Bordj-Bou-Arréridj, felt at Alger and Douéra (RK, GS).

April 3, many houses damaged at Izmir, Mytilíni, Aydin, Nymphio, Khíos (LC), I_0 = VII (MJD), epic.?

April 14, 00^h50^m, 42·8°N, 17·7°E, R.22a, I_0 = VIII—IX, I_0 = IX, r_{max} = 243 km (TD), April 21, I_0 = X (MF).
Five houses completely ruined, 50 partly destroyed at Ston near Dubrovnik, Dubrovnik VII, felt at Zadar and Kotor (LC).
Aftershock with I_0 < IX on April 29, 09^h, r_{max} = 42 km (TD), at Ston all houses rended uninhabitable.

Aug. 19, 07^h15^m UT, 42·6°N, 18·1°E, R.22a, I_0 = VIII? (TD), I_0 = VII (MJD).
The walls and one house collapsed at Mali Ston near Dubrovnik, Imotski heavily shaken.

Sept. 9, 07^h40^m, (52·4°N, 4·6°E), R.9, I_0 = VI—VII? local, h = sup?
A relatively strong shock felt at six localities around Haarlem, some walls fissured and some chimneys thrown down at Bloemendaal, Hillegom, Veenenburg and Zandvoort (TL, RU).

Nov. 11, 09^h UT, R.19, I_0 = VII? (MJD), local.
At S. Nicandro (Aquila) some chimneys fell down and all houses were fissured, rockfalls from Mt. Castello (BI).

Nov. 17, 16—17^h, 43·1°N, 0·2°W, R.17, I_0 = VI—VII? I_0 = VII (MJD).
Largely felt earthquake, the ceilings cracked in houses at Bagnères (RF), strongly felt at Lourdes, felt at Tarbes, Argelès, Cauterets, St. Pé, Lestelle, Bétharram, Loubajac (GS), violent shaking, no damage (GS).

Dec. 17, 12^h30^m, 36·5°N, 7·4^{h}E, R.15, I_0 = VI—VII? (GA, RK, GS), I_0 = VII (MJD).
Walls fissured at Guelma, strongly felt at Millesimo, Petit, Héliopolis, Bône (RK).

1851

Date? 40°N, 48·4°E, R.32, 34, I_0 = VII?, M. Kemalpaşa (TM).

Jan. 20, (41·2°N, 20·2°E), R.25, I_0 = VIII—IX?
A disastrous earthquake at Elbasan; Vlorë and Tepelenë heavily shaken (VII?) (GGK).

Febr. 5, a damaging shock in the Alps, in the region of Milano houses were fissured, at Bagano chimneys collapsed, felt in Tirol, in Switzerland (Zürich, Luzern, Bern, Basel, Grund) (LC), I_0 = VII (MJD), not mentioned by (MFS, TT), erroneous?

Febr. 28, 15^h, $36\frac{1}{2}$°N, $28\frac{3}{4}$°E, R.26c, I_0 = IX (GG), I_0 = VIII—IX (GGK), I_0 = IX—X (MJD), I_0 = X (CAM), I_0 = VI (EGU).
Destructive shock in Ródhos (Rhodos), Mákri and Kaya (Levisi) in Asia Minor suffered largely, earthslides on Buba-Dağ destroyed 14 villages (GG, SAM, LC, SF), tsunami at Fethiye much higher on April 3 (ANS). According to (LC) destructive aftershocks on March 5 (Mákri), April 3 (Mákri), April 4 (Ródhos).

May 15, 01^h47^m UT, 39·6°N, 2·8°E, R.15, I_0 = VIII (MU, MJD, GS), I_0 = IX (CAM), I_0 = IX—X (MF).

Violent shock in the Mallorca island caused serious damage in the epicentral area NE of Palma, slightly felt in the southern part of the island ($h = $ sup?), many severe aftershocks in 1851 and 1852, e. g. that on May 22, 04h, almost destroyed the church at San Marçial.

July 1, 21^h15^m UT, 47·8°N, 18·1°E, R.21a, $I_0 = $ VII—VIII, $I_0 = $ VII (CD, MJD), $I_0 = $ = VIII (KC), $I_0 = $ VIII—IX, $r_{max} = $ 95 km (RA).

A severe shock in Komárno where many chimneys collapsed and walls of all houses were fissured, felt at Budapest, Mór and Balassagyarmat (RA), Bratislava (KC).

Aug. 14. 01^h23^m UT, 41°N, 15·7°E, R.19, $I_0 = $ X(CAM), $I_0 = $ X—XI (MR), $I_0 = $ IX—X (MJD), $I_0 = $ XI (MF), isos. map in (BI).

Melfi completely ruined, 1—2 m cracks in the ground, similar disastrous effects also at Barile (partly ruined), Rionero, Rapolla, Atella, Venosa, Lavello, Monteverde, felt at Lecce, Ischia, Caserta, for details see (BI), 62 victims (LC).

Oct. 12, 07^h? ($40\frac{1}{2}$°N, $19\frac{3}{4}$°E), R.25, $I_0 = $ IX—X (GGK), $I_0 = $ VIII (MJD), $I_0 = $ X (MD, CAM, MF, MA), $P = $ 600 000 km².

A sequence of disastrous shocks ruined the towns Vlorë and Berat in Albania, all houses seriously damaged, some of them completely destroyed, 2000 victims, IX—X at Narta, VIII—IX at Smokthinë, Velçan, Berat, Karbunarë, Elbasan, Tepelenë, Dragot, Peshtani, Maricaj, Vasiari, Klisyra, Turani, Dukaj, VII — Delvinë, Tiranë, V — Ioánnina (Janina), felt in Italy at Lecce, Bari, Taranto, Barletta, Canosa, V — Cerignola and Napoli (MDC, MCA), tsunami at Vlorë (Valona) (MA, ANS).

Oct. 17, aftershock, R.25, $I_0 = $ IX?, $I_0 = $ IX—X (GGK), $I_0 = $ X (MD, MA).

A new disastrous shock almost completely destroyed Berat (40·7°N, 20·0°E), 400 victims, rockslide, VII — Vlorë, Elbasan (MD, MCA).

Oct. 20. 09^h30^m, aftershock, R.25, $I_0 = $ VIII? $I_0 = $ IX (MD).

A severe shock partly destroyed Elbasan, VI — Tepelenë, Vlorë (MD, MCA).

Nov. 22, 09^h30^m, 35·4°N, 0·1°E, R.25, $I_0 = $ VIII (GA), $I_0 = $ IX (MF).

A severe shock in the province of Oran, at Mascara various houses fissured or ruined, panic (GS); the French houses were less or more damaged, three collapsed (RK).

Dec. 29, 11^h, 17^h, 40·7°N, 20·0°E, aftershock, R.25, $I_0 = $ VIII—IX (GGK), $I_0 = $ IX (MD).

Two violent shocks ruined various houses at Berat and surroundings, casualties, large fissures in the ground (also in the mountains), at 11^h: VI — Cfiri, Klos, Velçan, Osman Zera, Balaj, Krute, Fier, Benja, V — Vlorë, Elbasan, at 17^h: VI — Vlorë, Elbasan (MD, MCA).

1852

May 12, Sept. 8, rather strong tsunamis at Smyrna (ANS).

July 24, 39·9°N, 41·3°E, R.33, $I_0 = $ IX (ANV, CAM), $I_0 = $ X, $M = $ 7·3 (ONB), $I_0 = $ IX to X (MJD).

Only Erzurum is mentioned in (PL) without additional information; 300 houses thrown down, the other seriously injured (PAT, LC), 17 victims (MO). Aftershock on Aug. 29, $I_0 = $ IX? $M = $ 6·7 (ONB), $I_0 = $ VII (ANV).

1852

Aug. 26, $(41\frac{1}{4}°N, 19\frac{1}{4}°E)$, R.25, $I_0 = $ VIII–IX (GGK), $I_0 = $ VIII (MJD), $I_0 = $ IX (MD, CAM), epic. in the sea?
The monastery of St. Antonio near Kep i Rodonit (Cap Rodoni) was destroyed by an earthquake with the epicentre near Shiak, VIII at Durrës (GGK).

Oct. 19, 03^h25^m, $(36.6°N, 29.1°E)$, R.26, $I_0 = $ VII? $M = 5.5$ (ONB).
Fethiye; at Çeşme a "terrible" shock (PL, PAT).

Dec. 9, $(41.7°N, 15.3°E)$, R.19, $I_0 = $ VII? (MR).
At Torre Maggiore an old arc collapsed, some small fissures observed at Apricena (BI, PAT).

1853

Jan. 6, 06^h50^m, $42.2°N, 0.9°E$, R.17, $I_0 = $ VII? (MU, GS).
A severe shock in Catalonia, strongly felt at Foix, Tremp; rockslides, large extension (GS).

Jan. 16, 01^h30^m UT, $45.9°N, 15.6°E$, $h = 5$ km, R.22b, $I_0 = $ VII, $r_7 = 6$ km, $r_5 = 16$ km, $r = 65$ km, $M = 4.2$ (RV).
At Zagreb chimneys collapsed (LC).

Jan. 18, 06^h, $40.6°N, 45.0°E$ (BZ), R.34, $I_0 = $ VIII (MJD).
A severe shock in Armenia at Cubuklu (Chubukhli) several houses and a wall of the church destroyed, 12 houses heavily damaged, at Dilizhan strong vibrations, felt at Tbilisi and in the island of Sevan (BZ, MO), no remark in (PL).

Febr. 19, all walls fissured at Amaro, Friuli, Carnia (MR), R.18b, $I_0 = $ VII? (MR, MJD).

April 1, 22^h45^m UT, $49°N, 2°W$, R.11, $I_{max} = $ VII (RF, MJD), $I_0 = $ VII–VIII, $r = 250$ km (PRF), $I_0 = $ VIII? (RF, ME).
Epicentre in the La Manche region between Jersey and Cotentin, damage observed at Coutances and Granville (RF), felt at Alderney, Sark, Chartres, etc., see (ME).

April 9, $40.9°N, 15.2°E$, R.19, $I_0 = $ VIII? (MJD), $I_0 = $ IX (MR, CAM), $I_0 = $ X (MF), isos. map in (BI).
At Teora, Caposele, Lioni, Calabritto several houses were partly ruined, the others largely fissured. Felt at Foggia, Melfi, Potenza, in the provinces of Napoli and Salerno (BI, MF, MR, CAM), many victims at Caposele, fissures in the ground, rockslides (LC).

Aug. 2, at Pomarance $(43.3°N, 10.8°E)$, R.19, several walls fissured and chimneys thrown down, violent at Volterra (BI), $I_0 = $ VII?, $I_0 = $ VII (MJD).

Aug. 11, R.18a, $I_0 = $ VII (MJD), $I_0 = $ VIII (MFS), $h = $ sup?
A strong, local shock at Soleure where 300 chimneys were thrown down and walls and ceilings fissured (MFS), some chimneys fell at Bechburg, walls damaged at Solothurn.

Aug. 18, 8^h30^m, $38\frac{1}{4}°N, 23\frac{1}{2}°E$, $h = n$, R.26a, $I_0 = $ IX–X (MJD), $I_0 = $ IX, $r = 530$ km (GGK, GG), $I_0 = $ X (CAM), $I_0 = $ XI (MF), isos. map in (SAM), p. 64.
At Thebes many houses and churches collapsed (SF), large landslides, tsunamis, casualties; according to the map of (SAM) $I = $ IX at Thebes, $I = $ VIII at Thespiaí, Khálki (Chalkis), $I = $ VII at Athínai, etc. Thebes partly ruined, great damage at Atalanti and Khálki, 17 victims, felt at Pátrai and Bursa (Brousse)(MF), strong

1853

tsunami in the Euboean Gulf (ANS).

A very strong aftershock on Sept. 29 (GG), $I_0 = $ VIII? $I_0 = $ IX (CAM).

Oct. 20, 15^h25^mUT, 43·4°N, 17·8°E, R.22a, $I_0 = $ VII? (MJD), $I_0 = $ VIII, $r = $ 82 km (TD).
At Mostar much damage, felt at Dubrovnik;
Oct. 10? (JM, MJD, LC).

Dec. 11, 15^h05^m UT, 42·4°N, 18·6°E, R.22a, $I_0 = $ VII? (TD).
At Dubrovnik walls fissured, at Hercegnovi, Baošič, Biela some houses damaged (BA).

1854

Febr. 12, (18^h30^m), 43°N, 12°E, R.19, $I_0 = $ VIII—IX, $I_0 = $ IX (MR, PET, CAM), $I_0 = $ X (MF), isos. map in (BI).
A violent shock in the region or Perugia, Foligno and Assisi, mainly at Bastia (many houses destroyed) and Spello, many houses fissured at Perugia (BI, PAT, MF).

Febr. 12,   39·3°N, 16·2°E, R.20, $I_0 = $ IX—X (MR, PET), $I_0 = $ VIII (MJD), $I_0 = $ X (CAM), $I_0 = $ X—XI (MF), $h = $ sup?, isos. map in (BI).
A disastrous shock in Calabria. At Cosenza some houses ruined, all were shattered and fissured, churches partly destroyed, 41 victims; at Donnici Soprano and Sottano (191 victims) all houses and 6 churches were destroyed, great damage at Torzano, Piane Crati, S. Ippolyto, Trenta, Spezzano Grande, Zumpano, S. Filli, etc. (BI), felt at Napoli and Reggio Calabria (BI, MF, MR, PAT, LC).

July 20, 02^h45^m UT, 43·0°N, 0·1°W, R.17, $I_0 = $ VII—VIII, $r = $ 380 km (RF, PRF), $I_0 = $ VII (MJD), $I_0 = $ IX (MF).
A severe shock in the Pyrenees, 100 houses fissured at Bagnères and Gripp, the church at Lourdes and many other churches injured, particularly at Saint-Savin, houses damaged at Argèles (RF, PAT), in Spain felt at San Sebastian, Vich and Navarra (GS), for details see (PAT).
(GS) and (MU) give also a shock on July 26 with $I_0 = $ VII—VIII in the Pyrenees, 42·7°N, 1·5°E, felt at Buerdos, Montpellier and Barcelona, the report probably corresponds to July 20, (MJD, PAT) give July 31.

July 30, 01^h30^m, $39\frac{1}{2}$°N, $20\frac{1}{2}$°E, R.25, $I_0 = $ VIII—IX (GGK), $I_0 = $ VII (MJD), $I_0 = $ IX (GG), $I_0 = $ X (MD, CAM), $I_0 = $ X—XI (MF), $P = $ 280 000 km².
Disastrous shock in the region of Delvinë (IX), Puli partly destroyed, casualties, Gziq VI—VII (MD, MCA).

Dec. 29, 02^h45^m, 43·8°N, 7·9°E, epic. in the sea?, R. 18a, $I_0 = $ IX, $r = $ 250 km (RF, PRF, RO), $I_0 = $ IX—X (MF), isos. map in (BI).
A disastrous shock in Liguria, in France at Cagnes the castle was injured, at Bar the gothic tower fell and several houses fissured, at Saint-Paul several houses damaged, at Grasse walls fissured, at Menton and Vintimille many houses thrown down, at the village at Comba all houses thrown down, considerable damage at Cannes, Grasse, Antibes (RO, PAT, LC), minor damage at Nice (walls and ceilings fissured) (RF), in Italy Oneglia, Poggio, Taggia, Bussana and Bordighera were much damaged, VIII (MR), felt also at Draguignan, Mandelieu, Toulon, Brignoles, Torino, Milano, Pavia, Porto-Maurizio, Rogliano in Corsica, Chambéry (RO, BI).

1855

Jan. 26, 37·7°N, 15·1°E, R.20, I_0 = VII (MR, MJD), h = sup?
At Fondo di Macchia the vault of a church and the walls of many houses fissured (PAT, BI).

Febr. 5, 15^h—16^h, 23 houses thrown down at Melouza, Algeria, after a violent shock (but also after a strong wind and a hail-strom), large cracks in the ground, felt in the plain and in the mountains, too (PAT), not given by (RK, GA), questionable.

Febr. 28, 03^h, 40·2°N, 29·1°E, R.27, I_0 = IX, M = 6·7 (ONB, CAM), I_0 = VIII (MJD), I_0 = X (EGU), isos. map in (SAM).
A destructive shock in NW Turkey, most severe az Bursa (Brussa), felt at Gelibolu (Gallipoli) (CAM), Izmir (Smyrna), Lüleburgaz (Loule-Boarga), Edirne (Adrinople); at Bursa many mosques and houses collapsed, thermal springs ceased for 6 hours, 300 victims, the ground was cracked (PAT), houses thrown down in the Ródhos (LC), 1600 victims, aftershocks till June (EGU).

March 2, 34·8°N, 27·2°E, R.26c, I_0 = VII? (ONB), I_0 = IX (CAM).
Felt at Izmir (Smyrna), the village Macri (Mákri?) sank by 105 feet (PAT).

April 11, 19^h40^m, (40·2°N, 29·1°E), R.27, I_0 = IX—X?, I_0 = X (TM).
All mosques, minarets and stone-built houses collapsed at Bursa, new thermal sources appeared, felt at Izmir (Smyrna), Nasildi, Metelin, Edirne (Adrianopole), Istanbul (PAT), 400—1300 victims, large cracks in the ground, felt in the large part of Anatolia (PL), many villages near Bursa destroyed (LC). The reports for April 11 and April 29 might be confused.

April 29, 40·2°N, 29·1°E, R.27, I_0 = X (CAM), I_0 = IX—X (MJD), I_0 = IX, M = 6·7 (ONB).
Large destructions at Bursa, 160 minarets and many houses collapsed, 1300 victims, many villages destroyed particularly between Bursa and Mouhalitch, e. g. Tepeidjík, Demirtaş (PAT).

July 3, (42°N, $19\frac{1}{2}$°E), R.25, I_0 = IX? (CAM, MA), I_0 = X (MD), I_0 = VIII (MJD).
Large destruction in the region of Shkodër (Skadar), all localities in the pleistoseismal region were almost destroyed (Shkodër, Buchati, Zadrimë, Jubani, Kosmaçi, Van-Dejës), strongly felt (VII?) in Albania at Karma, Koman, Dushman, Toplani, Xhuxhë and in Yugoslavia at Titograd, Tuzi, Mataguši (MD, MCA). Many houses fissured and three thrown down at Shkodër (PAT).

July 7, aftershock, R.25, I_0 = VII? (MD).

July 16, aftershock, R.25, I_0 = VIII?
Houses greatly injured, large fissures in the ground, VIII — Shkodër (Skadar), Bushati, Juban, Kosmaci, Van-Dejës (MD, MCA).

July 25, 13^h, 46·3°N, 7·9°E, R.18a, I_0 = VIII—IX, r = 320 km (RO, PRF), I_0 = VII (MJD), I_0 = IX (SW), I_0 = X—XI (MF).
One of the most violent earthquakes in Switzerland, damage in the region between Visp and Brig and in the Nikolaital (Saint Nicolas); the churches et Viège, Torbel, Saint Nicolas and partly that at Stalden collapsed, the tops of towers of several other churches fell as well as numerous houses, important damage in some Alpine valleys, e. g. at Macugnaga and San Carlo, at Belfort several walls fissured and many chimneys fell, mud streams, rockslides, cracks in the ground in the valleys of Rhône and Viège; large extension up to Paris, Clermont-Ferrand, Parma, Verona,

Dresden etc., see map in (RO). A detailed description of this shock in (MFS) where the monograph of O. Volger is quoted, and in (PAT, BI).

Aug. 14, $(42°N, 19\frac{1}{2}°E)$, R.25, $I_0 = VIII?$ (MJD), $I_0 = IX$ (MD, CAM), $I_0 = X$ (MF). Aftershock at Shkodër, serious damage at Bushati, Juban, Kosmaci (an old palace collapsed), Van-Dejës, Zadrimë, large fissures in the grouod (MD, MCA, PAT). Another shock on Aug. 11, $I_0 = VIII?$ (MD).

Nov. 23, 15^h15^m, 43·8°N, 6·4°E, R.18a, $I_0 = VIII$ (RF), $I_0 = IX$, $r = 40$ km (RO), $I_0 = VII$ (MJD).

Serious damage near Castellane, the church of Chasteuil thrown down, large fissures in the ground (RF), rockslide, some chimneys collapsed and in three villages one church and several houses fissured (PAT); a violent aftershock on Dec. 12 caused serious damage in the same three villages as on Nov. 23, roofs thrown down, several houses cracked, at Chasteuil the church collapsed, large fissure 10 km distant from Castellane (RO), $I_0 = VII-VIII?$, $I_0 = IX$ (CAM, MF).

Dec. 18. 01^h20^m, 38·4°N, 27·2°E, R.26, $I_0 = VII$, $M = 5.5$ (ONB).

1856

Jan. 12, $10-11^h$, (37·1°N, 8·0°W), R.12, $I_0 = VIII$ (FH, MJD, MU), $I_0 = VII-VIII$ (BRS), $I_0 = X$ (MF).

A severe shock at Loule where several houses were thrown down, the earth fissured, serious damage at Faro, Albufeira, Tavira, felt at Lisboa and Sevilla (GS, PAT).

Febr. 17,   40·2°N, 29·1°E, R·27, $I_0 = VII$, $M = 5.5$ (ONB), $I_0 = IX$ (CAM).

Febr. 22,   epicentre unknown, (41·3°N, 36·3°E), R.29?, $I_0 = VIII$, $M = 6.1$ (ONB), $I_0 = IX-X$ (MJD), $I_0 = X$ (CAM).

Several villages destroyed between Karpan (?) and Korgo (?), felt at Istanbul, Varna and Samsun (PAT), no report in (WS, EGU).

May 11−12, earthquake in the province of Bari, Italy, R.20, some churches and houses slightly fissured at Aequaviva, Canosa and Corato (BI), $I_0 = VII?$ (MJD), local?

June 5, 19^h, (43·7°N, 12°E), R.19, $I_0 = VII?$ (MJD), local? $I_0 = IX$ (MF).

Considerable damage at Pieve S. Stefano (Toscana), especially on soft, water saturated ground (PAT, BI).

June 22,   47·4°N, 18·2°E, R.21, $I_0 = VII?$ $I_0 = VII-VIII$ (RA), local?

At Mór two chimneys fell and several walls fissured (RA), isolated report, not included in (CD).

July 23,   40·6°N, 48·6°E (BZ, AV), R.32, $I_0 = VII$, $I_0 = IX-X$ (MJD).

At Shemakha all houses fissured or seriously injured (Asian type of houses), felt with lesser intensity in the surroundings of Shemakha, 1 victim (BZ, MO), several houses collapsed (PAT).

Aug. 21, 22, 21^h, 12^h, 37·1°N, 5·7°E, R.15, $I_0 = IX$ and X (second shock), epic. doubtful (GA, RK), epic. in the sea? $I_0 = VII$ (MJD).

A disastrous earthquake in Algeria, Djidjelli and Collo ruined, at Philippeville even some very stable constructions thrown down, at Bougie the light-house almost collapsed, at Sétif, Batna, Bône, Constantine, Gastonville, Robertville and Gu-elma the houses injured or cracked, fissures in the ground near Philippeville, tsunami, no damage at Alger and La Calle, felt also outside Algeria at Nice, Malta, Mahon (Baleares), Carloforte, Iglesias and San Pietro in Sardinia (RK, GS, BI).

Oct. 9, 00^h45^m, 38·0°N, 1·2°W, R.16, I_0 = VII? (MJD), I_0 = V (MU).
Walls fissured at Murcia and surrounding villages (GS, PAT), not mentioned by (BRS).

Oct. 12, 00^h45^m, $35\frac{1}{2}$°N, 26°E, $h = i$, R.26b, I_0 = X−XI (GGK), I_0 = XI, r = 1450 km (GG, MF, SAM), I_0 = X (CAM), I_0 = IX−X (PK, MJD), isoseismal maps in (SAM) pp. 46, 91.
An earthquake catastrophe in Kríti (Crete), at Iráklion (Heraklion), only 18 houses from the total of 3620 houses remained inhabitable, 538 victims, Sitia almost completely destroyed, heavy damage at Ierápetra (Hierapetra) and Golf of Mera-béllou (Mirabello), at Khaniá all houses injured but only some of them collapsed, disaster in the island of Kásos, in Kárpathos all houses damaged, in Rhódos 8 villages heavily injured, elevation of the coast, casualties, tsunami, slight damage observed in Thíra (Santorin), Malta, Gozo, Karien, Cyprus, S. Syria, N. Palestina, delta of the Nile, very large shaken area extending up to Palermo, Napoli, Ancona, Zara, Athínai, Bursa? (SAM, PAT, BI), disaster at Ioánnina (Jannina) (a relais shock?) (MCA), between Pentalidion and Kyparissía VI−VII (GGK), strongly felt in Cyprus (ANC).

Nov. 9, 22^h17^m, 45·9°N, 14·5°E, h = 8 km, R.22b, I_0 = VII, M = 4·5, r = 100 km (RV, MJD).
A strong shock in the Korinthia (Carinthia), at Ljubljana the walls fissured, felt at Trieste and San Magdalena (PAT).

Nov. 13, $38\frac{1}{4}$°N, $26\frac{1}{4}$°E, $h = n$, R.26c, I_0 = VIII−IX (GGK), I_0 = IX (GG).
At Khíos (Chios) many houses collapsed, allmost all were fissured, casualties (GG, GA), tsunami (ANS); error in date? it might correspond to 1865, Nov. 11−14 given by (SAM).

1857

Febr. 13, 00^h30^m, 38·4°N, 27·2°E, R.26c, I_0 = VII, M = 5·5 (ONB).

March 7, 02^h20^m UT, R.22b?, an earthquake near Ljubljana, many houses fissured at Klagenfurt, felt in all Slovenia, I_{max} = VII (RV), in Italy at Padova (BI), I_0 = VII? (MJD). Ribarič assumes an exchange of date for the event of Dec. 25, epicentre Rosegg, Austria.

April 9, 38·4°N, 42·1°E, R.33, I_0 = IX? M = 6·7 (ONB, CAM), I_0 = VIII (MJD), 39·0°N 41·6°E (ANV), not in (EGU).
Muş, district of Hinis, 4 villages in the plain of Bulanik destroyed (PAT).

Aug. 28, epicentre? R.18a, I_0 = VII? I_0 = VIII (MFS).
An earthquake in the Lower Engadine, tiles thrown down and walls fissured at Tarasp, Vulpera and Fetan (MFS).

Sept. 17, 22^h, (40·2°N, 29°E), R.27, I_0 = ? I_0 = VII (MJD), I_0 = IV (ONB), I_0 = VI (EGU).
A violent shock in Turkey, at Bursa (Brousse) 30 houses and two palaces thrown down (another reporter speaks only about the collapse of an old house), epic. near Yalova (PAT).

1857

Oct. 27,   according to a newspaper report a damaging shock in Azerbaydzhan, the little town Tesong(?) completely ruined, in Georgia the river Kura inondated several villages (PAT), very doubtful, not mentioned by (BZ), $I_0 =$ VIII (MJD).

Dec. 16, 21^b UT, $40\cdot3°$N, $16\cdot0°$E, R.20, $I_0 =$ X—XI, $I_0 =$ IX—X (MJD), $I_0 =$ X (CAM, PET), $I_0 =$ XI (MF, MU), isos. map in (BI).
A catastrophic shock in Italy, about $11\,000-24\,000(?)$ victims, most affected area (total destruction) in the triangle Montemurro—Sapponara—Viggiano, deformations of the ground, Tramutola (totally ruined), Marsico, Spinoso, Sarconi, Polla, Padula, Pertosa, Sassano, Montesano, Brienza, Calvello (total destruction), Sant' Angelo le Fratte, Puerno, Tito, Potenza, houses were thrown down at Laurenzana, Corleto, Perticara, Guardia, Aliano, Armento, Gallicchio, Missanello, S. Arcangelo, etc. (BI). Extended up to Reggio in Calabria, Brindisi, Vasto, Terracina, Napoli (PAT).

Dec. 25, 01^h30^m UT, $46\cdot6°$N, $14\cdot0°$E, R.18, $I_0 =$ VII.
Epic. near Rosegg (Kärnten) (TT), where allmost all walls were seriously damaged, heavy at Ossiach (PAT), see March 7, 1857.

1858

Jan. 15, 19^h15^m UT, $49\cdot2°$N, $18\cdot8°$E, R.21a, $I_0 =$ VIII—IX, $I_0 =$ IX (KC), $I_0 =$ IX—X (RA, MF), $I_0 =$ VII (MJD), $r = 145$ km about (KR, RA), $P = 78\,200$ km^2 (PAT), $P = 66\,070$ km^2 (SW).
A violent earthquake in the region of Žilina, seriously injured buildings at Divina, Višňové (the churches seriously damaged), Žilina (all houses cracked, stone houses in the centre of the town became uninhabitable, the chimneys fell), at Bánová, Teplička nad Váhom, Domaniža, Kotešévá, Predmér, Povážská Bystrica, Zigmundov, Púchov, Lieskovec, Klačno, Nitrianské Pravno, Martin, Párnica, Kláštor pod Znievom, the walls of houses or churches cracked. Detailed description of the shock in (RA), where intensity classifications are, however, overstimated by about one grade.

Febr. 21, 09^h, $38°$N, $23°$E, $h =$ sup? R.26a, $I_0 =$ IX—X (GGK), $I_0 =$ X (GG) $I_0 =$ XI (MF), $I_0 =$ VI (MJD).
Total destruction of the towns Kalamaki, Hexamilia, Perighiali and Old-Corinth, fissures in the ground at Kalamaki, felt (III—IV) at Athínai (Athens) (PAT), small shaken area, some victims (GGK).

March 8,   A shock caused new ruines at Tramutola near Potenza (Italy) (PAT), ($40\cdot3°$N, $15\cdot8°$E), R.19, $I_0 = ?$

April 5, $39\frac{3}{4}°$N, $20\frac{3}{4}°$E, $h = n$, R.26a, $I_0 =$ VII—IX (GGK), $I_0 =$ VIII (GG), $I_0 =$ IX (MF). A disastrous shock at Ioánnina (Jannina) (MCA), not mentioned by (MD).

Sept. 20, $40°$N, $20°$E, R.25, $I_0 \leq$ IX, $r = 400$ km (?) (GG), $I_0 =$ X (CAM, MA).
Serious destructions at Delvinë, felt strongly at Kuçi, Zhulat, Fushë-Bardhë, felt at Pontepigadia, Arta and Préveza, some destruction reported also from Sofia (MCA, MD, PAT), this report about Sofia corresponds probably to the following shock.

Sept. 30—Oct. 12, ($42\frac{1}{2}°$N, $23\frac{1}{4}°$E), R.24, $I_0 =$ IX (GGB), $I_0 =$ X—XI (MF).
A newspaper report indicates $35-40$ houses and 20 minarets collapsed, all houses injured and casualties at Sofia, another report says that the "djamiyas" and five of

seven churches were so seriously injured that religious services could not be performed for danger of collapse of some parts, 4 persons perished beneath the collapsed walls, changes in the ground water sources, long fissures in the mountains of Vitosha between the villages Bojana and Dragalevci, main shock on Sept. 30, the inhabitants lived outside the town for several days, rockfall on Vitosha (WS).

Oct. 10, 09^h30^m, 40°N, 20·1°E, R.25, $I_0 =$ IX−X (GGK), $I_0 =$ X(MD), $P =$ 310 000 km^2. A disastrous shock at Delvinë, Vlorë and Kanino, heavy injuries also at Smokthine, Velçan, Narta, Karbunarë, Tepelenë, Dragoti, Peshtani, Maricaj, Vasiari, Fushë-Bardhë, Këlcyrë, Turani, felt strongly in Italy at Lecce, Taranto, Bari, Canosa, Gioia (MD, BI, MCA), 1556 houses thrown down at 31 localities of the district of Delvinë (Kuç — 180 houses, Zhulat — 60 houses, Fushë-Bardhë — 11 houses, Gjirokastër — 6 houses), severe shocks every day between Sept. 20 and Oct. 10 (PAT). The reports on effects corresponding to the shocks from this period might be confused.

Oct. 11, 07^h05^m UT, 38·2°N, 9·0°W, R.12, $I_0 =$ IX, $M =$ 7·2 (FH, BRS), $I_0 =$ X (MP), 38·7°N, 9°W, $I_0 =$ VIII (MJD), isos. map (CMP), 38·7°N, 9·2°W, $I_0 =$ IX R. F. (MU).
A violent shock in Portugal, at Setúbal a part of the town ruined, many houses injured, IX — Azeitao, Sesimbra (Cezimbra), Alcácer-do-Sal, VIII — Sines, Santiago de Caceum, Evora, Montemor-o-Novo, Almada, Borba, Sintra, Sacavem, Leiria, Alquiedao, Sto. Amaro, Thomar, Lisboa, Algarve, Faro, Lagos, VII — Tavira, Olivarez, Mafra, Cartaxo, Santarém, Abrantes, Coimbra, VI — Aveiro, Oliveira-de-Arenas, Pôrto (Oporto), Braga, Caminha, Valenca de Douro, Vila Real, etc. (GS), damage also at Belem and Grandina, felt in Spain at Madrid (strong), Sevilla (violent), Cauceres, Huelva, Cadiz (PAT).

Oct. 25, 02^h42^m UT, 45·1°N, 7·8°E, R.18a, $I_0 =$?, $I_0 =$ VII (MJD), $I_0 =$ VIII (MR), $I_0 =$ IX−X (MF), local?
Houses destroyed at Abbadia, felt at Pinerolo (Pignerol), Cavour and surrounding localities (PAT), no damage reported by (RO).

Nov. 28,   44·4°N, 19·0°E, R.22b, $I_0 =$?, $I_0 =$ VIII (MJD), $I_0 =$ IX (CAM), $I_0 =$ IX−X (MF).
"At Tuzla (Bosnia) a terrible shock threw down the houses" (PAT, MJD). A disastrous shock at Ergheni (Albania), many houses collapsed (PAT, GS); no data in (TD, MD).

Nov. 29, 13^h, 43·3°N, 1·2°W, R.17, $I_0 =$ VII? (MJD, GS), $I_0 =$ VIII (BRS), $I_0 =$ VI−VII (RF).
A severe shock in the Pyrenees, one chimney and tiles from roofs fell at Saint-Jean-Pied-de-Port, old walls collapsed at Lecumberry, in the castles of M. Etcheverry the walls fissured, felt in the districts of Bayonne, Mauléon and Orthez, at Saint-Palais, Ostabat, Saliès, Dax, Pau (PAT, GS).

Dec. 24 08^h30^m UT, 45·9°N, 16·1°E, R.22, $I_0 =$ VII? (TD), not mentioned in (PAT), questionable; Kašina VII, no description of damage.

1859

Jan. 20 07^h55^m UT, 46·8°N, 12·1°E, R.18b, $I_0 =$ VII? (MR, MJD), $I_0 =$ IX (MF), isos. map in (BI).

A severe shock at Collalto, all houses and a church fissured, some chimneys thrown and some walls fissured in the surrounding villages (BI).

Jan. 21, 10^h (ONB) or 21^h30^m (PAT), 40·0°N, 41·7°E, R.33, $I_0 = $ VIII? (TM), $I_0 = $ IX, $M = 6·7$ (ONB).
Allegedly one village ruined near Erzurum (PAT); a heavy shock (MO).

March 9, 19^h, 37·2°N, 7·4°W, R.12, $I_0 = $?, $I_0 = $ VII (FH), $I_0 = $ IV (MU), not mentioned by (GS, BRS, PAT).

March 13, a disastrous shock at Vlone (Valona), (40·5°N, 19·5°E), R.25 (MCA), $I_0 = $ VIII to IX (GGK), $I_0 = $ IX (MD), local.

May 30, 16^h. Several houses thrown down by a shock at Tbilisi (41·7°N, 44·8°E) (PAT), no information in (BZ), doubtful.

June 2, 10^h30^m, 39·9°N, 41·3°E, R.33, $I_0 = $ IX, $I_0 = $ IX−X? (MJD), $I_0 = $ VIII (EGU, TM), $M = 6·1$ (ONB), $I_0 = $ X (CAM), $I_0 = $ VIII+, 39·8°N, 41·0°E (ANV).
At Erzurum half of mosques, bazars and houses thrown down, about 500 victims during the first shock, the second one ruined 4000 houses, 3000 houses partly ruined in total about 15 000 victims (PAT), heavy damage in Erzurum (PL), 1800 houses completely destroyed, 2400 houses heavily injured, strongest effects in the mountains of Palendzhukana and Yarli Dag (MO).

June 11, 13^h, 40·7°N, 48·4°E (BZ, AV), R.32, $I_0 = $ IX?, $r_5 = 30$ and 80 km about; $M = 5·5$ (AV), $I_0 = $ VII (MJD).
A violent shock in E. Caucasus, at Shemakha many houses ruined or seriously injured, more than 100 victims, the village Baskhal ruined, strongly felt at Marazy, Shusha and Jelizavetpol, felt from Kuba to the sea (BZ), (PAT) gives June 12, 740 houses completely ruined, 1464 seriously damaged, 100 persons killed, 186 injured, (MO) describes this second shock as less damaging and gives the total damage as follows: from 5412 houses 1181 totally destroyed, 1697 seriously injured.

June 26, 10^h, 39·9°N, 41·3°E, R.33, $I_0 = $?, $I_0 = $ VII (MJD).
Aftershock at Erzurum, new ruins (PAT, MO).

June 27, 10^h, 39·9°N, 41·3°E, R.33, $I_0 = $?
New aftershock at Erzurum threw down the injured houses (PAT).

July 13, 39·9°N, 41·3°E, R.33, $I_0 = $?, $I_0 = $ VII (MJD).
Another aftershock at Erzurum threw down the walls, the fortress and the houses having remained after the preceding shocks (PAT), strong at Tbilisi (MO, MJD).

July 16, 39·9°N, 41·3°E, R.33, $I_0 = $ VII? $M = 5·5$ (ONB), $I_0 = $ VI (ANV).

July 17, 39·9°N, 41·3°E, R.33, $I_0 = $?, $I_0 = $ IX, $M = 6·7$ (ONB), $I_0 = $ VI (ANV).
Complete destruction (PAT), underground shocks (MO).

July ? Much damage at Plovdiv (42·1°N, 24·7°E), R·23, $I_0 = $ VII−VIII (GGB), questionable, not reported by (WS).

Aug. 13, 39·9°N, 41·3°E, R.33, $I_0 = $ VIII, $M = 6·1$ (ONB); old style: July 31.

Aug. 13, 12^h30^m (40·1°N, 20·1°E), R.25, $I_0 = $ VII−VIII (GGK), $I_0 = $ VIII (MJD).
Serious damage at Gjirokastër and Turani (GGK).

Aug. 21, $02^h, 11^h55^m$, $40\frac{1}{4}$°N, 26°E, R.27, $I_0 = $ VIII−IX (GGK), $I_0 = $ IX, $r = 280$ km (GG), 40·2°N, 25·9°E, $h = 18$ km, $I_0 = $ VIII, $M = 6·1$, $r = 350$ km (ONB, MJD),

$I_0 = $ X (CAM); a sequence of strong shocks, main shock at 11^h55^m, the shock at 11^h20^m: $I_0 = $ VII? 11^h35^m: $I_0 = $ VII—VIII?

A disastrous shock in the Eastern Sporades region, at Iskinit and Panayia all houses were ruined or cracked, in other villages of the island Imroz (Imbros) almost all houses were fissured (at Pyrgos walls fell down, rockslides in other parts, cracks in the ground, at Kástron some mills thrown down), felt at Enez (Enos), Edirne (Adrianople), Istanbul, Gelibolu (Gallipoli) VII, Dardanelles VII, Mitilíni V, Aiwali V?, Izmir (Smyrna) IV, Bursa (Brousse), Büyükdere V?, Thessaloniki IV, Sofia, Plovdiv, Demotika (Dhidhimótikhon?), Límnos (two houses fell), Samothráki, Bozcada (Tenedos) (PAT).

Aug. 22, 18^h15^m, 42·8°N, 13·1°E, R.19, $I_0 = $ IX (CAM), $I_0 = $ VIII (MJD), $I_0 = $ X (MR), $I_0 = $ X—XI (MF).

At Norcia all poorly constructed houses were ruined, many others greatly damaged, 100 victims, Campi, Casali, Capo del Colle and Villa S. Angelo suffered almost in the same extent as Norcia, Abeto, Todiano; Villa d'Ancarano and Frascaro suffered partly, felt as far as Roma, Camerino and Pesaro (PAT, BI).

Sept. 12, 16^h06^m, (40·1°N, 19·8°E), R.25, $I_0 = $ VIII—IX? (GGK), $I_0 = $ IX (MJD).

A violent shock destroyed Himarë (MCA), VIII(?) at Dhërmi, Kudhësi, Vlorë and Kuçi (MD).

Oct. 17, 09^h30^m, 46·1°N, 20·9°E, R.21c, $I_0 = $ VI (RA), $I_0 = $ IX (PRA).

Earthquake felt at Periam and Lovrin (FA).

1860

April 10,   40·2°N, 20·0°E, R.25, $I_0 = $ IX (MD), $I_0 = $ IX—XI (GGK).

Tepelenë and Gjirokastër ruined, damage at Leskovik, Kónitsa and Ioánnina (Jannina), further Dragoti, Peshtani, Maricaj, Vasiari, Dukaj, Zhulat, Fushë-Bardhë, Këlcyrë, Turani, Vlorë, Kanino, Narta, Velçan and Karbunarë injured; aftershocks on April 15 at Leskovik, $I_0 = $ VII? $I_0 = $ VIII (MD), Kónitsa, destructive (MCA),

on April 16 at Leskovik, $I_0 = $ VIII? $I_0 = $ IX (MD), Kónitsa, destructive (MCA).

May 16, 17^h40^m, 41·3°N, 19·7°E, R.25, $I_0 = $ VIII—IX?, $I_0 = $ IX (MD), $I_0 = $ IX—X (MF), $I_0 = $ VIII (MA, MJD).

A violent shock near Tiranë (VI?) destroyed the village Ura Besirit and other localities (MCA, MD), Durrës VI?, felt at Vlorë (MD).

May 26 or 30, Norcia-Perugia (Italy), R.19, volcanic(?) shocks destroyed houses near Vesuvius (PAT, LC), $I_0 = $ VII (MJD).

June 4, 00^h30^m, 40·2°N, 29·1°E, R.27, $I_0 = $?, $I_0 = $ VIII? $M = 6·1$ (ONB), $I_0 = $ VI (EGU).

A shock felt at Bursa (PAT), Aug. 4 in (PL), a rockslide on the slope of Olimbos (Olympus?) (LC).

June 7, 18^h30^m, 40·2°N, 29·1°E, R.27, $I_0 = $?, $I_0 = $ VIII, $M = 6·1$ (ONB), $I_0 = $ VII (MDJ).

At Bursa only walls fell, no other damage, rockslides on Mt. Olimbos (PAT), the report might correspond to the previous shock.

July 19, 16^h38^m, (46$\frac{3}{4}$°N, 12°E), R.19, $I_0 = $ VII (MJD), $I_0 = $ VI—VII (MR).

A strong shock at Treviso and Valdobbiadene, chimneys fell, at Guia some walls were thrown down, felt at Venezia, Pieve di Soligo, Collalto, Verona, Trieste (PAT, BI).

Aug. 6, $40\frac{1}{2}°$N, $25\frac{1}{2}°$E, $h = n$, R.24, $I_0 =$ VII? (GG), $I_0 =$ VI−VII (GGK). Samothráki, fissures in the walls and in the ground, rockslides, changes in ground water level (GG, GGK); Aug. 17/22, $40.4°$N, $26.6°$E, $I_0 =$ VIII, Gelibolu, Sakiz, Edirne (EGU, TM), one shock?

Sept. 22, $34.7°$N, $2.5°$W, R.16, $I_0 =$?, $I_0 =$ V (MU), $I_0 <$ VII (BRS), $I_0 =$ IX−X (MF). A heavy shock awoke people at Almeria, felt at Lorca (GS, PAT); many houses thrown down at Vera, damage at Almanzora and Cartagena (MF, LC).

Sept. 27,   ($36.3°$N, $4.5°$E), R.16, $I_0 =$ VII−VIII (GA), $I_0 =$ VIII (MJD), $I_0 =$ IX (CAM). Houses collapsed at Oued-Sahel, Tensaout, Taourirt, felt at Bordj-Bou-Arréridj (RK, GS, PAT).

Dec. 2, 04^h. At Kütahya (Kutaiah), 280 km E of Izmir (Smyrna), R.26c, several houses were thrown down by an earthquake, felt at Izmir, Magnesia, in the island Khíos (Chios) V (PAT), not mentioned by (ONB), questionable.

Dec. 3, $39.9°$N, $41.3°$E, R.33, $I_0 =$?, $I_0 =$ VIII, $M = 6.1$ (ONB), $I_0 =$ VII (MJD), $I_0 =$ VI (ANV, EGU). Not destructive (MO).

1861

Febr. 9,   VII in Malta (MJD); Jan. 9, 00^h35^m, Malta, at Modica (Sicily) serious damage (LC), epicentre in the sea?, R.20.

March 16, 00^h30^m UT, $44.4°$N, $9.6°$E, R.18a, $I_0 =$ VII? (MR, MJD), local? At Varese Ligure chimneys thrown down, walls fissured (PAT, BI).

May 9, 01^h53^m UT, $42.9°$N, $12°$E, R.19, $I_0 =$ VIII (MR, MJD), $I_0 =$ IX−X (MF), isos. map in (BI). Several houses and a church destroyed at Citta della Pieve; at Catona and Chiusi buildings fissured (BI).

May 24, 13^h?, $39.5°$N, $47.9°$E (BZ), R.32, $I_0 =$ VII−VIII?, $I_0 =$ VII (MJD); $16^h−17^h$, $39.4°$N, $46.8°$E (MO). Earthquake in the E. Caucasus, at Shusha and Bozhie Kolodtsy and farther 10 km to the river Kura, chimneys fell and walls cracked, felt at Redkino, Bozhi Promysla, Shemakha, Nukha, Lenkoran, Baku, at Zurnabad and Sardob walls fissured (BZ), very strong (MO).

July 26, 17^h. A violent shock at Oran ($35.6°$N, $0.6°$W), R.16, the church Saint Louis fissured, panic (PAT, GS), $I_0 =$?, not mentioned by (RK, GA).

Oct. 16,   ($44\frac{1}{4}°$N, $12\frac{1}{4}°$E), R.19, $I_0 =$ VII (MR, MJD). A violent shock at Ravenna, Rimini and Forli, many chimneys fell and walls fissured, felt at Bologna, Venezia and Trieste (PAT, BI).

Dec. 9, 06^h, $40.8°$N, $14.4°$E, $h =$ sup, R.19, $I_0 =$ VII (MR, MJD), $I_0 =$ IX−X (MF), a volcanic earthquake. A very violent shock at Torre del Greco (Campania), houses and churches fissured, large cracks in the ground, accompanied by an eruption of Vesuvius (PAT).

Dec. 18, 9^h? UT, $45.8°$N, $16.4°$E, R.22, $I_0 =$ VII−VIII, $I_0 =$ VII (MJD), $I_0 =$ VIII (TD, JM), $I_0 =$ IX−X (MP), Dec. 8, $45.2°$N, $16.6°$E (JM).

Poor houses at Bešlinac, Kostajnica, Baliba, Dubica collapsed, many cracks in the walls, many chimneys thrown down (BA); felt at Zagreb, Vojnic, Petrinja VI, Karlovac VI and Trieste IV (MF).

Dec. 26, 06^h49^m UT, $38\frac{1}{4}°$N, $22\frac{1}{4}°$E, R.26, $h = n$, $I_0 = X-XI$ (GGK), $I_0 = XI$ (GG, MF), $I_0 = IX$ (CAM), $I_0 = VIII$ (MJD), $M = 7\frac{1}{2}$ (GGE), isos. map in (SAM).

An earthquakue catastrophe in Akhaïa, the plain between Aíyion (Aigion, Aeghion) and Kórinthos devastated, the highest damage at Valymitika and Trypia, Aíyion ruined, Kalamaki injured, at Kórinthos many houses injured (PAT), 15 km^2 large slump along the coast of Akhaïa, 13 km long crack, tsunami waves 2 m high in the Gulf of Kórinthos (ANS), fissures, sand craters, casualties (GG, GGT, GGK, LC). Dec. 16 (GGT).

1862

Date? $40\cdot2°$N, $30\cdot0°$E, R.27, $I_0 = VII$, $M = 5\cdot5$ (ONB). This shock at Canakkale might correspond to the shock at Bursa, Jan. 11, $40\cdot2°$N, $27\cdot8°$E, $I_0 = VI$ (EGU, PL).

Jan. 1, 09^h, $38\frac{1}{4}°$N, $22\frac{1}{4}°$E, R.26a, $I_0 = VIII-IX?$ (GG), $I_0 = VII$ (MJD).

A violent aftershock, at Pátrai houses injured, at Aíyion fifteen houses thrown down, upheaval of the coastline, houses at Kórinthos and Kalamaki fissured, tsunami (PAT).

March 14, 01^h30^m, $38\frac{1}{4}°$N, $20\frac{1}{2}°$E, $h = n$, R.26a, $I_0 = VIII?$, $I_0 = IX-X$ (GGK), $I_0 = X$ (GG), $I_0 = IX-X$ (MF). $M = 7\frac{1}{2}$ (GGE).

A shock in Zákinthos (Zante) (PAT), a few houses escaped considerable damage at Argostólion, less damage at Lixourion, little damage in Kerkíra (Korfu), none in Zákinthos (MF).

May 24, $36\cdot8°$N, $28\cdot3°$E, R.26c, $I_{max} = VII?$ (ONB, MJD), epic. in the sea?

Ródhos (Rhodos), some old walls ruined, felt at Marmaris and other localities along the Anatolian coast, in the islands Nísiros and Khálki (PAT).

May 26, 16^h30^m UT, an earthquake with the intensity of VIII occurred in the region of Pustertal and Lessachtal in Tirol, Austria—Germany, no description of damage (SW), R.18b, (TT) gives $I_0 = VI$ for an aftershock at 23^h20^m.

June 8, 12^h45^m, $(35\cdot7°$N, $0\cdot5°$E), R.15, $I_0 = VII?$ (MJD), $I_0 = VI-VII$ (GA).

A severe shock at Relizane, many houses less or more fissured, felt also at Mostaganem (PAT, RK, LC).

June 21, $36\frac{1}{2}°$N, $25°$E, $h = i$, R.26b, $I_{max} = VII-VIII$ (GGK), $I_{max} = VIII$, $r = 390$ km (GG), $M = 6\cdot8$ (GGE), $I_{max} = IX$ (MF), epic. in the sea.

Damage in Mílos, Antimelos, Sifnus, Folégandros and Thíra (Santorin), felt in Pelopónnisos, Zákinthos (Zante) and Kríti (Crete), allegedly felt also in Malta (SAM); fissures in the houses (LC).

Aug. 22, 17^h, $37\cdot2°$N, $5\cdot1°$W, R.16, $I_0 = ?$, $I_0 = VI$ (MU), $I_0 = VIII$ (BRS, MJD), $I_0 = IX$ (CAM), $h = sup?$

At Villanueva several houses thrown down, about twenty persons injured, strong at Algamitas (PAT, GS).

Oct. 4,   05h?, $40,5°$N, $19\cdot5°$E, R.25, $I_0 = IX?$, $I_0 = IX-X$ (MF, GGK), $I_0 = X$ (MD, MCA), $M = 6\frac{3}{4}$ (GGE).

Great destructions at Vlorë (Valona) and surrounding villages, serious damage in the region of Vijosë (Vojussa), felt with $I = $ IX? at Vlorë, Kanino and Narta, with VII? at Smotkhine, Velça, Karbunarë, Tepelanë, Dragot, Peshtani, Maricaj, Vasiari, Gjirokastër, Turani, Klisyra (MD, MCA).

Oct. 16, 01^h11^m UT, 45·6°N, 25·6°E, R.21d, $I_0 = $?, $I_0 = $ VIII (RA).
Allegedly a strong earthquake felt at Braşov, Bucureşti, Ruse, Grozeşti, Măgheruş (FA), at Braşov walls fissured and chimneys fell (RA), in the evening a terrible storm damaged houses (PAT).

Oct. 16, 38·8°N, 30·5°E, R.30, $I_0 = $ X, $I_0 = $ VIII, $M = $ 6·1 (ONB, MJD).
Afyon-Karahissar (MJD), the town Suhut was ruined (LC), (EGU) gives a shock in Bursa on Oct. 16, 1863, 40·2°N, 29·1°E, but also reports about destruction of Suhut in 1863 (date?), $I_0 = $ X, 38·5°N, 30·6°E, 800 people killed, cracks in the ground, strong at Dinar, Afyon and Konya.

Nov. 3, 03^h, (38·4°N, 27·7°E), R.26c, $I_0 = $ IX−X (MJD), $I_0 = $ IX, $M = $ 6·7, $r = $ 330 km (ONB), $I_0 = $ X (CAM).
At Turgutlu (Chekoud-Kassaba) the houses collapsed, 280 victims, less damage at six surrounding villages, felt at Afyonkarahissar and the district of Isparta (PAT). An aftershock(?) on Nov. 13 at Afyonkarahissar where houses cracked ($I_0 = $ VII?), felt also at Izmir (Smyrna), Aydin, Nazilli, Denizli, in the islands Khíos and Mit líni (Metelin) (PAT).

Nov. 30, 00^h25^m, (36·5°N, 5·3°E), R.15, $I_0 = $ VII (GA), not in (GS).

1863

Jan. 17, 03^h45^m UT, 47·8°N, 19·2°E, R.21b, $I_0 = $ VII? (RA, MJD), local, $I_0 \leq $ VII (CD).
Many chimneys thrown down at Penczen, felt at Budapest and Kecskemét (RA).

Jan. 19, 06^h UT, 41·5°N, 13·8°E, R.19, $I_0 = $ VII (MJD), $I_0 = $ VIII (MR), not mentioned by (PAT, MF).
At Montecassino cracks in the walls of the monastery, at S. Germano some old houses collapsed, the others fissured (BI).

Jan. 30, 11^h30^m, 40·7°N, 13·9°E, R.19, $I_0 = $ VII? (MR, MJD).
At Casamicciola some walls fissured, felt at Ischia, Barano, Testaccio (BI, PAT).

April 17, 07^h53^m, 36·8°N, 3·2°W, R.16, $I_0 = $ VII (BRS, MJD), $I_0 = $ VI (MU).
A severe shock (walls fissured) at Granada, at Alhendin cracks in the ground, some damage at Og jares and Gójar (GS).

April 22, 20^h30^m, $36\frac{1}{2}$°N, 28°E, $h = i$, R.26c, $I_0 = $ IX−XI (GGK), $I_0 = $ XI, $r = $ 1380 km (GG), $M = $ 8·5 (GGF), 36·3°N, 28·0°E, $I_0 = $ IX, $M = $ 6·7, $H = 10^h20^m$ (ONB), $I_0 = $ IX−X (MJD), $I_0 = $ X (CAM).
An earthquake catastrophe, thirteen villages in the Ródhos (Rhodes) were destroyed (Trianda, Bastida, Maritsa, Demetria, Salakos, Dimilia, Lardos, Katavia, Laerma, Pilona, Lachania, Istridos, Monotilos, Massari), casualties, other villages were partly destroyed, in total 2050 houses were thrown down, strongly felt at Izmir (Smyrna), felt at Aydin, Nazilli, Beirut (Beyrouth), Gelibolu (Gallipoli), Iráklion (Candia), Suez, Mersine, Cairo V, Jerusalem, Malta, Tripoli(?), several houses in the island Kos destroyed some others and the cathedral seriously injured, some houses destroyed in Mákri, Marmara and Khálki, no damage in the island Symi (PAT, SAM).

1863

June 9, 22^h, 44·0°N, 6·2°E, R.18a, $I_0 =$ VII? (MJD), $I_0 =$ VIII (RO).

At Beynes walls fissured and tiles fall down, rockslides, some minor damage at Mézel, felt at Digne (RO, PAT), not included in (RF).

June 10—Sept. 30, 37·0°N, 2·0°W, R.16, $I_0 =$ VIII (BRS, MU), $I_0 =$ VII (MJD), a sequence of shocks.

June 19: Subsidence, rockslides near Huércal-Obera, cracks in the ground, at Albox one house thrown down, fissures in old houses, serious damage at Jaroso, Vera and Cuevas, felt at Cantoria, Albox, Cartagena, Murcia, Lorca, Aguilas, Venta de los Cazadores, Almeria (PAT, GS).

July 9, 37·3°N, 2·6°W, $I_0 =$ VI (BRS), $I_0 =$ VII (MJD).

August 16 or 26, $38\frac{1}{4}$°N, $26\frac{1}{4}$°E, $h = n$, R.26c, $I_0 =$ VIII (GG), $I_0 =$ VII—VIII (GGK), $I_0 =$ VII (MJD), local?

A ruinous shock in Khadra (?) and Khíos (SAM), 30 000 inhabitants without shelter, two new hot springs originated (LC).

Oct. 6, 03^h22^m, (51°N, $2\frac{1}{2}$°W), R.10b, $I_0 =$ VII? (MJD), $I_0 =$ VIII (DCB).

Maximum intensities at Worcester, Hereford, Ross, Brecon, Monmouth and Newport, largely extended shaken area, epicentre near Galway where one wall fissured, one ceiling and bricks from one chimney fell (PAT).

Nov. 6, 10^h, 40·5°N, 29·1°E, R.27, $I_0 =$ IX, $M =$ 6·7 (ONB, CAM).

A severe shock, at Gelibolu (Gallipoli) one ceiling destroyed, at Gemlik several chimneys and a minaret thrown down, walls cracked, at Umurbey forty houses destroyed, two victims, felt at Bursa (Brousse) and Istanbul (PAT).

Oct. 16, 40·2°N, 29·1°E, $I_0 =$ VI, Bursa (EGU), or does it correspond to the report for Oct. 16, 1862?

Nov. 24, night, 40·7°N, 31·6°E, R.27, $I_0 =$ VII—VIII?, $I_0 =$ VII, $M =$ 5·5 (ONB, MJD), local?

At Bolu several houses thrown down (PAT). Not in (EGU).

Dec. 8, 03^h50^m, 43·9°N, 5·2°E, R.18, $I_0 =$ VII? (RF), $I_0 =$ VII—VIII (PRF, RO), $r = $ = 70 km (RO).

At Lagnes houses fissured, strongly felt at Vaucluse, felt at Avignon, Orgon, Carpentras, Graveson, Nimes and in the region of Durance (PAT).

1864

Jan. 3, the villages Ardebil, Lenkoran, Niar, Nouskhar, Gnert, Karabağ, etc. destroyed, 500 victims (MO, MJD), $I_0 =$ IX?, epicentre? R.34.

Jan. 10—11, Ródhos (Rhodes) IX (CAM), Mákri VIII (MJD), epicentre? R.26c, not mentioned by (ONB, GG, PAT), questionable.

Jan. 12, 05^h20^m, 37·8°N, 1·7°W, R.16, $I_0 =$ VII? (MJD), $I_0 =$ VII—VIII (BRS), 37·8°N, 1·9°W, $I_0 =$ V (MU).

A heavy shock in SE Spain, at Librilla one house ruined, at Alhama de Mureia the plaster cracked, felt at Murcia, Lorca and Vera (GS).

March 15, 02^h30^m, 44·3°N, 11·1°E, R.19, $I_0 =$ VII—VIII (MR), $I_0 =$ VII (MJD).

A severe earthquake with a most affected area between Vergato (many houses cracked, many chimneys were thrown down), Zocca (many chimneys fell, a belfry was thrown down), similar but smaller effects observed at Pracchia, Pocretta, Tole and Savignano, felt at Mantova, Granaglione, Firenze (Florencia), Bologna, Modena, Reggio, Parma and Milano (PAT, BI).

June 14, 04^h30^m, (41°N, 25°E), $h = n$, R.24, I_{max} = VII (GG, MJD), I_0 = IX—X (MF), $M = 5\frac{1}{2}$ (GGE), 40·3°N, 26·5°E, I_0 = VI (EGU), epicentre in the sea?
A series of violent shocks along the coast between Thessaloníki and Gelibelu (Gallipoli), at Thessaloníki several stone built houses injured, other buildings fissured, a part of the already poor and damaged ramparts collapsed, at Jenisea (bay of Lago) a custom house collapsed (PAT), I = VIII in the Struma valley in Bulgaria (GGB).

July 17, 0^h, $38\frac{1}{2}$°N, $23\frac{1}{2}$°E, $h = n$. R.26a, I_0 = VII (GG), I_0 = VI—VII (GGK), $M = 5\frac{1}{2}$ (GGE).
A damaging shock in the Evvoia (Euboea), particularly at Kastrovala(?) (SAE).

Aug. 21, 05^h15^m, 41·1°N, 20·1°E, R.25, I_0 = VII—VIII (GGK), I_0 = VIII (MD), I_0 = IX (MF), $M = 5\frac{1}{2}$ (GGE).
Extremely strong shock at Elbasan, Vlorë VI? (MD), "the third destruction of Elbasan" (MCA, MJA), tremors almost every day in 1864 (LC).

Aug.? Houses collapsed at Kalarash (region of Silistra 44·1°N, 27·3°E, R.21d), I_0 = ?, I_0 = VIII (GGB, MJD). Questionable, not reported by (WS, FA), Aug. 14, accompanied by a strong wind, I_0 = X (MF), I_0 = IX (CAM).

Oct. 2, 36·1°N 29·5°E, R.26c, I_0= ? I_0=VII, M=6.1 (ONB, EGU).
Cracks in the island of Meis, felt at Fethiye (PL).

1865

Febr. 23, 40·2°N, 26·4°E, I_0 = VIII (EGU), R.26c, 39·3°N, 26·2°E, I_0 = VII? M = 5·5 (ONB).
At Midili 8 victims and damage, felt at Izmir (Smyrna), Canakkale and Tekirdag (PL).

Febr. 25, 02^h. Some walls fissured by a shock at Bougie, Algeria, R.15 (PAT, GS), not mentioned by (RK, GA).

April 10? Slightly damaged houses and fissured walls at Kalámai (Greece, 37·0°N, 22·2°E?) (GGM), R.26a.

July 19, 01^h UT, 37·7°N, 15·1°E, h = sup, R.20, I_0 = IX—X?, I_0 = VIII (MJD), I_0 = IX (CAM), I_0 = XI (MR, MF), r = 20 km, isos. map in (BI).
A disastrous shock in the region of Etna, epicentre closely to a little village Fondo di Macchia which was completely ruined, alltogether 150—200 houses were destroyed and 52—64 persons perished, damage at Rondinella, Baglio, Scaronazzi, S. Venerina, Mangano, San Leonardello, etc., the pleistoseismal area was 7 km long and only 1 km wide, slight subsidence, fissures in the ground (BI). An aftershock on July 19, 07^h, damaged seriously Faggio, Linera, Mangano and Guardia (PAT).

July 23, 21^h30^m, 39·4°N, 26·2°E, $h = n$, R.26c, I_0 = IX—X (GGK, SAM), I_0 = X, r = 300 km (GG), M = 6·6 (GGE). Nov. 23, 40·2°N, 26·4°E, I_0 = VIII (TM).
In Lésvos, at Molivo and surrounding villages (radius of 2 miles about) 100 houses collapsed and the others were injured, several persons perished, thermal sources disappeared; felt at Mitilíni (Metelin), Istanbul, Gelibolu (Gallipoli), Rodosto (Tekirdağ), (Izmir (Smyrna) and in Dardanelles (PAT, SAM).

1865

Sept. 21, 20^h50^m UT, 43°N, 23°E), R.19, $I_0 =$ VII—VIII?, $I_0 =$ VIII (MJD), $I_0 =$ IX—X, Sept. 25 (MF, LC), Sept. 23 (PAT, FC).

At Perugia a violent shock threw down several roofs and many chimneys, at Citta di Castello several houses were thrown down (PAT, FC, BI).

Oct. 10,   40·7°N, 20·0°E, R 25, $I_0 =$ VIII—IX?, $I_0 =$ IX (MD, MF, MCA), $M = 6\frac{1}{4}$ (GGE).

A ruinous shock at Berat and surrounding villages Cfiri, Klos, Velçan, Osman Zeza, Krutë, Fiéri and Bënja (MD, MCA).

Oct. 11, 01^h15^m, $37\frac{3}{4}$°N, 27°E, $h = n$, R.26c, $I_0 =$ VII (GG), $I_0 =$ VI—VII (GGK). $M = 5\frac{1}{2}$ (GGE).

Houses injured in Sámos, strongly felt at Izmir (Smyrna) (SAM).

Nov. 11, $38\frac{1}{4}$°N, $26\frac{1}{4}$°E, $h = n$, R.26c, $I_0 =$ VIII—IX, $I_0 =$ VIII (GG), $I_0 =$ VII—IX (GGK), $I_0 =$ IX (MJD), $I_0 =$ X (CAM), $M = 6\frac{1}{4}$ (GGE).

A series of shocks in Khíos destroyed many houses and caused serious damage (GG), in the town and in the country the houses were injured in different extent, many walls collapsed, panic, all walls fissured, the storehouses very well resisted to all shocks (PAT).

1866

Jan. 2, 10^h, 40·4°N, 19·5°E, R.25, $I_0 =$ IX—X (GGK), $I_0 =$ X (MD), $I_0 =$ X—XI (MF), $M = 6\frac{3}{4}$ (GGE).

A catastrophic shock at Vlorë and along the coast, high tsunamis wasted the coast Narta—Vlorë—Kanino (ANS), houses destroyed, 60 victims, large fissures in the ground, strongly felt (VIII?) at Smotkhinë, Velça, Kudhesi, Karbunarë, Tepelenë, Dragot, Peshtani, Maricaj, Vasiari, with $I =$ VII? at Dukaj, Gjirokaster, Këlcyrë, Turani and Butrinto (MD, MCA). Strong tsunami also on Jan. 6 at Vlorë and Narta (ANS).

Jan. 19,   ($38\frac{1}{4}$°N, $26\frac{1}{4}$°E), R.26c, $I_0 =$ VII? (MJD).

Houses damaged in Khíos (SAM, GH), several walls fissured (PAT). Not mentioned by (GG), felt on the sea (GH), (GGK) gives an earthquake in Ródhos on Jan. 13, $M = 6·8$ (GGE), with the shaken area $P = 700\,000$ km^2.

Jan. 31,   36·4°N, 25·3°E, $h = n$?, R.26b, $I_{max} =$ VII, $r = 230$ km (GG), $I_0 =$ IX (MF), volcanic, $M = 6·1$ (GGE).

A heavy shock in Thíra (Santorin) followed on Febr. 1, by an eruption of the volcano Thíra (GG, PAT), 50 houses and two churches were damaged at Nea Kaimeni, the continuing subsidence (without shocks) splitted the houses (SAM).

Febr. 2, $38\frac{1}{4}$°N, $26\frac{1}{4}$°E, $h = n$, R.26c, $I_0 =$ VII—VIII (GGK), $I_0 =$ VII (MJD), $I_0 =$ VIII (GG).

In Khíos one house fissured and several others thrown down (PAT, LC), a series of disastrous shocks felt as far as Ródhos (GGE, LC, SAM), a destructive shock (GH); tsunami in Khios (ANS).

Febr. 6, 13^h45^m, 36°N, 24°E, $h = n$, R.26a, $I_{max} =$ VIII (GG), $I_{max} =$ VII—IX (GGK), $M = 6\frac{1}{4}$ (GGE).

At Pátrai two houses collapsed and some others injured, at Tripolis houses fissured, felt in the country as far as Árgos (but not felt there) and Kíthira (PAT). 8 m high tsunami at Avlemon (GGT, ANS), epicentre probably in Kíthira (GGK).

Febr. 28, 40·5°N, 19·5°E, R.25, I_0 = IX—X, I_0 = VIII—IX (GGK), I_0 = IX(MD, MCA), $M = 6\frac{1}{4}$ (GGE).

A disastrous shock in the region of Vlorë, 275 houses destroyed, 9 persons killed, 90 injured, most strongly felt at Vlorë, Kanino and Narta, further with I = VIII to IX at Smokthinë, Velçan, Karbunarë, Tepelenë, Dragot, Peshtani, Maricaj, Vasiare, Dukaj, I = VII (?) at Gjirokastër, Klisyra, Turani (MD, MCA, FC).

March 2, 11^h, 40·4°N, 19·5°E, R.25, I_0 = VIII—IX? (GGK), $M = 6\frac{1}{4}$ (GGE).

A further disastrous shock at Vlorë and Kanino, observed with I_0 = VIII (?) at Smokthinë, Velça, Dhermi, Butrinto (MD).

March 2, 15^h30^m, 40·4°N, 19·5°E, R.25, I_0 = IX?, I_0 = IX—X (GGK), I_0 = X (MD, CAM), $M = 6\frac{3}{4}$ (GGE).

Disaster at Vlorë, Kanino, Smothinë, Velça and Dhermi, I_0 = VIII(?) at Narta, Karbunarë, Tepelenë, I_0 = VII(?) at Gjirokastër, felt in Ipiros (Epiros) and very slightly in Kerkira (Korfu) (MD, MCA, PAT); 12 houses collapsed, 20 victims (LC); tsunami at Vlorë on March 3 (ANS).

Aftershocks (I_0 = VII): March 2, 20^h, I_0 = VII—VIII?, $M = 5\frac{1}{2}$ (GGE).

March 6, evening, I_0 = VIII—IX?, strong tsunami at Himara and Kanino (ANS), $M = 6\frac{1}{4}$ (GGE).

March 13, 18^h, I_0 = VII—VIII?, tsunami in the bay of Vlorë (MD, GGK, MCA, ANS). $M = 5\frac{1}{2}$ (GGE).

March 9, 01^h, 63·5°N, 9·5°E, R.3, I_0 = VIII, $M = 6·3$, $R = 450$ km (BMF), I_0 = VII (MJD).

A severe, largely felt shock in Scandinavia, epicentre near Kristiansund, felt with I = VII at Trondheim (parts of chimneys thrown down), felt as far as Bodö, Solleftea, Söderhamn, Felsund, Langesund, Bergen, Shetland islands (PAT, KRS).

May 19, 09^h12^m, 44·2°N, 6·1°E, R.18a, I_0 = VIII (RF), I_0 = VIII—IX, $r = 150$ km (PRF).

Houses damaged at Thoard, Saint Geniez, Volonne, La Motte, Gigors, etc., largely felt as far as Toulon, Nice, Marseilles, Grenoble, Chambéry, Genève (RO, RF).

June 20, 39·9°N, 41·3°E, R.33, I_0 = ?, I_0 = IX, $M = 6·7$ (ONB), I_0 = VIII, 39·8°N, 41·1°E (ANV), I_0 = VI (EGU).

Not reported by (PAT, FC). In (PL) only Erzurum is mentioned without indication of any damage.

July 22, (38·5°N, 40·1°E), R.33, I_0 = ?, I_0 = VII (ANV), $M = 5·5$, 38·7°N, 39·2°E (ONB). I_0 = IX—X (MJD).

"A terrible earthquake" between Euphrates (Eufrat) and Tigris, many fissures in the ground, near Diyarbakir 16 villages allegedly swallowed by the earth (FC, PAT), epicentre questionable, probably on the line Lice and Kulp (PL).

Aug. 11, midnight, epicentre? R.19, I_0 = VII (MJD), I_0 = VII—VIII (MR).

At Monte Baldo (Verona) chimneys, balconies and plaster fell down, strong at Baretta and Cassone (BI, PAT), rockfalls (FC).

Aug. 18, 39·0°N, 41·3°E, R.33, I_0 = VII (ANV).

Sept. 14, 05^h, (47$\frac{1}{2}$°N, 2°E), R.11, I_0 = VII? (RF).

Largely felt in the Bassin de Paris and in Plateau Central, highest intensity at Indre-et-Loire and Loire-et-Cher (RF), felt at la Châtre, Arcachon, Vendôme, Angoulême, Clermont, Dijon, Ouges, Vitteaux, Montbard (PAT).

1866

Nov. 4, 12^h. A strong shock in Bessarabia and Moldavia, R.21d, the ceilings and the framework of houses fissured (FA, MO), $I_0 = ?$, $I_0 =$ VII (MJD).

Dec.? A heavy shock damaged some old houses and poor walls at Sofia (WS), (GGB) gives $I =$ VIII for the region of the Struma valley, incomplete information; R.24.

Dec. 4, 40·1°N, 20·1°E, R.25, $I_0 =$ IX−X? (GGK), $I_0 =$ X (MD), $M = 6\frac{3}{4}$ (GGE). A destructive shock at Gjirokastër, great damage in the direction Vlorë-Himara, extremely strong at Ioánnina (Janina), $I_0 =$ IX (?) at Vlorë, Kanino, Himara and Tepelenë, $I_0 =$ VIII(?) at Smokhtinë, Velçan, Kudhesi, Karbunarë, Dhermi, Zhulat and Fushë-Bardhë (MD, MCA).

1867

Jan. 2, 07^h13^m, 36·4°N, 2·7°E, R.15, $I_0 =$ IX−X (MJD), $I_0 =$ X−XI (GA), $I_0 =$ X (CAM). The village Mouzaïaville almost completely destroyed (48 persons perished, 100 injured), Bou Roumi, La Chiffa and El Affroun were also seriously injured in the same extent, strong at Ameur el Aïn (3 killed), Chancelade, Blida (many houses fissured, public buildings seriously injured), Medea (some damage), Marengo, Duperré, Tipasa (slight damage), houses damaged at Dalmatie, Douéra and Cherchel, felt at Miliana, Boghari, Teniet el Haad, Aumala, Dellys, Tizi Ouzan, Dra al Mizan, Fort National, Orléansville, Djidjelli.
The majority of collapsed houses was built of pebbles poorly connected with earth, brick built buildings at Blida suffered very little (RK), a detailed report in (PAT).

Jan. 7? A letter from Sofia reports about a shock which shook heavily the town but only poor, old houses and fences were injured (WS).

Jan. 27, 39$\frac{3}{4}$°N, 20$\frac{3}{4}$°E, $h = n$, R.25, $I_0 =$ VII?, $I =$ IX−X (GGK), $I_0 =$ X (MD, GG), $M = 6\frac{3}{4}$ (GGE). Felt at Gjirokastër (VII), Vlorë, Berat (VI) (MLD), (FC) gives only "an earthquake at Ioánnina"; the nineth destruction of Ioánnina, a disastrous shock also on Jan. 14 (MCA).

Febr. 3, 20^h30^m, 38·4°N, 0·7°W, R.16, $I_0 = ?$, $I_0 =$ VII−VIII (BRS), $I_0 =$ V (MU). Cartagena and Murcia were shaken (FC).

Febr. 4, 04^h15^m, 38$\frac{1}{4}$°N, 20$\frac{1}{4}$°E, $h = i$, R.26a, $I_0 =$ X, $r = 810$ km, (GG, DM, CAM), $I_0 =$ IX−X (GGK), $I_0 =$ XI−XII (MF), 06^h20^m (PAT, FC), $M = 7·9$ (GGE). A disastrous shock in the Ionian islands, most damaging in the Keffalinía's peninsula Paliki, Lixoúrion (Lixuri) and the small villages Anoge, Katokhí (Katoge), Mesokhóra (Mesochorites) and Thinia almost completely destroyed, Argostólion heavily damaged, one third of houses collapsed, "the villages in Keffalínia thrown down, rockslides at the coast, large and deep fissures in the ground", about 200 victims, in total 2642 houses destroyed, in the district of Livathos four villages completely ruined, felt in Zákinthos (severe), Levkás (walls fissured, some houses ruined), Cap Skinari (some houses collapsed), the village Eugerea, Apolionia and Leuke (Lefki) seriously injured, in Itháki severe, in Kérkira (Corfu), at Misolongion (one house collapsed), at Pátrai, Athínai (Athens), Tripolis, in the provinces Mantinea, Kaspakas, Acarnania, Lamía, Agrafa, etc. (PAT, FC, MCA), felt with $I_0 =$ VII(?), at Vlorë, Elbasan and Durrës (MD), (GGM) gives Febr. 7 and many houses collapsed in some villages of Levkás, (FC) gives Febr. 6 as the date of the greatest catastrophe in Greece since 1817, epicentre Paliki.

1867

March 7, 16^h, 39·1°N, 26·6°E, $h = n$, R.26c, $I_0 = $ X—XI (GGK), $I_0 = $ XI, $r = $ 300 km (GG), $I_0 = $ VIII (EGU), $I_0 = $ X (CAM, MJD), $M = $ 6·7 (GGE), 05^h58^m, 38·4°N, 27·2°E, $I_0 = $ VIII, $M = $ 6·1 (ONB).

At Mitilíni 1500—2500 houses destroyed, large cracks in the ground, 500 victims, other localities even more seriously destroyed, several houses collapsed in Khíos and Lemnos, felt at Izmir, Magnisía, Kavalla, Edremit (Adramanti), Gelibolu (Gallipoli), Istanbul V, Turgutlu (Cassaba), Aydin (damage), Aivalik, Edremit, Kydonies (Cydonies) (walls fissured) (PAT, FC, SAM, KLC).

Aftershock on March 19, 09^h, $I_0 = $ VII (ONB, PAT).

March 29 and 30, 41°N 24°E, R 24, $I_0 = $ VIII(?) (GG), $I_0 = $ IX—X (MF), $M = $ 6¼ (GGE).

A violent shock at Drama, several walls and minarets thrown down, felt at Thessaloníki and Filippaívi (Philippoi) (PAT, FC).

July 22, 03^h, 39·3°N, 26·2°E, R.26c, $I_0 = $ VIII?, $I_0 = $ VII, $M = $ 5·5 (ONB), $I_0 = $ VI (EGU).

At Mitilíni (Lésvos) houses collapsed, particularly at Ipios (FC, PAT, LC), not mentioned by (GG, GGK).

July 22, 37°55′N, 40°18′E, R.35 "Sixteen villages near Diyarbakir (Turkey) were swallowed up by the earth after violent shocks" (PAT), not mentioned by (ONB), probably a wrong date, the report might correspond to July 22, 1866.

July 23, 12^h, 40·6°N, 46·3°E, R.32, $I_0 = $ VII (MJD).

At Zurnabad(?) and Kirovabad (Elizavetpol) (BZ) the walls of many houses injured, felt at Zakataly, Telavi, Shemakha, Mukharavani, Tbilisi, Bely Klyuch, Vanka, Geokchal, etc. (BZ, PAT, FC, MO).

Sept. 20, 03^h15^m, 36½°N, 22¼°E, $h = i$, R.26a, $I_{max} = $ VIII—X (GGK), $I_0 = $ VIII (MJD), $I_{max} = $ IX, $r = $ 700 km (GG, CAM), $I_{max} = $ X (MF), 36°N, 23°E (SAM, LC).

A disastrous shock in the province of Mani, Laconia, tsunami along the south coast of Pelopónnisos (ANS), felt in Kérkira (Corfu), Zákinthos, at Filiatrá, Avia, Kalámai (Calamata), Návpliion (severe), Khálki, Pátrai (slight), Tripolis, Athínai (slight), Spárti, Kíthira (Cerigo), Khaniá (Canea); Gythion destroyed by tsunami; damage at Oesylos, at Kythnos tiles fell from roofs, Areópolis suffered much, violent shaking at the cape Drosos where houses collapsed and some persons perished, at Izetzino thirty houses thrown down, at Petrína walls of a church fissured, the belfry of the monastery of Zarbitza fell, new springs originated, felt also in Sicily, Kríti, Malta, tsunami observed at Catania (PAT, FC, GGT, LC), felt at Brindisi (MCA). (FC) gives a damaging foreshock on Sept. 19, 17—18^h, with a very large shaken area.

Oct. 22, 20^h30^m UT, 39¼°N, 23½°E, $h = n$, R.26a, $I_0 = $ VII, $r = $ 200 km (GG), $I_0 = $ VI to VII (GGK), $M = $ 5·9 (GGE).

A severe shock felt at Tripolis, Pátrai and in the islands Skiathos and Skópelos (PAT, FC).

Dec. 31, "Some houses destroyed on the north and north-east coast of Iceland at Akureyri, Husavik and Vopnafjörd", $I_0 = $ VIII (MJD), $I_0 = $ IX (MF).

1868

Febr. 18, 17^h, 41·2°N, 43·5°E (BZ), R.33, $I_0 = $ VII?, $M = $ 5·5 (ONB, MJD).

A severe earthquake in Armenia (Caucasus), felt at Leninakan (Aleksandropol) (VI), Akhaltsikhe, Kvirila (strong). Akhalkalaki (some old houses damaged),

Chatakh, Tbilisi (IV), Manglisi, Sharopan, Zurnabad, Telavi (III), Gori (III), Surami (III), Borzhomi, Kars (IV), Atskhuri, Ardagan; epicentre probably near the village Spasskoe, which suffered mostly (BZ, PAT, MO).

Febr. 25, 11^h10^m, $39\cdot9°$N, $41\cdot3°$E, R.33, $I_0 = $ VII—VIII?, $I_0 = $ VIII (MJD), $I_0 = $ IX, $M = 6\cdot7$ (ONB), $I_0 = $ VII, $40°$N, $42°$E (ANV, EGU).

Earthquake in the region of Erzurum, felt in the Caucasus, poorly constructed houses damaged at Leninakan and Akhalkalaki (BZ, MO), Kars, many strong aftershocks till April (PL).

March 18, 17^h05^m, $39\cdot4°$N, $47\cdot4°$E (BZ), R.34, $I_0 = $ VII (MJD), 14^h05^m (BZ).

A severe earthquake in the E. Caucasus, felt at Zurnabad (two chimneys fell, fissures in the walls of some houses), Shusha (large cracks in some walls), Belyasuvar (fissures in walls, plaster fell), Telavi, Zakataly, Dilizhan, Dzhebrail (plaster fell), Shemakha (V), Geokchai, Chatakh, Tsarkie Kolodtsy, Tbilisi; the strongest effects observed at the frontier localities along the Araxes and in the province of Karadag where the epicenter should be located (BZ, MO).

April 23, 02^h37^m, $40\cdot0°$N, $41\cdot7°$E, R.33, $I_0 = $ IX—X?, $I_0 = $ IX, April 11, $M = 6\cdot7$ (ONB), $I_0 = $ VIII (MJD), $I_0 = $ VII (ANV, EGU).

A disastrous shock in the region of Erzurum, villages destroyed, felt at Kars and Tbilisi (weak), (BZ, PAT, MO, FC), epicentre near Pasinler. (ANV) gives also Apr. 23, 04^h19^m, $I_0 = $ VIII, which should correspond to Apr. 12, according to the new style.

May 3, $37\frac{3}{4}°$N, $27°$E, $h = n$, R.26c, $I_0 = $ VII (GG), $I_0 = $ VI—VII (GGK). $M = 5\frac{1}{2}$ (GGE).

A heavy shock in the Sámos, heavily felt at Pagóndhas (Pagonda) where 100 houses were damaged, some poor ones collapsed (GGK), $I_0 = $ VIII—IX, $I_0 = $ VII (MJD).

June 21, 05^h10^m, $45\cdot7°$N, $20\cdot0°$E, R.21b, $I_0 = $ VIII—IX, $I_0 = $ VII (MJD), $I_0 = $ IX (RA, CD, MF), isos. map. $P_5 = 19\ 000\ km^2$ (RA).

A damaging shock in Hungary, at Jászberény churches, monasteries and many houses were seriously injured, two houses collapsed, many chimneys fell down, large fissures in the ground (RA, FC, PAT).

Aftershocks with $I_0 \geqq$ VII: Aug. 20, 19^h, $I_0 = $ VII—VIII? $r = 45$ km, $I_0 = $ VIII—
          X (RA).
         Sept. 17, 18^h, $I_0 = $ VII? (RA).
         Dec. 15 and 26, $I_0 = $ VII? (RA, MJD).

Oct. 3, 23^h30^m, $39\frac{1}{4}°$N, $23\frac{1}{2}°$E, $h = n$, R.26a, $I_0 = $ VII—VIII (GGK), $I_0 = $ VIII, $r = $
= 140 km (GG), $I_0 = $ IX—X (MF), $M = 5\cdot6$ (GGE).

At Skíathos in N. Sporades 150 houses damaged, serious damage also in Skópelos and at Vólos (houses injured) (SAM, FC, LC), felt in Athínai (GGK), damaging aftershocks on Oct. 9 and 11 (LC).

Nov. 1,   (MF) gives an earthquake with $I_0 = $ IX at Reykjavík ($64\cdot1°$N, $21\cdot9°$W, R.2) and Börgarfjord, $I_0 = $ VII (MJD), (FC, PAT, MF) do not report about any damage, violent shocks were probably connected with a volcanic eruption.

Nov. 13, 07^h45^m UT, $45\cdot8°$N, $26\cdot6°$E, $h = i$, R.21d, $I_0 = $?, $I_0 = $ VIII—IX, $M = 6\cdot9$ (PRA), $I_0 = $ IX—X (RA).

A damaging shock in the region of Vrincioaia, felt over large area in Rumania, Bulgaria, as far as Ruse (Ruscuk), Bessarabia and Odessa, felt at Bucureşti, Braşov, Bodoc, Chervonoarmeyskoie (Cubei), Cetatea Albă, etc. (FA, PAT, MO).

1868

Nov. 27, 20^h35^m, 45·7°N, 25·6°E, $h = i$? R.21d, $I_0 = $?, $I_0 = $ IX (PRA).

A heavy shock felt over the same area as the preceding one, felt at Bucureşti, Ruse, Constanţa, Odessa, etc. (FA, MO).

1869

Jan. 10, 10^h37^m UT, 42·6°N, 18·1°E, R.22, $I_0 = $ VII? (MJD), $I_0 = $ VIII (TD).

Many shocks at the beginning of 1869, most severe at Dubrovnik, where many houses were damaged (FC), the series of severe shocks lasted until June 1869. No description in (BA), not mentioned by (PAT).

Jan. 10, 20^h30^m, epicentre in the Black Sea?

A largerly extended shock felt in Rumania, Moldavia and at Odessa (FA, FC).

Febr. 7, 05^h UT, (43$\frac{1}{2}$°N, 11$\frac{1}{2}$°E), R.19, $I_0 = $ VII (MR, MJD).

Epic. N-NE of Siena (MR). At Siena many chimneys fell down, violent at Fontebecchi and Monteliscai, felt at Arezzo, Firenze and Perugia (PAT, FC, BI).

March 18, (40·4°N, 19·5°E), R.25, $I_0 = $ VII−VIII (GGK), $I_0 = $ VIII (MD).

A series of shocks in Albania, Vlorë VIII(?), Durrës VII, Elbasan VII (MD), not mentioned in (MCA), felt in Kérkira (Corfu) (GGK).

April 18, 04^h, 36$\frac{1}{2}$°N, 27$\frac{1}{2}$°E, $h = i$, R.26c, $I_0 = $ VIII (GG, MJD), $I_0 = $ VII−IX (GGK), $r = 400$ km (GG), 05^h45^m, 36·3°N, 28·0°E, $I_0 = $ VIII (ONB), $I_0 = $ IX (AM), $M = 6·9$ (GGE).

A disastrous shock in the island Symé, 75 houses thrown down, most of the houses seriously injured, all houses damaged, several victims, rockslides, the villages in the islands Niséros, Ródhos and Kálimnos suffered also much, felt over Sporades and on the coast at Izmir and Bursa (SAM, PAT, FC), a strong aftershock on April 22 (SAM, GG).

June 25, 14^h UT, 44·3°N, 11·2°E, R.19, $I_0 = $ VII? (MJD), $I_0 = $ VIII (MR).

At Vergato, Marzabotto, Zocca and Caprarola (Capraro) some old cottages collapsed, et Casio some chimneys and parts of walls fell down, felt at Bologna, Siena, Forli, Firenze, Vicence, Urbino (FC, PAT, BI).

Aug. 14, 40·3°N, 19·6°E, R.25, $I_0 = $ IX−X? (GGK, MF), $I_0 = $ X (MD, MCA), $M = $ $= 6\frac{3}{4}$ (GGE).

A catastrophic earthquake along the coast Vlorë-Himara, highest intensity X (?) at Vlorë, Kanino, Narta, Himarë, IX(?) at Smokthinë, Velça, Kudhesi, Karbanarë, Dhermi, Kuç and Durrës (MD), no information about the observed damage.

Sept. 1, a shock ruined Durrës (41·3°N, 19·4°E), R.25, and surrounding localities, $I_0 = $?, $I_0 = $ IX (MD, MCA), $I_0 = $ VIII−IX (GGK), $M = 6\frac{1}{4}$ (GGE), local?

Sept. 2, 12^h, 40·6°N, 48·7°E (AV, BZ), R.32, $I_0 = $ IX, $P > 5000$ km^2, $M = 5·5$ (AV).

A damaging earthquake near Shemakha, epic. to the north from Shaldash and SE from Chaili, in total 227 houses destroyed, 987 seriously injured, 3608 slightly damaged, at Sundi all houses destroyed, Chobani, Khilmili and Chaili suffered much, felt at Shemakha and as far as Bezovand, Dzhengi, Alty-Agach, Kuba, changes in hot springs (BZ, MO, LC, PAT, FC).

Sept. 11, 38·4°N, 42·2°E, R.35, $I_0 = $ VII (ANV).

Sept. 20−25, an earthquake felt over the Aegean islands, particularly in Skíathos (39·2°N, 23·5°E) and Aedipsos, at Lamía houses were damaged (FC, LC), $I_0 = $ VII?

1869

Sept. 26, 21^h UT, 43·5°N, 11·1°E, R.19, $I_0 = VII-VIII$, $I_0 = VIII$ (MR, MJD).
A severe shock in Italy, at San Gimignano (Toscana), most houses were damaged in different extent, two of them collapsed, heavily felt at Certaldo and Poggibonsi, felt at Colle di Val d'Elsa, Firenze, Napoli, Colle, Siena, Castelfiorentino, Volterra (FC, LC, MR, BI); (PAT) gives 21^h45^m, San Germignano, Etna, houses collapsed. Sept. 24 (LC).

Oct. 11, 13^h, $44\frac{3}{4}°N$, 35°E, R.28, $I_{max} = VII-VIII$ (PV), $I_{max} = VII$ (MJD), $I_{max} = IX$ (MF), $r = 250$ km ca.
A very strong earthquake felt throughout the Crimea, at Sudak church walls fissured, the region Sudak-Taraktash suffered much (PV), at Feodosiya and Yalta houses were damaged, felt at Sevastopol and Livadia (FC, PAT, MO).

Oct. 13, 02^h30^m, 46·3°N, 14·2°E, R.22, $I_0 = VII$, $r = 90$ km?, $h = 7$ km (RV, TD, MJD), $H = 12^h37^m$ (TD).
Many chimneys fell at Rádovljica (Radmannsdorf) (FC, LC), fissures in walls and damaged chimneys (RV), no other data.

Oct. 31, 15^h25^m, 17^h26^m, Nov. 1, 04^h, 07^m, Nov. 2, 21^h26^m, 49·9°N, 8·5°E, R.8, $I_0 = VII$, isos. map, $r = 40$, 125, 160, 170 km (SW, RSF).
A swarm of damaging shocks at Gross-Gerau and Rüsselsheim where chimneys were thrown down, felt as far as Giessen, Hanau, Heidelberg, Meisenheim and Remagen, a detailed list of shaken localities in (SW); a foreshock with $I_0 = VII$ on Oct. 31, 14^h25^m UT, $r = 30$ km and 59 km, other shocks with $I_0 = VII$ on Nov. 1, 03^h07^m UT, $r = 160$ km, Nov. 2, 20^h26^m UT, $r = 170$ km.
During the period of maximum activity about 20 shocks per hour were felt at Gross-Gerau, the swarm continued till 1872 (SW).

Nov. 16, 12^h45^m, 34·9°N, 5·9°E, R.15, $I_0 = IX$ (RK, GA), $I_0 = IX-X$ (MJD), $I_0 = X$ (CAM).
A disastrous shock in the region of Biskra, the oases near Biskra suffered very much, at Biskra the military buildings were fissured, at Seriana and El Hebbab many houses, at Sidi Okba 45 houses and at Gurta one third of houses thrown down, houses damaged at Tchouda, Droh, Mechonnek, about 40 victims, felt at Batna strongly and at Setif slightly, landslides? (RK, PAT, LC).

Nov. 28, 18^h30^m, 38·6°N, 16·1°E, R.20, $I_0 = VII$ (MR, MJD), $I_0 = IX$ (MF), Nov. 26 (MR).
A severe shock in Calabria, some poor houses collapsed at Monteleone, strongly felt at Mileto, Nicotera, Soriano, Arena, Monterossa, Pizzoni, Laureana, Polistena, Oppido, Serrata (houses damaged), felt at Reggio, Castrovillari, Messina and Tiriola (PAT, FC, BI).

Dec. 1, 18^h, 36·8°N, 28·3°E, $h = n$, R.26c, $I_0 = VIII$, $r = 220$ km, $M = 6·1$ (ONB, EGU), 37°N, 28°E, $I_0 = X$ (GG, CAM), $I_0 = IX-X$ (GGK, MJD), $M = 7\frac{1}{2}$ (GGE).
A damaging shock with the focus near Menteçe (Menteche), changes in springs, Izmir was violently shaken, the small town Ula completely destroyed, at Marmaris cracks in the ground and walls fissured, felt also in Ródhos and Mákri (FC, PAT), Budrun, Simi, Mitilíni (LC), smaller damage at Ula and Mugla (GGK).

Dec. 26, 19^h, (40·1°N, 44·2°E), R.32, $I_0 = VII-VIII$, $I_0 = IX-X$? (MJD).
Damaging earthquake in Armenia, 6 houses injured at Malye Dzhamzhili, 7 houses at Dzhangigan (?) and 200 m long fissures in the ground near the village on a hill, at Yerevan several walls fissured, felt at Tbilisi and Leninakan (BZ, FC, MO, LC).

1869

Dec. 28, $38\frac{3}{4}°$N, $30\frac{3}{4}°$E, $h = n$, R.26a, $I_0 = $ X, $I_0 = $ XI (GGK, MF), $I_0 = $ IX−X (MJD), $r = 370$ km, $I_0 = $ XI (GG, MD), $I_0 = $ X (CAM).

The town Levkás almost completely ruined as well as the village Tsoukalades, 15 victims, damage reported from other Ionian islands and the coast of Akarnania (Acarnia), felt strongly at Vlorë, Durrës, Preveza and as far as Otranto (PAT, FC, MAC, MD, GGM, LC), tsunami in front of Vlorë (GGT, ANS).

1870

Febr. 8, 16^h30^m UT, 43·6°N, 13·5°E, R.19, $I_0 = $ VII, epic. in the sea? (MR), $h = $ sup? A heavy shock at Ancona with relatively small shaken area, at Ancona chimneys thrown down, all houses fissured, felt strongly at Osimo, Gallignano, felt at Camerino, San Benedetto, Fano, etc. (FC, BI), Loreto, Sinigaglia, etc. (LC).

Febr. 22,   (36·6°N, 29·1°E), R.26c, $I_{max} = $ VIII?, $I_0 = $ VII, $M = 5·1$ (ONB), $I_0 = $ IX (CAM). At Fethiye strong, uplift of the shoreline (PL), at Mákri several houses collapsed, felt in Rhódos, at Amfissa and in the Golf of Kórinthos (SAM, FC, GGK).

March 1, 19^h08^m UT, 45·4°N, 14·4°E, R.22a, 5−6 m subsidence near Novokracine in the alluvial ground, $I_0 = $ IX (TD, CAM), $I_0 = $ X (MF), $r = 180$ km, $I_0 = $ VIII−IX (RV).

At Klana 40 houses destroyed and the others damaged, serious damage (VII−VIII), reported also from surrounding villages Studena, Lipik, Zabice, VI−VII at Susak, Bakar, Trnovo, Lisac, Sežana, Podgraje, felt at Pula, Trnovo (Dornegg), Ilirska Bistrica (Feistritz), Gorizia (Görz), Schwarzenberg, Ljubljana (fissures in walls), St. Georgen, Vigann, Stein, Trieste, Novo Mesto (Rudolphswerth), etc., as far as Venezia, Urbino and Dubrovnik (FC), with $I = $ VII at Trnovo, Podgraje; other localities in (LC); Gomance, Ilirska, Trnovo (RV).

Aftershocks: $I_0 = $ VI−VII? on March 14 UT (RV),

    $I_0 = $ VII? on May 10, 16^h56^m UT, (TD, FC), strong at Klana, Skalica, strong wind at the same time (BA).

    $I_0 = $ VII? on May 11, 01^h50^m UT (TD, FC), strong shaking at Klana (BA).

April 14, 10^h, (41·1°N, 21·1°E), R.25, $I_0 = $ VI−VII (GGK), $I = $ VII (MD), $I_0 = $ VIII (MA). A strong earthquake at Elbasan (MCA).

June 24, 18^h, epicentre? R.38, $I_0 = $?, $h = i$?, $I_0 = $ X−XI and isos. map in (SAM). A heavy shock largely felt, felt at Il Iskandarîya (Alexandria) VII (few damage, tsunami), Ismâilîya (Ismaila), El Qâhira (Cairo), Beirut, Esti Sham (Damascus), Zebdani, Cypruṣ (ANC), Crete (?), Aden, Dardanelles, Greece, Albania, S. Italy, Malta, Benghazi (FC, MCA, ANS), epicentre probably in the eastern part of the Mediterranean Sea near the delta of the Nile; (ANC) assumes Anti-Lebanon as the focal zone.

July 7,   noon, epicentre? R.32, $I_0 = $ VII? A severe shock felt along the Caucasian coast of the Black Sea, at Chutorsk (Kubansk) several chimneys fell down, at Lesnoye several houses were damaged, felt at Dakhorsk, Adler, Lazarevsk, Gelovinsk, Veljaminovsk, Bozhi Vody and Prochnookap (MO, FC), not mentioned in (BZ, GC).

July 11, 01^h30^m, $39\frac{1}{4}°$N, $26\frac{1}{2}°$E, $h = n$, R.26c, $I_0 = $ VI−VII (GGK), $I_0 = $ VI, $r = 100$ km (GG). Minor damage at Mytilíni (Lésvos), felt at Izmir (Smyrna) (GGK).

1870

Aug. 1, 00^h40^m, $38\frac{1}{2}°N$, $22\frac{1}{2}°E$, $h = n$, R.26a, $I_0 = X-XI$, $I_0 = X$ (GG, GGK, CAM), $I_0 = XI-XII$ (MF), $M = 7\frac{1}{2}$ (GGE). Isoseismal map in (SAM), p. 64.

An earthquake disaster in Greece in the Gulf of Kórinthos, fissures, sand craters, new fountains, large rock slides on Parnassós, Kovax and Kirphis, many foreshocks and aftershocks (70 000 within 3 years), 100 victims (GG), the towns Amfissa, Galaxídhion, Arákhova, Kastri (Kastrítsi), Itea, Chrysos, Pyrna, Cassotis and others were ruined, fissures in the ground 5—6 feet wide (PAT, FC, LC).

Sept. 28, 05^h, ($41·3°N$, $19·4°E$), R.25, $I_0 = VIII-IX$ (GGK), $I_0 = IX$ (MD), $P = 150\,000$ km^2 (GGK), $M = 6·2$ (GGE).

A disastrous shock at Durrës, Elbasan VIII(?) (MD); Sept. 28, 1870 — April 18, 1871 earthquakes with different intensity in the region of Ioánnina, Preveza, Kérkira and Elbasan (MCA).

Oct. 4, 17^h UT, $39·2°N$, $16·3°E$, R.20, $I_0 = X$ (MF, FC, PET, CAM).

A disastrous earthquake in Calabria, new springs originated, mud craters, a total disaster in the region Cellara—S. Stefano—Longobucco—Figline—Vegliaturo, 1600 houses damaged, 250 victims, serious damage at Cosenza, Celico, Rovella, Zumpano, Motta, etc., felt slightly at Tropea, Palmi, Reggio, Napoli, Messina and Palermo (FC, PAT, BI).

Oct. 19, 23^h45^m, $39·2°N$, $16·3°E$, R.20, $I_0 = VIII$?

A strong aftershock at Cosenza, at Conegliano and Rossano houses were injured and some collapsed (FC, PAT).

Oct. 25, 19^h, Many shocks in the provinces of Amfissa and Ftiotis (Phthiatis) caused damage, felt strongly in Athínai (FC), ($38\frac{1}{2}°N$, $22\frac{1}{2}°E$), R.26a, $I_0 = ?$, $I_0 = XI$ (MF), not mentioned in (GG), Amfissa ruined, an aftershock of Oct. 5 (GGK).

Oct. 30, 18^h30^m UT, $44·1°N$, $12·1°E$, R.19, $I_0 = VIII-IX$?, $I_0 = X$ (MR), $I_0 = IX$ (CAM), $I_0 = VIII$ (MJD).

Epic. area Dogheria—La Caminate (?) (MR), destruction at Bertinoro (walls collapsed), Meldola (houses collapsed), Predappio, Castrocaro (some victims), Forlimpopoli (large fissures), minor damage at Cesena, Ravenna, felt also at Firenze, Siena, Bologna, Urbino, Perugia, Iesi (FC, PAT, BI).

1871

Jan. 22, $37°N$, $24°E$, R.26b, $I_0 = ?$, $I_0 = VII-VIII$ (GGK), $M = 6·4$ (GGE).

A sequence of shocks in Milos, the strongest effects (no damage) observed in Pylonos, Kímolos and Sérifos, near Kastro the ground fissured (GGK).

Febr. 10, 04^h32^m UT, Febr. 16, $49·8°N$, $8·6°E$, R.8, $I_0 = VII$, $r = 150$ km (RSF), $I_0 = IX$ (MF), isos. map. in (SW).

A heavy shock in the province of Hessen (Germany), allegedly "serious devastation at Lorsch" but no other reliable information is available; felt as far as Bonn, Marburg, Nürnberg, Esslingen, Strasbourg and Saarbrücken. About 42 chimneys were thrown down at Lorsch during the aftershock on Febr. 16 (SW, FC, PAT, LC).

Febr. 28, 07^h30^m, $40·5°N$, $19·5°E$, R.25, $I_0 = ?$, $I_0 = VII$ (MD).

A heavy shock at Vlorë, felt at Berat (MD, FC, MCA).

March 17, 23^h05^m, epicentre? R.10, $I_0 = ?$, $I_0 = VII$ (DCB).
A series of shocks in NW England and Wales, the area heavily shaken comprises Scarborough, York, New Malton, Leeds, Bradford, Preston, Longridge, Kendal, Penrith, Carlisle, Newcastle and Sunderland, there are, however, no reports on damage available (PAT, FC).

March 25?, $38·5°N$, $43·9°E$, R.33, $I_0 = VII$, $M = 5·5$ (ONB, EGU).
Region of Van.

April 8, 23^h30^m, $39\frac{1}{2}°N$, $20°E$, $h = n$, R.25, $I_0 = VII-VIII$ (GGK), $I_0 = VIII$ (GG), $I_0 = IX$ (MF), $M = 6\frac{1}{4}$ (GGE).
A heavy shock accompanied by destruction was felt in Kérkira (Korfu) (GGK). Not mentioned in (FC, PAT).

June 7, $36·8°N$, $28·3°E$, R.26c, $I_0 = VII$, $M = 5·5$ (ONB, EGU), $M = 6\frac{1}{4}$ (GGE).
A damaging shock in the Sporades, little damage also at Marmaris (SAM, PAT, FC, GGK, LC), epicentre in the Sporades, felt over the nearby coast of Anatolia (PL).

July 29, 21^h35^m, ($43·3°N$, $10·6°E$), R.19, $I_0 = ?$, $I_0 = VIII$ (MJD), $I_0 = IX$ (MR, CAM), $I_0 = IX-X$ (MF).
A severe shock in Central Italy, felt violently at Livorno, Guardistallo (epic.?), Montescudaio, Casale, Bibona, further at Pisa, Grotta, Ferrata, Rocca di Papa, Siena and Porto Ferrajo (FC, PAT), there are, however, no reports on damaged houses; not mentioned in (BI)!

Oct. 8, 11^h, ($38·4°N$, $26·1°E$), R.26c, $I_0 = VIII$, $M = 6·1$, $r = 370$ km (ONB), $M = 6·9$ (GGE).
A heavy shock in Khíos (Chios), felt also in Canakkale (Dardanelles), at Radosto, Tekirdag, Burgas and Istanbul (SAM, FC, PAT, GH, PL); two shocks? one in Khíos and another one in the Marmara sea?

Oct. 11, $40·4°N$, $26·7°E$, R.27, $I_0 = VII$, $M = 5·5$ (ONB, EGU).
Some damage at Gelibolu (PL), not mentioned in (PAT, FC).

Oct. 22, 13^h UT, R.19, $I_0 = VII?$ (MR, MJD).
At Vagliagli serious damage, one wall fissured by a shock which was felt also at Siena (BI, PAT, MR, FC).

Nov. 2, 03^h20^m, $45·8°N$, $27·4°E$, R.21d, $I_0 = ?$ (FA), local?
A severe shock "destroyed all large houses at Tecuci", "in some houses the chimneys fell", very differing observations, no reports from other localities (FA).

Dec. 2, 22^h UT, $45·9°N$, $15·0°E$, R.22b, $I_0 \leq VII?$, $I_0 = VIII$, $r = 120$ km, $h = 5$ km, $M = 4·8$ (RV). A severe shock in Slovenija, no reports on damage corresponding to $I_0 = VIII$, the localities quoted are Trebnje (RV, NL), Mokronog (Nassenfuss), Tressen (V−VI), Dec. 3 (FC, PAT). At Trebnje windows broken from frames (BA).

Dec. 3, 22^h UT, $46·1°N$, $14·6°E$, R.22b or 18b, $I_0 = ?$, $I_0 = VII$, $r = 54$ km (TD).
Aftershock? No data in (TT, RV, PAT).

Dec. 11, 14^h, $39·9°N$, $43·6°E$ (BZ), R.34?, $I_0 = VII-VIII?$
A violent shock in Armenia, at Incesu (Indzha) 4 houses destroyed, 3 houses with injured walls, at Kalacha-Parchinis a rockslide, the old fortress damaged, at Molla-Gaspar a collapsed wall killed a man, felt at Echmiadzinskiy Uyezd (BZ, MO, FC, PAT).

1872

Jan. 28, 15^h, 36·6°N, 3·7°W, R.16, $I_0 = ?$, $I_0 = $ VIII (BRS), $I_0 = $ III (MU).
An earthquake felt at Malaga, Velez-Malaga, Alhama and Granada, at Motril various houses were damaged (BRS).

Jan. 28, 07^h, 40·6°N, 48·7°E, R.32, $I_0 = $ IX—X (MJD), 40·3°N, 48·2°E, $M = 5·5$ (AV).
A disastrous shock in the E. Caucasus region, Shemakha was almost completely destroyed, only 20 houses remained undamaged, the heavily shaken area comprised Dzhabani, Marazy, Khilmili, Sundi, Chaili, Alty-Agach, stations Dzhengi and Arbat, Infikheran, Adzhidara, Dzhafarkhan, Molgan, Chukhur-Yurt, Kilva, Meisary, Kelakhen, Charagan, Chukhanli, Matrasy, station Akhsu, Kush-Engidzha, 118 victims, the radii of pleistoseismal area are 45 and 20 km, respectively; felt as far as Kuba, Baku, Sabirabad, Karamarjan (BZ, FC), Jan. 16, 17 (LC).

Febr. 11, 20^h, $39\frac{3}{4}$°N, $20\frac{1}{4}$°E, $h = n$, R. 25, $I_0 = $ IX—X (GGK), $I_0 = $ X (GG), $I_0 = $ = X—XI (MF), Febr. 10 (MD).
Almost complete destruction of the villages Sagiada and Konispol on the coast facing the Kérkira, some victims, fissures in the ground at Sagiada, felt in Albania on Febr. 10(?) at Permet VI, Vlorë VI, Durrës V (MD).

March 6, 14^h55^m, 50·8°N, 12·3°E, R.6, $I_0 = $ VIII, $r = 290$ km (SW, KC), $I_0 = $ VII (MJD), $I_0 = $ IX (MF), isos. map in (SW), p. 76.
A damaging shock in the province of Thüringen (Germany), largely felt as far as Wroclaw, Praha, Passau, Hechingen, Heidelberg, Marburg, Braunschweig, epicentre near Posterstein, where walls and ceilings cracked, one wall collapsed, rockfalls, localities with $I = $ VII: Gössnitz, Schmölln, Crimmitschau, Gera, Glauchau, Greiz, Niederwinkel i. Sa., Nöbdenitz, Waldenburg i. Sa., a detailed list of other shaken localities given in (SW, FC, KC).

April 3, (18^h30^m), 44·9°N, 35·4°E, R.28, $I_0 = $ VII?, $I_0 = $ VI $+$ (MSK), $I_0 = $ VI—VII (PV), $r_5 = 40$ km.
A strong earthquake at Feodosiya (Crimea), many houses damaged, destruction of some structures, a landslide in the mountains (PV).

Apr. 2, 07^h45^m, 36·2°N, 36·1°E, R.35, $I_0 = $ X, $M = 7·3$ (ONB, PL, EGU), Apr. 3 (CAM, PL, MJD).
A disastrous earthquake at Antakya (Antiochia) one third of which was totally ruined, 500—1800 victims, the other houses were seriously damaged except the 150 wooden ones, also one half of the town Samandag (Süveydiye) destroyed, Altinözü was damaged, damage at Fatikli, felt at Halab (Aleppo), Beirut, Esh Sham (Damascus), Diyarbakir, Iskenderun (Alexandretta) and Trípolis (FC, LC, PK), Tel Aviv-Jaffa? (KAD). Aftershock of the same intensity $I = $ X? on April 10 (ONB), (FC) gives only "an earthquake at Antiochia"; another aftershock with $I_0 = $ VII in the night on May 15 (ONB).

April 18,   (64·7°N, 13·6°W), R.2, $I_0 = $ VIII? (MJD), $I_0 = $ IX—X (MF).
A shock destroyed 20 houses at Husavik (Iceland) (FC, LC), many houses and farms destroyed (MJD), tsunami took away houses in the island of Flatey (MF).

May 19—23, 39·3°N, 0·5°W, $h = $ sup, R.16, $I_0 = $ VIII (MU), 39·2°N, 1·5°W, May 20 (BRS), $I_0 = $ IX—X (MU).
A violent shock in the province of Valencia, at Carlet several houses were destroyed, felt strongly at Benimodo, Alginet and Alcudia de Carlet (GS).

1872

July 29,   08^h, 35·9°N, 0·1°E, R.14, $I_0 =$ VII (GA).
Several houses fissured at Mostaganem, the belfry at Mazagan injured, felt at Oran and Alger (RK).

Dec. 26,   12^{h}40^m, 48·4°N, 23·3°E, R.21a, $I_0 =$ VII (RA, EU), $h =$ sup?
A severe shock at Dolgoe (Dolha) where chimneys and plaster fell down, cracks in the walls of a tower, felt also in the surroundings (RA, EU, FA), local.

1873

Date?   40·5°N, 37·8°E, R.29, $I_0 =$ VII, Niksar and Şebinskarahisar (TM).

Date?   A severe shock in Crimea, one half of the mosque at Bakhchisarai (44,8°N, 33·9°E) collapsed (PV), R.28, $I_0 = $?

Jan. 31,   23^{h}15^m, 37$\frac{3}{4}$°N, 27°E, $h = n$, R.26c, $I_0 =$ VIII−IX (GGK), $I_0 =$ IX, $r =$ 320 km (GG, CAM), $M =$ 6·6 (GGE), 38·4°N, 27·2°E, $I_0 =$ VII (ONB), $I_0 =$ VIII (MJD).
A very severe shock in Samos, particularly its eastern part suffered very much, at Wathy and Chora many houses became uninhabitable, felt in the Mikale peninsula and at Izmir, Afyonkarahisar (SAM, FC), Thessaloníki (LC); a strong aftershock on Febr. 29 (GG).

March 12,   20^h UT, (43$\frac{1}{4}$°N, 13°E), R.19, $I_{max} =$ VII (MR), $I_0 =$ IX (CAM, MF), isos map in (BI).
A ruinous shock at Camerino, Fabriano and S. Ginesio, damage at Mondavio, Cagli, Urbino, Perugia, Orvieto, Todi, Spoleto, etc. (BI), felt along the Adriatic coast from Venezia to the south, severe at Ancona, strong at Roma, Spoleto, in Umbria and the Albano mountains, felt also in the Alps at Belluno and at Dubrovnik, in the surroundings, in the province of Macerata (MF), in Austria.

April 19,   15^{h}15^m UT, 45·0°N, 14·9°E, R.22a, $I <$ VII, $I_0 =$ VII, $r_{max} =$ 75 km (TD).
Not mentioned in (RV, FC), at Senj many walls fissured, felt at Karlovac (BA).

June 29,   04^h UT, (46·1°N, 12·3°E), R.18b, $I_0 =$ X (CAM, MF), $I_0 =$ XI (MR, MC).
A disastrous shock near Belluno which was partly ruined (according to (MF) 10% of buildings destroyed), the cupola of the cathedral collapsed, at S. Pietro di Felletto a church was thrown down and killed 38 persons, further victims at Pnos (11 persons), Curanga (4), Visone (2), several hunderd meters long cracks in the ground, large landslides. 52 villages seriously damaged, for details see (BI); 410 buildings completely destroyed, 1376 had to be pulled down later, felt as far as Augsburg, Venezia, Verona (FC), epicentre east of the S. Croce lake (MR). A heavy aftershock on July 5, the cupola of the cathedral at Belluno which had been fissured on June 29 collapsed (FC, BI).
On July 11, 01^h UT at Farra d'Alpago and Pnos another strong aftershock (VII?) (FC, BI).

July 12,   06^h UT, 41·7°N, 13·8°E, R.19, $I_0 =$ VIII, $I_0 =$ IX−X (MF), $I_0 =$ VII−VIII (MR).
Some houses ruined at Alvito and S. Donato, serious damage at Atina, Sora, Settefrati, Belmonte, etc., felt at Napolis, Caserta, Tivoli (BI). Aftershock(?) on July 21 at "S. Donato several houses were destroyed", felt at Alvito (FC), exchange of date?

July 14,   02^{h}55^m, 44·5°N, 4·7°E, R.11, $I_0 =$ VIII, $r =$ 190 km? (PRF, RF), $I_0 =$ IX−X (MF).
A severe shock which caused damage at Donzère, Châteauneuf(?) and Pierrelatte

where houses were fissured and some fell, the church at Châteauneuf fissured and was closed, changes in the springs, felt at Montélimar (one chimney fell), Rac, Allan, Goulardes, Viviers (one house fissured), Le Teil, Rochemaure, Meyné, La Garde (PRF, RF, FC, RO). Another violent shock on Aug. 8, 04^h35^m, with epicentre at Châteauneuf, heavily felt at Donzère, Bourg-Saint-Andéol and Viviers, felt as far as Clermont-Ferrand, the houses which had been fissured by the preceding shock were again injured more seriously (RO), $I_0 = $ VII?

July 25, 09^h30^m, $37\frac{3}{4}°$N, $23\frac{1}{4}°$E, $h = n$, R.26a, $I_0 = $ VIII (GG), $I_0 = $ VII−IX (GGK), $M = 6\frac{1}{4}$ (GGE).
Earthquake at Athínai and Demoskleonas (FC), a disastrous shock in Korinthia particularly at Epidauros (Epidharos?) and Solygia(?) (GGK).

Sept. 11, 9^h UT, a strong shock at Cosenza (39·3°N, 16·3°E) (FC), R.20, $I \leqq$ VII? (MR), houses fissured (BI), local.

Oct. 22, 08^h45^m, 50·9°N, 6·0°E, R.9, $I_0 = $ VII−VIII (SW, RU, FC), isos. map in (SW) p. 83.
A severe shock in the region of Herzogenrath where houses were seriously fissured, the pavement also cracked, many chimneys were thrown down, similar effects observed at Heinsberg, at Waubach (the Netherlands), the church walls cracked from the top to the basement, some houses just under construction were seriously injured, etc., the description of lower intensity observations is given in (SW, RU), felt along the valleys of the Rhine from Neuwied to Kleve and of the Maas from Liège to Gennep and also at Bruxelles, Giessen, Stavelot, Cleve, Münster, Bochum (SW, RU, LC).

Oct. 25, 22^h, 38°N, $21\frac{1}{4}°$E, $h = n$, R.26a, $I_0 = $ VIII, $r = 220$ km (GG), $I_0 = $ VII−VIII (GGK), $M = 6·1$ (GGE), $I_0 = $ IX (MF).
An earthquake was felt over a large part of Peloponnísos (FC), a destructive earthquake in Zákinthos and Elide (MF), largely damaged houses at Vartholomión, 25% of houses at Kyllini was rended uninhabitable, serious damage at Vlacherna, felt at Pyrgos, Pátrai, Kórinthos, Khálki, Kérkira (GGK).

Nov. 9,   40·5°N, 25·6°E, R.24, $I_0 = $ VII (EGU).

Nov. 26, 04^h33^m, 43·1°N, O°, R.17, $I_0 = $ VII, $r = 200$ km? (RF, PRF).
Chimneys fell at Bagnères, felt from Bordeaux to Barcelona (RF, FC).

Dec. 15,   41·4°N, 14°E, R.19, $I_0 = $ VII? (MR).
All houses at S. Pietro Infine and Mignano were seriously damaged (BI).

1874

Jan. 17, 01^h46^m, $38\frac{1}{4}°$N, $23\frac{3}{4}°$E, $h = n$, R.26a, $I_0 = $ VI−VII (GGK), $I_0 = $ VII (GG), $M = 5\frac{1}{2}$ (GGE).
A damaging earthquake in Athínai (FC, GGK).

March 18, 03^h08^m, $38\frac{1}{2}°$N, $23\frac{1}{2}°$E, $h = n$, R.26a, $I_0 = $ VII (GG), $I_0 = $ VI−VII (GGK), $I_0 = $ IX (MF), $M = 5\frac{1}{2}$ (GGF).
Houses fissured and one thrown down at Erithrai (Eritria) (GGK), damage at Khálki (Chalkis), felt at Kími, Vasilikón and on the continent in the region of Lamía (MF).

March 28, 11^h, 36·6°N, 2·2°E, R.15, $I_0 = $ VII (GA, MJD).
Almost all houses fissured at Cherchel, felt at Alger (RK), March 29 (MJD).

1874

May 4,   38·8°N, 38·8°E, R.35, $I_0 \leqq$ VIII?, $I_0 =$ VIII, $M = 6·1$ (ONB), $I_0 =$ IX (CAM). In the region of Maden and Diyarbakir(?) one village destroyed and many other damaged (FC, LC), near Keban-Maden some damage and cracks in the ground (PL).

May 5,   a shock threw down the stone tiles from the roofs of the Rilski monastery (Bulgaria, R.24) (WS).

June 28,   37·8°N, 26·8°E, R.26c, $I_0 =$ VII (EGU).

July 5,   39·2°N, 26·3°E, R.26c, $I_0 =$ VII (EGU).

Sept. 17, 07^h30^m UT, A heavy shock in N. Italy (epicentre unknown) threw down chimneys at Lucca, Ferrara, Baura, Parma and Cento, fissured some houses at S. Pier d'Arena, Lerici, strongly felt at Milano, Nice, Padova, Mantova and Livorno, etc. (BI), $I_0 =$ VII?

Sept. 26,   37·9°N, 15°E, R.20, $I_0 =$ VII? (MR), local.
A heavy volcanic shock at Randazzo in Sicily, many houses fissured (FC, BI, LC).

Oct. 7, 16^h UT, epicentre? R.19, $I_0 =$ VII, isos. map in (BI).
Chimneys thrown down and walls fissured at Tossignano, Tirli, Palazzuolo, Valnera etc., felt slightly at Modena, Prato, Firenze, Pergola (BI).

Nov. 16,   36°N, 28°E, R.26c, $I_0 =$ VII−IX? (GGK). $M = 7·3$ (GGE).
A heavy shock in Ródhos, felt as far as Istanbul (GGK), this report might correspond to Nov. 18.

Nov. 18, 05^h, 39·1°N, 26·9°E, R.26c, $I_0 =$ VII, $M = 5·5$ (ONB).

Dec. 1,   (46°N, $7\frac{3}{4}$°E), R.18, $I_0 =$ VII?, $I_0 =$ VIII (MFS).
At Zermatt the chimneys were thrown down (MFS).

Dec. 6, $15^h 45^m$ UT, $41\frac{3}{4}$°N, 14°E, R.19, $I_0 =$ VIII (MR), $I_0 =$ VII (MJD), isos. map in (BI), $h =$ sup?
Ruinous at Posta, Acquafondata, Gallinaro, S. Donato, Scapoli, etc., felt at Alvito and Castel di Sangro, Roma, Ancona, Arpino, Pesaro, Firenze, Genova, Napoli, Viterbo (FC, BI).
An aftershock(?) on Dec. 9, at Pasta (region of Sora) several houses collapsed and some persons were killed, felt at Isola del Liri (FC), not in (BI).

1875

Jan. 7, 23^h45^m UT, (37·6°N, 15·2°E), R.20, $I_0 =$ VIII? $I_0 =$ VII (MJD), $I_0 =$ IX (MR), local, isos. map in (BI).
At Acireale a series of shocks thrown down several local houses and walls, cracks in the ground, 8 victims (FC, BI).

Febr. 26,   a strong earthquake at Varna, Shumen, Ruse (Bulgaria, R.23), many chimneys and old houses collapsed at Shumen (WS).

March ?,   40·2°N, 26·4°E, R.27, $I_0 =$ VII (EGU), local?

March 18, 00^h ($44\frac{1}{4}$°N, $12\frac{1}{2}$°E), R.19, $I_{max} =$ VII−VIII (MR), $I_0 =$ VII (MJD), isos. map in (BI), epic. in the sea?
At Rimini, Cesenatico and Servia houses were fissured, some walls and chimneys thrown down, felt also at Belluno, Pola, Trieste, Parma, Padua, Firenze, Aquila, Bologna, Roma (BI, FC).

March 23, 06^h, 36·5°N, 2·6°E, R.15, $I_0 = $ VII (GA).
Houses fissured at El Affroun, Oued Djer VI, Mouzaiaville VI, felt at Alger and Medea (RK).

March 27,   39·3°N, 41·0°E, R.33, $I_0 = $?, $I_0 = $ VIII, $M = 6\cdot1$ (ONB), not in (EGU).
Epicentre between Palu, Sivrice and Kiği-Karliova (PL). No additional information available.

April 20, 08^h, (44·2°N, 12·4°E), R.19, $I_0 = $ VII?
Several houses were damaged by a shock at Cesenatico in the region of Ravena, felt at Rimini, Forli, Camerino, Cervia, in Romagna and Toscana (FC).

April 24, $37\frac{1}{4}$°N, $21\frac{3}{4}$°E, $h = n$, R.26a, $I_0 = $ VII−VIII (GGK), $I_0 = $ VIII (MJD), $I_0 = $ VII (GG), $I_0 = $ IX (MF), local? $M = 5\frac{1}{2}$ (GGE).
Kyparissia, (FC, CAM) give April 29 for a "terrible" shock at Kyparissía where collapsed church walls killed 47 persons (GGK, LC).

May 3−5,   38·1°N, 30·2°E, R.30 or 26c, $I_0 = $ X?, $M = 7\cdot3$ (ONB, FC), $I_0 = $ IX (EGU).
A disastrous shock in the Asia Minor, the villages Civril and Yaka(?) were completely destroyed, an devastation at Işikli, allegedly more than one thousand houses destroyed and several thousands persons killed, near Civril the ground fissured and hot springs originated (FC, LC), exaggerated?; epic. probably in the region of Dinar, heavy damage in the region of Dinar and Civril, at Uşak and Afyon small damage, 1300 victims (PL, EGU).

May 11,   38·7°N, 29·4°E, R.30 or 26c, $I_0 = $ VIII? $M = 6\cdot1$ (ONB), $I_0 = $ VI (EGU), May 12 (CAM, FC).
Another disastrous earthquake in the region of Uşak, villages were again demolished and people killed, felt at Izmir (FC), small damage at Uşak (PL), houses fissured at Izmir and in Ródhos (LC).

July 7, $37\frac{3}{4}$°N, 27°E, $h = n$, R.26c, $I_0 = $ VIII−IX? (GGK), $I_0 = $ IX (GG), $M = 6\frac{3}{4}$ (GGE).
An earthquake destroyed 150 houses in Sámos (SAM, LC).

July 25, 6^h30^m, 44·6°N, 33·6°E, R.28, $I_0 = $ VII (PV), $r \leq 100$ km.
A severe shock in Crimea near Sevastopol where walls fissured and plaster fell, two lighthouses were fissured, in the Georgievski monastery the cupola cracked, more damage done at Kadykovka, Karani and near Balaklava (MO, PV), chimneys were thrown down and some damage was done (FC).

Aug. 7, 16^h, ($40\frac{3}{4}$°N, $48\frac{1}{2}$°E), R.32, $I_0 = $ VIII−IX?
A damaging shock in the region of Shemakha, 15 villages NE of Shemakha suffered by the shock, 136 houses destroyed, 160 damaged, felt at Nukha (BZ, MO).

Aug. 8,   (13^h), 45°N, 35·4°E, R.28, $I_0 = $ VII?, $r \leq 150$ km. Feodosiya VI−VII; confusion with April 3, 1872? (MSK).

Aug. 17, 16^h45^m, ($50\frac{1}{2}$°N, $24\frac{1}{2}$°E), R.21a, $I_0 = $ VII (MJD), Aug. 19 (FC).
An earthquake in W. Ukraine-Galitzia in the region of Lvov, most strongly observed at Velikiye Mosty (some chimneys fell), Chervonograd-Krystynopol (walls of the church fissured and some chimneys destroyed), Sokal (many chimneys destroyed, fissures in some walls), felt heavily at Lopatino, Selec, Konotopy, Ugnov, Nesterov, further at Lvov, Radekhov, etc., see (EU, MO), at Dolgobychevo two houses destroyed, an impact earthquake? (EU).

1875

Aug. 21, 36·2°N, 36·1°E, R.35, $I_0 =$ VII?, $M = 5·5$ (ONB). Not mentioned in (FC, PK, EGU).

Oct. ? 40·2°N, 26·4°E, R.27, $I_0 =$ IX?, $M = 6·7$ (ONB), questionable, $I_0 =$ VI (EGU). Canakkale. No details available.

Oct. 21, 00^h, 39·9°N, 41·3°E, R.33, $I_0 =$ X?, $M = 7·3$ (ONB), $I_0 =$ IX (ANV), questionable, not in (EGU).
Erzurum. No details available.
An aftershock on Oct. 22, $I_0 =$ VII (ANV).

Nov. 1, 38·6°N, 26·5°E, R.26c, $I_0 =$ VII, $M = 5·5$ (ONB), $I_0 =$ VI (EGU).
Not mentioned in (FC), some damage at Karaburun and Mordogan (PL).

Nov. 1, 10^h, 39·9°N, 41·3°E, R.33, $I_0 =$ IX?, $I_0 =$ X, $M = 7·3$ (ONB), $I_0 =$ VIII (ANV), 38·5°N, 26·4°E, $I_0 =$ VI (EGU).
Not mentioned in (FC), much damage at Erzurum, 125 houses collapsed at Meldovassi, destructive effects at Karaburun and Mordogan (LC).

Dec. 6, 02^h20^m UT, 41·7°N, 15·7°E, R.19, $I_0 =$ VIII (MJD), $I_0 =$ IX (MF), isos. map in (BI).
At S. Marco in Lamis houses were demolished, at S. Giovani Rotondo two houses collapsed, 150 houses were seriously injured, 4 victims, houses fissured at Foggia, Manfredonia, Biccari, Bovino and Troia, felt at Roma, Urbino, Aquila, Matera, Brindisi and Cosenza (FC, BI).

1876

May 13, 06^h, 38·8°N, 30·5°E, R.30, $I_0 =$ IX?, $M = 6·7$ (ONB), questionable, $I_0 =$ VI (EGU).
At Afyonkarahisar allegedly many inhabitants killed (LC).

June 4, 23^h50^m UT, 42·4°N, 19·3°E, R.25, $I_0 =$ VII?, $I_0 =$ VIII (TD, MD), $I_0 =$ IX (MF).
A heavy shock with much damage at Titograd (Podgorica), Vrelo, Tuzi, no damage reported in (FC), Shkodër VII (MD).

May 24—June 11, 37·8°N, 13·3°E, R.20, $I_0 =$ VII (MR).
Some old walls collapsed and houses fissured at Corleone during the severest shocks of a sequence in May—June, felt also at Palermo, Palazzo Adriano and Misilmeri (BI, FC).

June 26, 16^h, $37\frac{3}{4}$°N, $22\frac{3}{4}$°E, $h = n$, R.26a, $I_0 =$ VIII (GG), $I_0 =$ VII—VIII (GGK), $I_0 =$ IX (CAM), $I_0 =$ IX—X (MF), $M = 6\frac{1}{4}$ (GGE).
Seven villages near Kórinthos, particularly Agios Georgios, were much damaged, severe at Kórinthos, felt in Athínai, Évvoia (Euboea, Avvia), Kefallinía, Vólos (FC), a disastrous shock with a small shaken area at Neméa (some small houses collapsed), in the Demos Pellene fissures and rock slides (SAE, GG), 02^h10^m, felt over all Pelopónnisos (LC). (FC) and (CAM) give another shock at Kórinthos on July 9 which threw down several houses and rocks, not mentioned in (GG) where, however, July 2 is given as the date of a strong aftershock which threw down some houses at Neméa and Sikyon (Sikiá?).

July 17, 12^h17^m, 48·0°N, 15·2°E, R.18b, $I_0 =$ VII—VIII, $r = 240$ km (NW) and 160 km (NE), (TT, RA, KC).
In Wien houses fissured and many chimneys fell down, felt over Bohemia ($I_{max} =$ VI) and Moravia ($I_{max} =$ V) as far as Prosmyky near Mělník, Cheb and Přerov,

1876

the pleistoseismal area involves Scheibbs,.Gaming, Euratsfeld, Wieselberg, Melk, Oberwölbling, Traism, Tulln, St. Pölten, Rabenstein, Frankenfels, Puchenstuben, M. Zell, Weichselboden, Aflenz and Kindberg (KC, LC, TT, FC), a special study published in Mitt. der Erdb.-Kom. der kais. Ak. Wiss. in Wien, N. F., XL, 1911.

July 19, 00^h?, 45·4°N, 15·3°E, R.22b, $I_0 \leq$ VII?, $I_0 =$ VII, $r =$ 16 km (TD), local.
A heavy shock at Bosiljevo (VII), Ogulin V (FC, BA), no damage reported.

Aug. 6, $38\frac{1}{4}$°N, $21\frac{3}{4}$°E, R.26a, $I_0 =$ VI(GG), $I_0 =$ V−VI (GGK), Aug. 12, $I_0 =$ VIII (MJD), $I_0 =$ IX (CAM).
A severe shock threw down several houses at Pátrai in August (FC), some houses were damaged at Pátrai (GGK).

Sept. 13,   (38·1°N, 15·7°E), R.20, $I <$ VII? local.
Slight damage at Reggio Calabria (BI), at Reggio several houses collapsed, strongly felt at Messina (FC, LC).

Oct. 26, 02^h20^m UT, (41·8°N, 12·9°E), R.19, $I_0 =$ VII? isos. map. in (BI).
At Palestrina many walls fissured, some chimneys thrown down, general damage to all buildings, strong at Poli and Gallicano, felt at Roma (BI).

Nov. 30, 09^h, 46·3°N, 17·1°E, R.21b, $I_0 =$ VII (CD), $I_0 =$ VII−VIII (RA), $I_0 =$ VIII−IX, isos. map (RAB).
The strongest shock in a sequence observed at Surd, chimneys thrown down, walls fissured at Surd, Bükkösd, slight damage at Murakeresztúr, Porrog, houses damaged also during two heavy foreshocks on Oct. 21, 14^h, walls fissured (CD, RA) and July 6, 06^h, $I_0 =$ VII (RA), $I_0 <$ VII (CD).

Dec. 6, 08^h, 46·0°N, 18·7°E, R.21b, $I_0 =$ VII?,$I_0 =$ VIII (RA), $I_0 <$ VII (CD).
At Mohács 29 chimneys fell (RA), some fissures in walls (LC).

1877

March 30, 00^h30^m UT, 45·0°N, 14·9°E, R.22a, $I_0 =$?, $I_0 =$ VII (TD).
Not mentioned in (FC, RV). At Senj very violent (BA).

April 4, 19^h45^m UT, 46·2°N, 15·2°E, R.22b, $I_0 =$ VII, $M =$ 4·0, $r_{max} =$ 160 km, $h =$ 4 km (RV), $I_0 =$ VIII (TD).
Three severe shocks in Slovenia, epic. near Lasko (much damage), Celje VII (bricks and chimneys thrown down), felt at Gradec, Šiška, Pula, Ljubljana, Zagreb, Maribor, Landsberg, Ehrenberg, Sevnica (Lichtenwald) (FC, RV, BA, LC).

June 24, 07^h53^m, 50·8°N, 6·0°E, R.9, $I_0 =$ VII−VIII, $I_0 =$ VIII (SW), isos. map in (SW), p. 93, $r =$ 60 km ca.
A damaging shock in the region of Herzogenrath felt between the Maas and the Rhine, in the focal area many chimneys were thrown down and the walls of many buildings were fissured, Herzogenrath VII−VIII, Bardenberg VII, Kerkrade VII, Pannesheide VII, Richterich, etc., for details see (SW, LC, FC, RU).

July 2, 9^h45^m, 38°N, $22\frac{3}{4}$°E, $h = n$, R.26a, $I_0 =$ VII (GG), $M = 5\frac{1}{2}$ (GGE).

Aug. 24, 02^h54^m UT, (41·7°N, 13·4°E), R.19, $I_0 =$ VII−VIII?, $I_0 =$ VII (MJD), $I_0 =$ VIII (MR), local, isos. map in (BI).
Ruinous at Veroli, very strong at Alatri, Frosinone, Bauco, Arce, Anagni, Jenne, felt at Roma, Caserta, Benevento (BI).

1877

Oct. 8, 04^h12^m UT, epicentre?, R.18a, $I_0 =$ VII?, $I_0 =$ VIII (MFS), $I_0 =$ VI?, $r = 250$ km? (RO).

A damaging shock in Switzerland (Mont Blanc massiv?) with the epicentral area surrounded by Genève VI—VII, Bonneville, Annecy and Bellay, the chimneys were thrown down at Roche-sur-Foron, houses fissured at Bellay and Genève, felt strongly at Chambéry VI, Léman, felt as far as Neuchâtel, Chaux de Fonds (MFS), Mühlhousen, Besançon, Bern, Lyons (FC, MFS, LC), Piccolo S. Bernardo VI—VII (BI), Malesine (Italy) seriously damaged (RO).

Oct. 13, $40.6°N$, $27.6°E$, R.27, $I_0 =$ VIII—IX?, $I_0 =$ VIII, $M = 6.1$, $r = 200$ km (ONB, EGU), $37\frac{3}{4}°N$, $27°E$, $I_0 =$ VIII—IX (GG, GGK), $M = 6\frac{1}{4}$ (GGE).

In the island Marmara 128 houses were thrown down in two villages during a sequence of shocks lasting from Oct. 13 to Nov. 1 (FC), heavy damage in the Marmara island, strongly felt at Ezine, Canakkale, Edirne and Istanbul (PL). (GG) gives a different epicèntre in the island Sámos, Avlakion-Kokkarion (45 houses collapsed, 70 houses damaged), very probably *another shock*; not mentioned in (SAM).

Nov. 1, $40.6°N$, $27.4°E$, R.27, $I_0 =$ VIII?, Marmara Islands (TM).

1878

Jan. 27, 10^h06^m, $46.1°N$, $14.9°E$, R.18b, $I_0 = $?, $I_0 =$ VII (TD).

A heavy shock in Steiermark, felt strongly in Judenburg, St. Lambrecht, Neumarkt and Knappenberg (FC), no damage reported, not in (BA, RV, TT).

Jan. 28, $11^h(35^m)$, epic. Jersey? R.11, $I_0 =$ VII? $r = 300$ km? (RPF).

A shock in Normandy, strongly shaken Elbeuf, St. Sever, Rivière-Thilon, felt at Caudebec, Le Havre, Caën, Dieppe, Argences, Paris (FC).

March 12, 21^h26^m, $(44.4°N, 11.5°E)$, R.19, $I_0 =$ VIII? $h = $ sup? isos. map in (BI).

The wault of a church at Quaderna d'Ozzano fissured, some chimneys fell at Castel S. Pietro and Varignana, felt slightly at Venezia and Firenze (BI).

April 19, 09^h, $40.8°N$, $29.0°E$, R.27, $I_0 =$ IX, $M = 6.7$ (ONB), $I_0 =$ VIII (EGU), $I_0 =$ X (CAM), May? (TM).

A disastrous shock in the Marmara region, the town Izmit (Esme) was destroyed, 40 victims, at Izmit several stone-built houses were destroyed, the others were damaged, 4 mosques collapsed, houses were thrown down also at Sapanca, felt at Bursa and also on the European side of the Bosporus (FC), rather strong tsunami at Izmit (ANS); damage in the region of Bursa-Izmit and Sapanca, houses fissured in Istanbul, felt over all the Marmara region (PL).

June 7, 10^h UT, $(44\frac{1}{2}°N, 7\frac{1}{4}°E)$, R.18a, $I_0 = $?, $I_0 =$ VII (MR, MJD).

Some roofs collapsed at Cartignano, strongly felt at S. Stefano Belbo (BI).

Aug. 21, 06^h UT, $46.0°N$, $15.2°E$, R.22b, $I_0 =$ VI—VII, $I_0 =$ V (RV), $I_0 =$ VII (TD, MJD), $M = 3.6$, $r = 50$ km, $h = 16$ km (RV).

At Mokronog (Nassenfuss) many tiles thrown down (FC, BA).

Aug. 26, 08^h, $50.9°N$, $6.4°E$, R.9, $I_0 =$ VIII, $P = 175\,000$ km^2, isos. map in (SW), p. 97.

A severe shock in the Lower Rhine region, the highest intensity $I =$ VIII was reached at Elsdorf, Etzweiler, Oberempt (Ober Emth) and Tollhausen where many chimneys were thrown down, even thick walls were cracked and no house remained undamaged, $I = $ VII—VIII at Buir, Distelrath, Ellen, Golzheim, Höllen, Königshoven, Morschenich, $I = $ VII at Aachen, Bergheim, Butzheim, Düren, Giesendorf, Hilden, Horrem, Jüngersdorf, Linnich, Nettesheim, Norf, Richardshofen, Volmerswerth, $I_0 = $ VI, etc., for detailed list of localities see (SW), felt as far as Basel, Brugge, Amsterdam, Hannover (FC, SW, RU).

Sept. 10, 01^h32^m UT, (44·3°N, 10·1°E), R.19, $I_0 = $ VII (MR, MJD).
An earthquake felt at Livorno, Bedonia, Genoa, Chiavari (FC), at Fivizzano chimneys fell down, some fissures occurred, strong at Bagnone (BI).

Sept. 15, 7^h20^m UT, (42·8°N, 12·6°E), R.19, $I_0 = $ VIII (MR, MJD), $I_0 = $ IX (CAM), $I_0 = $ IX—X (MF), isos. map in (BI).
A damaging shock in Umbria with the epicentre near Montefalco-Castel Ritaldi (MR), about 15 houses collapsed and 40 were rended uninhabitable at Bruna, Marcatello, Fratte and Turritta; Spoleto, Trevi and Foligno were heavily shaken, the shaken area involves Roma, Bologna, Perugia and Viterbo (FC, BI), the church at Montefalco ruined (LC).

Sept. 23, 20^h20^m UT, 45·0°N, 14·9°E, R.22a, $I_0 = $ VII? $I_0 = $ VIII (TD).
A strong shock at Senj (FC), some houses fissured, three chimneys collapsed (BA).

Oct. 4, 00^h46^m UT, 37·3°N, 14·7°E, R.20, $I_0 = $ VII? (MR).
Serious damage at Mineo, felt at Palagonia, Vizzini, Scordia, Militello, slightly at Catania, Acireale, Giarre, Riposto, Piedimonte (FC, BI, LC).

Oct. 8—9, 00^h UT, 45·8°N, 17·0°E, R.22b, $I_0 < $ VII, $I_0 = $ VIII (TD), local?
At Severin a "terrible explosive" shock (BA); not mentioned in (FC).

Dec. 31, 05^h30^m, 47·8°N, 19·9°E, R.21b, $I_0 = $ VII (CD).

1879

Jan. 8, (43·0°N, 44·2°E), R.32, $I_0 = $ VII? (PV).
Serious damage in many houses at Alagir during a severe shock (PV, MO), Jan.1 (PV).

March 22—23, $38\frac{1}{4}$°N, $47\frac{1}{2}$°E, R.34, $I_0 = $ IX—X?
NW Iran, two villages destroyed, several localities from Tabriz to Zendjan and Mianeh suffered much (LC), Ardebil, Kara-Shiran, Meshkidzhik and Dashanly destroyed, other villages near the mountain Sawalan (Nir, Khadili, etc.) partly destroyed (MO).

April 16—17, 22^h15^m, 46·1°N, 20·6°E, R.21c, $I_0 = $ VI—VII (RA), $I_0 < $ VII (CD).
Several chimneys fell down at Sînnicolaul Mare, felt at Kiskomlo (RA).

April 27, 04^h UT, 44·2°N, 11·6°E, R.19, $I_0 = $ VII—VIII, isos. map in (BI), $h = $ sup?
The strongest shocks of a sequence in the region of Romagna reached the intensity VII—VIII at Palazzuolo and Casola Valsenio, slight at Firenze and Forli (BI), a castle and a church damaged, three houses and many chimneys collapsed at Castel del Rio (LC).

May 2, at the village Salka (Schabka) in the district Soroki (Szoroki, 48·1°N, 28·2°E), Bessarabia, R.21d?, 27 houses were thrown down and the ground fissured during a severe shock (FC, FA, LC), $I_0 = $ VIII (MJD), $I_0 = $ IX—X (MF), local?

1879

May 2, $15-16^h$ UT, (48·8°N, 23·2°E), $I_0 \leq$ VII (RA), local.
At Zawadka, district of Volovec, small fissures originated in stone-built houses during a sequence of shocks (EU), at Babuliska one stone built house fissured, other effects only $I = $ IV—V (RA).

May 26, $37\frac{1}{2}$°N, 15°E, $h = $ sup, R.20, $I_0 = $ VII, volcanic.
A volcanic shock during an eruption of Etna caused fissures in walls and damaged the church at Riposto (FC, PAT).

June 17, $37\frac{1}{2}$°N, 15°E, $h = $ sup, R.20, $I_0 = $ IX—X?, $I_0 = $ X—XI (MF), main volcanic shock.
Between Monochilo and Zafferana 600 houses and many churches were destroyed in an area $2 \text{ km} \times 4 \text{ km}$, 10 victims, Bongiardo and S. Venerina suffered less (FC, BI, LC), (MR) gives Aug. 7—8 and $I_0 = $ IX at Carico, Aci Platani and Linguaglossa.

July 3, 14^h15^m, 38·2°N, 22·6°E, R.26a, $I_0 = $ VII, $r = 100$ km (GG), $I_0 = $ VII—VIII (GGK).
Epicentre near Xilókastron, all houses fissured, felt at Kórinthos, Argos, Amfissa, Chrysso, Pátrai, Athínai (GGC, FC).

Sept. 28, 15^h30^m UT, 44·8°N, 21·5°E, R.22b, $I_0 = $ VII, $r = 12$ km (TD), local, not mentioned in (FC).
Veliko Gradište, a foreshock of Oct. 10?, no data on damage in (BA).

Oct. 9,   in the evening, in the region of Kurgansk (44·9°N, 40·2°E), Varenikovsk, Troitsk (45·1°N, 38·1°E), Gostagaievsk (45·0°N, 37·4°E), Temryuk (R.32) houses were fissured and some corners collapsed (MO), $I_0 = $ VII (MJD).

Oct. 10, 15^h45^m UT, 44·9°N, 21·4°E, $h = n$, R.22b, $I_0 = $ VIII (PRG, MJD), 44·6°N, 21·6°E, $I_0 = $ IX (TD, MF, RA), Oct. 11 (RA), 45.7°N 21·7°E (NR).
A strong earthquake in Banat; at Golubac, Drenkova chimneys and ceilings collapsed, VII? at Veliko Gradište, Majdanpek, Bela Crkva, Kruševac, etc., felt at Oršova, Moldova, Beograd, Timişoara, in Moldavia and S. Hungary (FC, RA, BA). According to (FC, FA) a relatively stronger shock occurred on Oct. 11, in the same area, $I_0 = $ VIII? (TD).
Aftershocks with $I_0 = $ VII on Oct. 10, 18^h20^m UT, 21^h30^m UT,
               Oct. 11, 01^h00^m UT, 11^h45^m UT,
               Oct. 17, 02^h53^m UT,
               Oct. 20, 10^h45^m UT (TD).

Oct. 31, 18^h30^m UT, 46·1°N, 20·6°E, R.21b, $I_0 = $ VII (RA), 45·9°N, 20·4°E, $I_0 = $ VIII (TD), $I_0 < $ VII (CD), 46·9°N, 20·4°E (RN).
Fissures in the walls at Sînnicolaul Mare (Nagyszentmiklos) (RA), several houses collapsed at Nagyszentmiklos, felt at Cărpinis and Periam (FC), VIII at Mokrin (BA).
Aftershocks with $I_0 = $ VII: Oct. 31, 18^h31^m UT,
               Nov. 1, 07^h UT (TD).

Nov. 19, 23^h10^m, 45·6°N, 21·5°E, R.21c, $I_0 = $ VII?, $I_0 = $ VIII (RA).
An earthquake in the region of Timişoara, fissures in the walls, chimneys fell, felt at Oršova, Ghioroc (Gyorok), Lipa, etc. (FA, RA).

Dec. 7, 20^h05^m, 47·8°N, 19·9°E, R.21b, $I_0 = $ VI?, $I_0 = $ VII (RA), not mentioned in (CD).
Fissures in some walls at Gyöngyös (RA).

1879

Dec. 22, 04^h03^m UT, 44·6°N, 21·6°E, R.22b, $I_0 = ?$, $I_0 = $ VII, $r = 33$ km (TD).
A strong shock at Golubac, Moldova (FC, BA).

Dec. 30,   46·1°N, 6·8°E, R.18a, $I_0 = $ VII, $r = 140$ km (RPF, RF, MJD).
The area of $I = $ VII was surrounded by the villages Sixt, Samoëns, Montriond
and Saint Jean d'Aulph and covered the valleys of the rivers Arve and Dranse,
felt as far as Lyon, Lucerne, Chamonix, Annecy, etc. (RO).

1880

Jan. 27, 15^h30^m UT, 43·3°N, 18·1°E, R.22a, $I_0 = ?$, $I_0 = $ VII (TD).
A very strong shock at Nevesinje in Bosnia (FC, BA), no report on damage.

Febr. 23, 21^h30^m UT, 44·6°N, 21·6°E, R.22b, $I_0 = ?$, $I_0 = $ VII (TD), not mentioned in
(FC).
Moldava, Golubac and Gradište (V) (BA).

March ?,   38·4°N, 26·1°E, R.26c, $I_0 = ?$, $I_0 = $ IX−X (ONB), $I_0 = $ IX (CAM, EGU),
erroneous?
Heavy damage in the Khíos (Chios) island, 4000 victims (PL), *very probably an
exchange for the catastrophic shock on April 1881.*

March 28,   42·0°N, 35·2°E, R.29, $I_0 = ?$, $I_0 = $ VII, $M = 5·5$ (ONB), $I_0 = $ VIII (MJD),
$I_0 = $ IX (CAM), not mentioned in (FC).
The village Heleddi near Sinop, i. e. its 60 houses, collapsed into the sea during an
earthquake (FC).

April 13, 12^h20^m, 44·6°N, 21·6°E, R.22b, $I_0 = ?$, $I_0 = $ VII, $r_{max} = 153$ km (TD), not mentioned in (FC).
Golubac, Moldova (BA).

May 29,   42·0°N, 1·1°W, R.17, $I_0 = ?$, $I_0 = $ VII (BRS), $I_0 = $ III (MU).

July 4,   46·3°N, 7·9°E, R.18a, $I_0 = $ VII (MJD), $I_0 = $ VI−VII (RO), $r = 140$ km, $I_0 = $
$ = $ VII−VIII (MR), $I_0 = $ VIII (MFS, BI).
At Brigue, Bérisal, Fiesch and Viège large fissures in the walls and damage inside
the houses, strongly felt at Simplon, Andermatt, Gondo, rockslides (MFS), felt
at Milano and Poschiavo (BI).

July 22,   38·1°N, 27·8°E, R.26c, $I_0 = $ VII (EGU), 38·4°N, 27·2°E, $I_0 = $ VIII, $M = 6·1$
(ONB, MJD), $I_0 = $ IX (CAM).
A severe shock at Izmir and surroundings, many houses were destroyed and many
people killed (FC), not given by (PL).

July 24, 25, $40\frac{3}{4}$°N, $13\frac{3}{4}$°E, R.19, $I_0 = $ VII?
Two shocks fissured houses, the first one felt at Ventotene on July 24 was very
local, the other one on July 25, felt strongly in the island of Ischia, had probably
a deeper focus being felt up to the distance of 100 km about (BI).

July 29, 04^h40^m, 38·6°N, 27·1°E, R.26c, $I_0 = $ IX, $M = 6·7$ (ONB, EGU), June 29, $I_0 = $ X
(CAM), $I_0 = $ VII−IX (GGK).
In the region between Izmir and Gediz about 100 houses were destroyed, the
others were damaged, 30 victims, the village in the plain of Gediz Nehri (Haer-
mus) as far as Magnisia suffered more than Izmir, particularly at Menemen, where
all houses were rended uninhabitable, further destruction at Magnisia, Giaurköy,

Horoskiö, Kordelio, Tomaso; cracks in the ground and new springs interrupted the rails between Izmir and Turgutlu, felt also in Khíos (Chios), Lésvos and Sámos and Istanbul (GGK, FC). Menemen, Izmir heavily damaged, at Turgutlu, Manisa and Alasehir little damage (PL).

Sept. 21, (46·8°N, 7·1°E), R.18a, I_0 = VII?, I_0 = VIII (MFS), local.
A strong local shock threw down tiles and chimneys at Fribourg, walls were fissured (MFS, LC).

Oct. 3, 04^h46^m, 46·4°N, 23·8°E, $h = n$, R.21d, I_0 = VIII (PRG), I_0 = VII (MJD), I_0 = = IX, P = 68 000 km^2, h = 10 km (RA, FA).
A severe earthquake in Transsylvania with a large area of perceptibility; houses collapsed at Luduş (Marmaros-Ludas) (FC), houses in the epicentral area rended uninhabitable (RA), felt at Debrecen, Baia-Mare, Borsec, Ditrău, etc. (FA).

Oct. 22, an earthquake damaged several houses and killed two persons at Shemakka (40·7°N, 48·6°E), E. Caucasus, R.32 (MO, BZ), I_0 = VII (MJD), local? Dec. 2, I_0 = VIII (FC).

Nov. 9, 06^h34^m UT, 45·8°N, 16·0°E, R.22b, I_0 = IX, r = 435 km (TD, CAM, RV, MF), I_0 = VII (MJD), I_0 = IX−X, P = 330 000 km^2 (RA), details in the report of F. V. Hochstetter in Monatsblättern des Wiss. Club in Wien, Ausserord. Beil. zu Nr 3, Jg II (1881?), 1−14.
The disastrous earthquake of Zagreb (Agram) with epicentre closely to the town, at Zagreb about 500 houses were seriously injured and became partly uninhabitable, chimneys and parts of roofs were thrown down, 3000 houses damaged, some churches suffered much (towers of the Marcus and Maria churches were so fissured that there was a danger of collapse), casualties, very severe also in the neighbouring country, some castles were partly thrown down, the churches at Remete and Granešina were destroyed, the church at Ščitarjevo (Ščitarje) had to be closed for serious damage as well as those at Šestina, Zelina, Prozor and Bosjakovina, at Vrabec the church tower collapsed, the chimneys and roofs were thrown down and walls fissured, etc., near Resnick and Drenje mud erupted in the valley of Sava along many cracks in the alluvial ground, subsidence in the forest near Stubica (FC, BA, Hochstetter) felt as far as Ancona, Belluno, Bologna (RV, BI), in S. Bohemia at Č. Budějovice, in Moravia at Prostějov etc., in SW Slovakia at Bratislava, Komárno (KC), in Banat at Timişoara (FA), Hungary (RA).

Aftershocks: Nov. 10, 07^h UT, 46·0°N, 16·0°E, I_0 = VIII? (TD), a strong shock at Zagreb and surroundings (FC).
Nov. 11, 10^h UT, 46·0°N, 15·9°E, I_0 = VIII? (TD), panic, bells sounded (FC),
Nov. 12, 09^h UT, 45·8°N, 15·3°E, I_0 = VII? (TD), several walls collapsed at Zagreb (FC),
Nov. 13 22^h UT, 46·1°N, 16·0°E, I_0 = VII? (TD).
Nov. 15, 03^h UT, 45·2°N, 16·1°E, I_0 = VII? (TD), the shock caused destruction in neighbouring villages and in the mountains,
Nov. 15, 23^h UT, 45·8°N, 16·0°E, I_0 = VIII? (TD), 2 heavy shocks (FC),
Nov. 16, 00^h30^m, 02^h UT, 45·8°N, 16·0°E, I_0 = VII? (TD), heavy shocks (FC),
Nov. 25, 21^h UT, 46·0°N, 16·2°E, I_0 = VII? (TD),

Dec. 7, 23^h UT, 45·9°N, 16·2°E, $I_0 = $ VII? (TD), a strong shock (FC),

Dec. 16, 22^h UT, 45·6°N, 15·3°E, $I_0 = $ VIII? (TD), a largely felt, severe shock (TD).

Dec. 18,   an earthquake at Somaki (Shemakha?) in Armenia where several persons were killed bellow the ruined houses (FC), $I_0 = $?, $I_0 = $ VIII (MJD), not mentioned in (BZ, MD).

Dec. 25, 14^h30^m, 45·8°N, 26·6°E, R.21d, $h = i$, $I_0 = $ VII–VIII?, $I_0 = $ VIII–IX, $M = 6·9$, $P = 92\,000\ km^2$ (PRG, RA), $P = 118\,000\ km^2$ (FA), $I_0 = $ VII (MJD).
A severe shock in the Vrincioaia (Vrancea) focal region, at Silistra the walls cracked, a violent shaking observed at Sibiu, Tușnad, Vărghiș, Praid, Tulcea, Vaslui, etc., strongly felt at Ruse, Kishinev, București, Izmail, Odessa, etc. (FA, FC).

1881

Jan. ?,   "The hospital at Makhachkala (43·0°N, 47·5ᵛE, R.32) was damaged by a strong earthquake, $I = $ VI" (PV).

Jan. 27,   (46·9°N, 7·5°E), R.18a, $I_0 = $ VII, $I_0 = $ VI–VII (SW), $I_0 = $ VII (MJD), $I_0 = $ VIII (MF).
A severe shock near Bern where about 20–100 chimneys were thrown down and walls of many houses were fissured, the same intensity at Köniz, Münchenbuchsee and Munzingen, outer limits of the shaken area touched Genève, Freiberg i. Ü., Luzern, St. Gallen, Schaffhausen, Mulhouse i. E., Biel (FC, LC).

Febr. 26 or 27, At Glavnica near Zagreb and at St. Ivan Zelina chimneys allegedly collapsed, $I_0 = $ VII? (MJD, FC, LC).

March 4, 12^h UT, 40·7°N, 13·9°E, R.19, $h = $ sup, $I_0 = $ IX–X, $I_0 = $ VIII (MJD), $I_0 = $ IX (CAM), $I_0 = $ X (MR, MF), local.
A destructive shock in the Ischia, a part of Casamicciola completely destroyed (250 houses), 126 victims, houses fissured at Forio, Monterone and Fontana, limited only to the island Ischia (BI, FC, LC).

March 24, 17^h45^m UT, 45·3°N, 15·3°E, R.22a, $I_0 = $ V?, $I_0 = $ VII (TD).
At Zagreb the dishes on the table were jumping (BA).

April 3, 11^h30^m, 11^h40^m, $38\frac{1}{4}$°N, $26\frac{1}{4}$ E, $h = $ sup?, R.26c, $I_0 = $ X–XI (GGK), $I_0 = $ XI, $h = n$, $r = 220$ km (GG), 38·4°N, 26·1°E, $I_0 = $ X, $M = 7·3$ (ONB, CAM), isos. map in (SAM), p. 52 and in (GGC), pp. 224.
An earthquake catastrophe in the island Khíos, the villages in its SE part were completely destroyed, i.e. Neueta, Kalamoti, Kallimasia, Tholopotamion, Myrminghion, Neochori, Vunon, Phlatsia, Koeni, Didymoi, Kataraktis, Kato to Panagia, in total of the 64 villages of the island 25 were destroyed and in 15 villages serious damage was observed, 4181 inhabitants were killed, according to (SAM) the northern part of the island was not damaged, i. e. the radius of the pleistoseismal area was small; felt at Izmir; subsidence of the ground by 0·8 m near Myrminghion, cracks, sand craters, landslides, changes in fountains, disaster reported also from the Çeşme peninsula and Çeşme district where 3460 houses collapsed and 3685 persons were killed, felt as far as Chalandrion (N), Foça (E), Athínai (W) (GGC, GH, FC, SAM). Many strong *aftershocks*, according to (FC) from April 3 to April 9 about 250

aftershocks, among them 30—40 damaging or destructive ones. The most severe shocks were on

April 5, further destruction of houses (GGC).

April 11, 06^h, collapse of ruins (GGC).

April 11, 17^h, destruction of the still standing part of the town Khíos and of many other localities, two shocks of the same intensity (GGC), (I_0 = IX?).

April 12, 16^{h}30^m, collapse of many damaged houses at Khíos (GGC).

April 13, accomplishment of the destruction (GGC).

April 18, new ruins at Khíos (LC).

May 20, I_0 = VII (ONB), I_0 = IX (CAM), some houses collapsed at Khíos (FC).

Juni 9, 07^h, some damaged houses collapsed at Khíos (GGC).

Aug. 26, I_0 = IX (ONB), I_0 = X (CAM), terrible (GGC), I = VI at Çeşme, Lésvos (EGU).

May 23,   19^{h}20^m UT, 42·9°N, 17·4°E, R.22a, I_0 = VII, r = 30 km (TD), not in (FC).

At Pijavičina in Korčula VII, Ston V, Metkovič IV (BA), no description of damage.

May 30,   38·5°N, 43·3°E, R.33, I_0 = IX (MJD), I_0 = X, M = 7·3 (ONB, CAM), I_0 = VIII (ANV), not in (PL).

At Van and Tegut 180 houses completely destroyed (MO).

June 7,   38·7°N, 42·4°E, R.33, I_0 = VII (ANV, EGU), 38·5°N, 43·3°E (EGU), identical with the previous report?

District of Van and Nemrut, 400 houses demolished, 95 victims in the vicinity of Van, a village destroyed by a landslide (PL).

July 22,   02^{h}45^m, (45·4°N, 6·1°E), R.18a, I_0 = VIII?, r = 180 km (RF, RO, PRF), I_0 = = VII (MJD, LC).

Exact epicentre unknown, intensity VIII between Saint Jean de Maurienne, Moutiers, La Chambre, Allevard and Chambéry, in the region of Lyons and in Savoie houses were fissured and chimneys thrown down (RF, RO), no other information on damage; felt in SE France, in Switzerland at Basel, Mulhouse, Domodossola and in NW Italy in Savoia (RO).

Sept. 10,   07^h UT, 42·2°N, 14·3°E, R.19, I_0 = VIII (MR, MJD), I_0 = IX (FC), I_0 = IX—X (MF).

A damaging shock in the Abruzzi mountains, at Orsogna all houses seriously damaged, the old ones partly destroyed, similar effects at Lanciano, serious damage at Castelfrentano, Canosa, Arielli, Guardiagrele, Crecchio and Ortona, felt at Napoli, Roma and Rocca di Papa (FC), Orsogna almost destroyed (LC).

A strong aftershock on Febr. 12, 1882, 02^h UT, I_0 = VII? (BI) at Chieti, Castelfrentano, Orsogna (FC).

Sept. 28,   40·6°N, 33·6°E, R.29, I_0 = VIII, M = 6·1 (ONB, MJD), I_0 = VII (EGU), not in (FC).

Many houses destroyed in the region of Cankiri, 12 victims (PL).

Sept. 28,   05^{h}38^m UT, 44·1°N, 12·1°E, R.19, I_0 = VII?

At Bertinoro many chimneys fell down, at Prati an old house ruined, strong at Forlimpopoli, Palazzolo, Modigliana, Forli, felt at Bologna, Urbino, Firenze, Cesena (FC, BI).

Oct. 23,   09^{h}11^m UT, 45·8°N, 15·3°E, R.22a, I_0 < VII?, I_0 = VII (TD).

Fissures in some houses at Zagreb (FC).

1881

Nov. 18, epicentre? R.18a, $I_0 \leqq$ VII?, $I_0 =$ VII (MFS).
Houses were fissured at Sax, Gams, Appenzell, Stein, Ebnat, Coire, chimneys were thrown down at Gams (MFS).

Nov. 25, 18^h, 46·3°N, 7·0°E, R.18a, $I_0 <$ VII?, $I_0 =$ VII−VIII, $r =$ 50 km (PRF), not in (RF).
An earthquake in the Alps, strongly felt at Chamonix, Mégève and Montriond, epicentre near Aigle where the intensity reached allegedly VII−VIII (RO), no exact data on the damage.

Dec. 29, 40·6°N, 33·6°E, R.29, $I_0 =$ VII, $M =$ 5·5 (ONB), $I_0 =$ VIII (MJD), $I_0 =$ IX (CAM), not mentioned in (FC). Aftershock of Sept. 28, Dec. 30, 40·2°N, 29·1°E, $I_0 =$ VI (EGU).

1882

Jan.?, A series of shocks in the region of Viterbo (42·4°N, 12·1°E), R.19, culminated at the end of January 1882 by a severe shock which ruined 5 houses at Latera and fissured all houses, at Rocca Respampani some walls collapsed (BI), $I_0 =$ VII to VIII?, $h =$ sup?

Jan. 31, 04^h30^m, $39\frac{1}{4}$°N, 23°E, R.26a, $I_0 =$?, $I_0 =$ VII (FC).
A very strong earthquake in Vólos particularly in the coastal villages (GGC), not given in (GG).

Febr. 10, 12^h, an earthquake with $I_0 =$ VII at Kórinthos (37·9°N, 22·9°E) (FC), R.26a, two shocks at Aetolikon (GGC).

Febr. 15, 06^h, ($44\frac{3}{4}$°N, $9\frac{1}{4}$°E), R.19, $I_0 <$ VII (BI), $I_0 =$ VII (MJD).
An earthquake in Italy, epicentre in the valley of Trebbia, $I =$ VII(?) at Bobbio Coli, Cabella, Albena, Tortona, Casone, Carega, $I =$ VI at Volpeglino, Parma, Piacenza (FC), slight damage (BI).

Febr. 27, 06^h30^m UT, (46°N, $10\frac{1}{2}$E), R.18b, $I_0 =$ VII? (MR, MJD, FC), isos. map in (BI).
At Castione della Presolanana and at Rovetta several chimneys fell down, felt heavily at Vilminore, Sondrio, Tirano, felt at Bergamo, Brescia, Ornavasso, Pallanza, in Switzerland at Castasegna, Olivone, etc. (FC, MR, BI).

March 21, 38·4°N, 26·1°E, R.26c, $I_0 =$ VII? $M =$ 5·5 (ONB, MJD).
Three strong shocks at Khíos (GGC), not in (GH).

June 6, 06^h UT, 41·5°N, 14·2°E, R.19, $I_0 =$ VII−VIII, $I_0 =$ VII (MR), $I_0 =$ VIII (MJD), $I_0 =$ IX−X (MF), isos. map in (BI).
At Monteroduni and Isernia some rural houses were destroyed, other houses damaged, some damage at Longano and Cantalupo, felt at Napoli, Caserta, etc. (BI).

July 17, 07^h51^m, 46·0°N, 14·3°E, R.22b, $I_0 =$ VII, $M =$ 4·8, $h =$ 12 km, $r_7 =$ 12 km, $r_6 =$ 25 km, $r_5 =$ 45 km (RV, MJD, RA), $r_{max} =$ 120 km (TD).
Damage at Vrhnika, Borovnica, Sinja Gorica, Bistra, Polhov Gradec, Žiri (RV); at Ljubljana tiles were thrown down from the roofs, the cupola of a church was fissured, felt at Aurisina (Nabrežina), at Klagenfurt (heavily), Pötschach, Bled (Veldes), Trieste, Capo d'Istria, Miramar, Monfalcone, etc. (FC), rockslide (BA).

1882

Aug. 16, 03^h, (43·0°N, 13·9°E), R.19, $I_0 = ?$, isos. map in (BI), epic. in the sea?
A severe shock on the Adriatic coast of Italy, $I = VIII$ (?) at San Benedetto de Tronto, Grottammare, Cupramarittima, $I = VII$ (?) at Monte Brandone (FC, BI), no description of damage available.

Sept. 18, epicentre?, R.19, $I_0 = ?$, $I_0 = VII$ (MR), $I_0 = IX$ (FC), doubtful.
An earthquake at the coast of the Garda lake and on Monte Baldo (VIII), Casone IX, Verona VI (FC), not in (BI).

Sept. 20, (43·4°N, 13·7°E), R.19, $I_0 = ?$, local?
A shock at Porto Recanati with $I = VII$ (FC), not in (BI).

Nov. 18, (38·8°N, 15·2°E), R.20, $I_0 = ?$, a volcanic shock.
A severe shock with $I = VII$ accompanying an eruption of Stromboli (FC), not in (BI).

1883

Jan. 24, 07^h UT, 43·1°N, 18·0°E, R.22a, $I_0 = ?$, $I_0 = VII$, $r = 51$ km (TD), not in (FC).
VII at Stolac, Nevesinje and Gacko, felt at Dubrovnik, Mostar and in all Hercegovina (BA).

Febr. 11, 09^h UT, 45·3°N, 16·1°E, R.22b, $I_0 = ?$, $I_0 = VII$, $r = 100$ km (TD).
An earthquake with epicentre at Bos. Krupa in Bosnia (FC), VII at Maja and Viduševac, felt at Topusko, etc., a bell was ringing (BA).

March 21, $(37\frac{1}{2}°N, 15°E)$, R.20, $I_0 = VII$ (MR, MJD), $I_0 = IX-X$ (MF), $h = $ sup, volcanic.
During a numerous sequence of volcanic shocks in the region of Etna from March 20 to March 30 some rural houses were ruined at Zafferana, Nicolosi, Milo, more solid houses and churches were damaged (BI, FC, MF).

May 3, serious damage, all houses fissured at Tabríz (38·1°N, 46·3°E), NW Iran, R.34 (MO), many victims (LC).

July 28, 20^h25^m UT, 40·7°N, 13·9°E (PET), R.19, $h = $ sup, $I_0 = X-XI$ (MF), $r = 1·5$ km, $I_0 = IX-X$ (MJD), $I_0 = X$ (CAM), $I_0 = XI-XII$ (MR), isos. map in (BI).
A disastrous shock over a smaller part of the island of Ischia, at Casamicciola 2278 houses were partly or completely ruined, also Lacco and Forio were destroyed, 2313 victims, not felt at the village Ischia and over a part of the island, not recorded by penduli of the Vesuvian Observatory (FC, LC, BI).
Aftershock: Aug. 3, 13^h UT, Casamicciola, Foria, Barano VII?

Aug. 5, 01^h30^m, $37\frac{1}{2}°N$, $24\frac{1}{2}°E$, $h = n$, R.26a, $I_0 = VII?$, $I_0 = VI$ (GG), $I_0 = V-VI$ (GGK), Aug. 4 (GG), Aug. 5 (GGC, FC, CAM, MF), $I_0 = IX-X$ (MF), $I_0 = IX$ (CAM).
A sequence of shocks which slightly damaged houses at Kythnos (GGC). On Aug. 4, 01^h35^m a largely felt shock in the Aegean Sea area, at Athínai the tiles of most roofs were thrown down as well as the ruins of older houses (GGC).

Sept. 2, 01^h30^m UT, 41·8°N, 12·7°E, $h = $ sup, R.19, $I_0 = VII$, $I_0 = VIII$ R. F. (BI).
Heavy shock in the region of Monti Albani, maximum intensity observed at Frascati, Grottaferrata, Rocca di Papa and Monte Cavo (BI).

Sept. 2, at Grumevano near Napoli two houses and at Pomigliano (40·9°N, 14·4°E), R.20, six houses collapsed during a heavy shock, 11 victims (FC), not in (BI), $h = $ sup?, $I_0 = ?$

1883

Oct. 15, 15^h30^m, 38·4°N, 26·1°E, R.26c, $I_0 = $ X, $M = $ 7·3 (ONB, FC, CAM), $I_0 = $ IX—X, $P = $ 210 000 km^2 (GGK), $M = $ 6·4 (GGE), $I_0 = $ IX (EGU).

A disastrous shock in the region of Izmir, at Izmir and Ayvalik several houses collapsed, particularly severe at the coast between Urla and Çeşme where all houses were destroyed, casualties, at Çeşme many houses were thrown down, destructive also in the villages Kritha, Alazata, Balcik, Serandumo, Korentzik, more than 120 victims, the earth cracked near Alazata, felt as far as Canakkale, Athínai, Syra, Andros, Thíra (FC). Houses heavily damaged along the western coast of the Çeşme peninsula, small damage at Izmir, epic. probably between Çeşme and Khíos (PL, EGU), at Khíos only old houses collapsed (LC).

Aftershocks: Oct. 22, a damaging shock at Çeşme.

Nov. 1, 38·3°N, 26·3°E, $I_0 = $ VII (ONB), $I_0 = $ VIII (EGU), $I_0 = $ X (CAM), cracks in the ground near Pyrgi, a rockslide near Safdere, Karaköy ruined (FC), a strong shock at Çeşme and in the bay of Izmir (PL).

Nov. 3, 06^h?, 40·6°N, 43·1°E, R.33, $I_0 = $ VIII, $M = $ 6·1 (ONB, MJD), $I_0 = $ IX (CAM), not in (EGU).

Yerevan (CAM), a shock felt at Karakoyunlu (BZ), one old church destroyed by the earthquake (MO).

Nov. 15, 02^h45^m UT, 45·3°N, 18·4°E, R.22b, $I_0 = $?, $I_0 = $ VII, $r_{max} = $ 10 km (TD), local? not in (FC).

At Hrkanovci VII, Bizovac VI, Bocunavac VI (BA), no description of damage available.

Dec. 20, 21^h, 46·2°N, 16·6°E, R.22b, $I_0 = $?, $I_0 = $ VII (TD).

A strong shock at Zagreb, Križevci, Zukany, Pécs (Fünfkirchen), Barcs (FC), VII at Ludberg, Koprivnica, Virje, etc. (BA), no description of damage.

Dec. 22, 03^h30^m, (38·7°N, 9·2°W), R.12, $I_{max} = $ VI? (FH), $I_0 = $ VII (MJD), $I_0 = $ VI to VIII (BRS), epic. in the sea?

In Lisboa panic and damaged walls, felt at Ferreira do Zezere, Oporto, Guarda (violent), Alemtejo (GS).

1884

Jan. 10, 16^h30^m UT, (42·6°N, 14·4°E), R.19, $I_0 = $ VII? (MR), isos. map in (BI), epic. in the sea?

An earthquake in the Abruzzi with maximum intensity VII(?) at Notaresco, Giulanova, Mosciano, Atri, Montepagano, felt at Fermo, Chieti (BI, FC).

Jan. 23, (41·4°N, 33·8°E), R.29, $I_0 = $ VII?, $M = $ 5·5 (ONB, MJD), 39·8°N, 26·3°E, R.26c, $I_0 = $ VI (EGU). A series of shocks at Kalodjik, province Kostambul (Kastamonu?), threw down several minarets (FC, LC), Káadjik? (MJD).

Febr. 10, 38·4°N, 42·1°E, R.33, $I_0 = $ VIII?, $M = $ 6·1 (ONB, MJD), $I_0 = $ IX (CAM), 37·8°N, 42·6°E, $I_0 = $ VI (EGU).

Many houses collapsed during a severe shock in Birvari (Perivari?), province Bitlis (FC, LC). Unusual epicentre.

March 24, 20^h UT, 45·3°N, 18·4°E, R.22b, $I_0 = $ VII—VIII, $I_0 = $ VIII, $P = $ 21 600 km^2 (RA), $I_0 = $ IX (TD), $I_0 = $ VII (MJD).

A damaging shock with the epicentre near Djakovo (Djakovar) where the municipal

building was seriously ruined and almost no house remained undamaged, many roofs and chimneys collapsed, 6 churches damaged, VI—VII at Kondurič, Piško- revci, Strizivojno, Budrovac (30 chimneys collapsed), Dragotinja (2 chimneys collapsed), etc, (BA)., strong at Vinkovci, felt at Pozega, Pécs (Fünfkirchen), Osijek, Zagreb (FC), in Hungary at Baju (RA,LC).
Aftershock on March 27, 21^h45^m UT, with the same intensity, $r_{max} = 87$ km (TD, FC, LC).

April 22, 09^h15^m, (51·9°N, 0·9°E), R.10b, $I_0 =$ VIII, $I_0 =$ IX (DCB, MF), $I_0 =$ VII (MJD).
A destructive earthquake in Essex, at Colchester a church tower collapsed, at Wivenhoe and Langhoe all houses were injured and the churches partly or com- pletely collapsed, at Ipswich many chimneys fell, at Alberton and Teldon one house was thrown down, felt strongly at Southend, Shrewsbury, Mannigton, Chelmsford, Bury, felt at Cambridge, Northampton, Harwich, Woolwich and London (FC), somewhat smaller effects described by (LC).

May 13, 40·4°N, 27·8°E, R.27, $I_0 =$ VII?, $M = 5·5$, $r = 130$ km (ONB), $I_0 =$ VI (EGU), $I_0 =$ VIII (MJD), $I_0 =$ IX (CAM).
A severe shock at Crevassa (?) where a Greek church was destroyed, at Bandirma several houses were damaged, strong at Erdek, felt slightly at Istanbul (FC, PL).

Nov. 27, 22^h UT, 44·7°N, 6·7°E, R.18a, $I_0 =$ VII—VIII, $r = 180$ km (PRF, RF), $I_0 =$ VII (MJD).
At Briançon ceilings were deformed, chimneys thrown down, walls fissured, Guillestre VI, Embrun V, violent at Savines, Chorges, Gap, felt at Antibes, Biot, Cagnes, Draguignan, Grenoble, Torino, Voiron, Saint Marcellin, Wien, Chambéry, Zürich, Interlaken, Marseille, Nice, Mulhouse (RO).

Dec. 25, 20^h55^m, 37·0°N, 4·0°W, R.16, $I_0 =$ XI (BRS, SAE, MF), $I_0 =$ X F. M. (GS, MU, CAM), isos. map in (SAE).
An earthquake catastrophe in Andalusia, mostly affected were Alhama (973 houses destroyed, 307 victims), Albuñuelas (362 h., 102 v.), Arenas (160 h., 135 v.), Tayensa (100 h., 10 v.), Loja (47 v.), Zafarraya (371 h., 50 v.), Murchas (805 h.), at Malaga all churches were fissured, Granada VI?, Madrid V, at Sevilla a part of the monastery collapsed, near Alhama and Guerifar lanslides, large cracks in the ground, horizontal movements in the region of Periana, in total 17 178 houses destroyed or damaged in the provinces Granada and Malaga (FC, GS, LC), felt in Morocco (RG).

Aftershocks: Dec. 26, Granada $I =$ VII? (FC).

Dec. 27, $I_0 =$ VII—IX?, at Antequera and Velez several houses collapsed, Albuñuelas almost totally destroyed, houses de- stroyed also in other localities (FC).

Dec. 29, Houses destroyed at Rio Gordon, Vinneda, Allarnatajo, Albuquerra, $I_0 =$ VIII—IX? (FC, LC).

Dec. 30, Houses injured at Granada, Jayena destroyed, several at Archidona and Periana (FC), $I_0 =$ VIII?, $I_0 =$ VII (MJD).

Dec. 31, Several houses destroyed at Velez, Torrox destroyed, at Albuquerque" the earth opened and swallowed a church and four houses", felt at Nerja (FC), houses collapsed at Archi- dona, Murches (LC), $I_0 =$ IX?, $I_0 =$ VIII (MJD).

Jan. 1, Velez, Torrox, Nerja, $I_0 = $ VIII (MJD).

Jan. 5, Nerja, Trujillano, $I_0 = $ IX (CAM), $I_0 = $ VIII (MJD), $I_0 = $
$= $ II (GS).

Jan. 12, Malaga, Granada, $I_0 = $ VII (MJD).

Jan. 16—17, at night, 35·5°N, 5·7°E, R.15, $I_0 = $ VIII? (GA, RK).
Three native houses collapsed at the village Oued Bou Adjam and a 100 m long
stripe of ground subsided, felt at Bougie, Sétif, Bou Saáda and Bordj Bou Arréridj
(RK); Jan. 31, M'sila, $I_0 = $ VIII (MJD), $I_0 = $ IX (CAM).

Jan. 25, epicentre?, R.2, $I_0 = $?, $I_0 = $ IX (MF, MJD).
Iceland, many houses damaged at Kelduhverf and at Vikingvatn (MF, MJD).

Jan. 27. $I_0 = $ VIII?, Alhama, Granada (GS, BRS, MJD).

Febr. 18, 14^h30^m, $38\frac{1}{2}$°N, $21\frac{3}{4}$°E, $h = n$, R.26a, $I_0 = $ VII—VIII, $I_0 = $ VII (GG), $M = 5\frac{1}{2}$
(GGE).
Two or three workshops and the school building collapsed at Návpaktos, almost
all houses were injured in different scale, many fissures in the walls (GGC).

Febr. 29, 18^h30^m, 37·2°N, 22·2°E, $h = n$, R.26a, $I_0 = $ IX, $r = $ 180 km (GG, GGC, CAM),
$I_0 = $ IX—X (MF, MJD, GGK), $M = $ 6·0 (GGE).
At Messíni, Loï, Filiatrá, Jannitsanika, Agios Joannes, Meligalá, Katsaroú,
Karyai, Kalámai several houses collapsed, the others were uninhabitable or serious-
ly injured, serious damage to buildings at Kalámai, Megalópolis, etc., for details on
shaken localities see (GGC, GGM), rockslides, casualties.

April 13, 11^h25^m, epic.?, R.18a, $I_0 = $ VII (MJD), $I_0 = $ VIII—IX (MFS), local?
A weak tectonic shock released allegedly an "impact earthquake" which caused
cracks in the walls and threw down chimneys at Zweisimmen, Mannried, Gruben-
wald, Boltingen, rockslides (MFS), some damage at Simmen, Grossheid, etc. (LC).

April 30, 23^h15^m UT, 47·5°N, 15·4°E, R.18b, $I_0 = $ VIII (KC), $I_0 = $ VII—VIII (TT), $I_0 = $
$= $ VII (MJD), $I_0 = $ IX (RF, Heritsch), a detailed report of F. Heritsch in Mitt.
d. Erdbeben-Kom. der Kais. Ak. d. Wiss. in Wien, N. F., No XXXII, 1908, with
an isoseismal map, $r = $ 260 and 150 km.
A damaging shock in the E. Alps, at Kindberg all houses were injured, most
walls were fissured, chimneys overthrown or shifted, mortar fell down, in some
houses walls and vaults collapsed, the school had to be closed for safety reasons,
at Aumühl, Allerheiligen, Wartberg the chimneys and tiles thrown down, short
fissures in the valley of Mürz, at Krieglach, Langenwang, Turnau, Aflenz, Veitsch
some chimneys fell and some walls were injured, some damage was done at Hohenwang, Grassnitz, Stanz, etc., details on other localities given in the report of F.
Heritsch. The earthquake was felt as far as Bratislava, Jihlava, Praha, Aš, Nürnberg, Klagenfurt and Rann.

May 26, 08^h30^m, 47·3°N, 23·3°E, $h = n$, R.21d, $I_0 = $ VII—VIII? (FA, RA), $I_0 = $ VIII
(PRA), $h = $ 5 km, $r_3 = $ 62 km, $r_4 = $ 50 km, $r_5 = $ 32 km, $r_6 = $ 27 km (RA).
A shock heavily felt in the district of Sălaj, Sătmarei, Cluj, Somes, at Apahida,
Huedin, Beclean, etc. (FA, RA).

July 1, 06^h15^m UT, At Vernante (Piemonte) walls were fissured and some roofs damaged,
felt at Valdieri (BI), R.18a, $I_0 = $ VII (MR, MJD), local.

1885

Aug. 22, 20^h30^m, $38\frac{3}{4}°N$, $23\frac{1}{2}°E$, $h = n$, R.26a, $I_0 = VII$, $r = 110$ km (GG, GGC), $I_0 = VI-VII$, $P = 40\ 000$ km^2 (GGK).

Two houses collapsed and many others were fissured at Agia Anna, Skíathos VI—VII, felt at Skópelos, Xerochori, Khálvi, Vólos, Athínai (GGC).

Sept.—Oct., $37·5°N$, $15°E$, $h = $ sup, R.20, $I_0 = VII$ (MR).

Volcanic shocks at Nicolosia, where walls and ceilings were damaged, felt at Acireale, Belpasso (BI).

Dec. 3, 20^h30^m, $36·1°N$, $4·6°E$, R.15, $I_0 = VII-VIII$, $I_0 = VII$ (RK, GA), $I_0 = IX$ (MJD), $I_0 = X$ (CAM).

A largely felt earthquake which caused considerable damage at M'sila, Bou Saada, Sétif, Batna, Bordj Bou Arréridj, felt as far as Nemours, Collo and Ghardaia, houses collapsed at Boghari (LC).

Dec. 14, $38\frac{3}{4}°N$, $20\frac{1}{2}°E$, $h = n$, R.26a, $I_0 = VII-VIII$ (GGK), $I_0 = VII$ (GG), $M = 5\frac{1}{2}$ (GGE).

Earthquakes in the island Levkás, many houses at Kariá were seriously damaged, heavily felt at Enkluvi, rockslides (GGL).

Dec. 26, 02^h UT, $41·6°N$, $14·7°E$, R.19, $I_0 = VII?$ (MJD), $I_0 = VII-VIII$ (MR), isos. map in (BI).

A heavy damaging shock at Baranello, Vinchiaturo, Oratino, Campobasso, etc., felt at Napoli (BI).

1886

Febr. 6, 06^h UT, ($37·5°N$, $15·1°E$), R.20, $I = VII?$, volcanic?

Monteleone and Limpidi, felt at Pizzo, Tropea and Gerrace (BI), a series of shocks connected with the eruption of Fossa di Vulcano and Etna.

March 6, 06^h30^m UT, $39·3°N$, $16·2°E$, R.20, $I_0 = VII-VIII$, $I_0 = IX-X$ (MF), isos. map in (BI).

Cracks in houses at Montalto, Marano Marchesato (some houses collapsed), Uffugo and Rende (a church destroyed) (BI).

March 14, 23^h, $37·2°N$, $4·2°W$, R.16, $I_0 = VII-VIII?$, $I_0 = VII$ (MJD), $I_0 = VIII-IX$ R. F. (BRS, GS), $I_0 = IX$ R. F. (MU).

Slight damage at Granada, more serious damage in the country between Loja, Albuñuelas and Talará (GS).

May-July $37.6°N$ $14.8°E$, R.20, $h = $ sup, $I_0 = VII-VIII$ (MR), volcanic, isos. map in (BI).

A series of volcanic shocks in the region of Etna, the most violent ones had damaging effects at Zafferana, Acireale, Biancavilla, Belpasso, Linera (BI).

July 1, 09^h45^m, $35·5°N$, $5·3°E$, R.15, $I_0 = VII?$ (GA, RK).

Houses damaged at Takitount, felt strongly at Setif, slightly at Djidjelli and Bougie (RK).

July 6, 11^h23^m, 20^h30^m, $36·7°N$, $4·4°W$, R.16, $I_0 = VII$ (MJD), $I_0 = VIII$ R. F. (BRS), $37·2°N$, $4·1°W$ (MU).

The second shock damaged some walls at Malaga, felt at Velez Malaga, Canillas de Aceituno, Ríogordo, Periana, Alcaucín and other localities of the province of Granada (GS).

Aug. 27, 21^h30^m, 37°N, $21\frac{1}{4}$°E, $h = i$, R.26a, $I_0 =$ X–XI (GGC, GGK), $I_0 =$ XI, $r =$ = 1250 km (GG, MF), $I_0 =$ X (CAM), $I_0 =$ IX–X (MJD), $M = 8\cdot4$ (GGE). An earthquake catastrophe in Messinia, three towns Filiatra, Ligudista and Koróni and about 123 villages were almost completely destroyed, 7 towns and 37 villages were very seriously damaged, 65 villages were injured, in total 6000 houses were destroyed or rended uninhabitable, about 326 victims; (GGC) gives the foll013owing localities with $I >$ X; Ligudista, Marathopulis, Válta, Kanalupu, Chalazoni, Christiana, Agia Kyriaki, Agorelitsa, Flota (Phloka), Muzusta, Raches, Stasion, Spilia, Koróni, Patalidhion, Jalova, Iklaena, Karamanoli, Kukunara, Lezaga, Papulia, Pisaski, Platanos, Romanu, Osmanaga, Pyla, Chandrinon, Kremmydia, Dauti, Kalaphati, Chomatades, Mimaghia, Kastelion, Kobö, Tsaïzi, Chomateron, Petriades, Tzapherogli, Charakopion, Saratzas, Loga, Gamvria, Agios Trias, Kandirogli, Katiniades, Vigla, Panyperi, Pasalina, Sgrapa, Vlachopulon, Sgurokampos, Grustesi, Dara, Ammusta, Karakasilion, Kurtaga, Miska, Pelekanada, Pispisia, Skarmiga, Phurtzi, Chalabreza, Armeni, Meligala, Ismaïla, Philippaki, Chastemi, Loï, Polaena, Vytina, Makraena, Manesi, Lezi, Skala, Bala, Solaki, Alitselepi, Tsausi, Toskesi, Malta, Sandani, Sterna, Kambos, Katakolon, Strofades; damaged area over all Pelopónnisos and the Ionian islands. Casualties, fissures, slump, changes in ground water level, tsunami, cable breaks (GG, GGC, GGM, GGT, ANS), felt as far as Albania, Syria, Malta and over almost all Italy (MCA).

A strong aftershock on Sept. 6 completed the destruction (GGM).

Sept. 4, $39\frac{1}{2}$°N, $26\frac{1}{2}$°E, $h = n$, R.26c, $I_0 =$ VII (GG), $I_0 =$ VII–VIII (GGK), $M = 5\frac{1}{2}$ (GGE). Some walls collapsed and several houses fissured in Lésvos and Kydonies, felt at Izmir (GGC).

Sept. 5, 20^h50^m UT, 45°N, $7\frac{1}{4}$°E, R.18a, $I_0 =$ VII (MR, MJD), isos. map in (BI). At Coazze, Volvera and Pinasca all houses cracked and many chimneys collapsed, some damage at Condove, S. Antonio, Vayes, Chiavrie, etc., felt at Aosta, Oropa, Milano, Genova, Como (BI).

Sept. 9, 11^h15^m, 36·2°N, 3·6°E, R.15, $I_0 =$ VII? (GA). Fifteen native houses, already in poor condition, were destroyed at Ouled Mériem and Ouled Boussaaf (RK).

Oct. 6, 39·6°N, 29·0°E, $I_0 =$ VII (EGU), 37·7°N, 27·2°E, R.26c, $I_0 =$ VII, $M = 5\cdot5$ (ONB). Some damage at Balat, Sagir, Bigadic, Aydin, Koycegiz, Marmaris, highest damage at Gökcedäg (PL), Tavsanli (EGU).

Oct. 9, 18^h10^m, 48·5°N, 7·8°E, R.8, $I_0 =$ VI (SW), $I_0 =$ VII, $r = 25$ km (RSF), $h =$ sup. Chimneys thrown down at Schutterwald (SW).

Nov. 27, 08^h, $38\frac{1}{4}$°N, $26\frac{1}{4}$°E, $h = n$, R.26c, $I_0 =$ VII (GG, GGC), $I_0 =$ VI–VII (GGK), $M = 5\frac{1}{2}$ (GGE). At Khíos some old houses collapsed and many new ones were fissured, Izmir VI, felt at Çeşme (GGC, GH), in Sámos and Lésvos (GGK).

November, 38·3°N, 29·3°E, R.26c, $I_0 =$ VII (EGU), Denizli, Usak.

Nov. 28, 22^h30^m UT, 47·3°N, 10·8°E, R.18b, $I_0 =$ VII (TT), $P = 376\,000$ km^2 (SW). A strong shock in Tirol, felt in S. Bavaria in München, Baden and E. Switzerland, at Nassereith almost all houses were fissured, several chimneys and some ceilings fell down, at Imst the chimneys were damaged, detailed information on intensities in S. Germany are given in (SW).

Dec. 11,   38·4°N, 26·1°E, R.26c, I_0 = VIII, M = 6·1 (MJD, ONB), I_0 = IX (CAM). Izmir, Khíos. No data in (PL).

1887

Date?   An earthquake in Egypt, houses were damaged at Cairo, felt in the Nile valley as far as Suakin in Sudan (SAM).

Jan. 6,   epicentre?, R.15, I_0 = VII? (ANT), I_0 = VIII (MJD), I_0 = IX (CAM), local?
An earthquake in Tunis which caused light damage to a village near Ejenel(?) (ANT); Jan. 5, catastrophe at "Djemel—Tunis", houses destroyed, Jan. 6. at Jemel(?) several houses collapsed (LC).

Jan. 8, 20^h, 36·1°N, 4·6°E, R.15, I_0 = VIII (RK, GA).
The walls collapsed at Medjana and Achir, at Mansourah forty native houses were thrown down, houses were fissured at Bordj Bou Arréridj, felt at Sétif, M'Sila and Azazga (RK).

Febr. 23, 05^h23^m UT, 43·9°N, 8·1°E, R.18a, I_0 = X, r = 300—400 km (RF, PRF, MF, CAM, MR), epic. in the sea?, P = 568 000 km^2, isos. map in (BI).
A disastrous shock in the Provence with the epicentre near the Ligurian coast, Diano Marina and Oneglia almost completely destroyed, Savona, Albenga, Porto Maurizio, Alassio, Taggia, Nice, Menton, Noli Albissola, Bussana were seriously damaged, etc., see (RO); the outer limits of the shaken area passed through Livorno, Basel, Luzern, Chamonix, about 640—1000 victims, changes in springs, fissures 100 m long at Valdo with eruptions of mud (RO). Detailed description in (RO, BI).

April 14,   Asola (45·7°N, 11·9°E), Italy, R.18b, I_0 = VII (MR).

May 14, 05^h30^m, 40°N, $25\frac{1}{2}$°E, h = n, R.27, I_{max} = VII—VIII, I_{max} = VII, r = 340 km (GG, GGC), M = 6·5 (GGE).
Three houses collapsed in the Límnos and the others were fissured, felt at Raedestos, Edirne, Mytilíni, Istanbul, Gelibolu and Izmir (GGC), no data in (PL, EGU).

May 26, 01^h35^m UT, (41·5°N, 14·8°E), R.19, I_0 = VII (MR).
At Ielsi some chimneys were thrown down and some houses fissured, felt strongly along the Adriatic coast from Civitanova Marche to Fano (BI).

July 10, 02^h56^m, 46·1°N, 21·1°E, h = n, R.21c, I_0 = VII, P = 8000 km^2 (PRA, RA).
An earthquake in Banat, at Vinga chimneys were thrown down (RA), felt at Arad, Timişoara, Glogovat, Periam, Kikinda Mare (FA).

July 17, 07^h45^m, 36°N, 26°E, h = i, R.26b, I_{max} = VII (GGC, MJD), I_{max} = VI—VII (GGC), I_{max} = VI (GG), P = 250 000 km^2, M = 7·7 (GGE).
Houses fissured at Iraklion and in Ródhos, largely at El Iskandarîya (Alexandria), Izmir, Khíos, Míkonos, Neápolis, Zákinthos, Kalámai, Trípolis, Mesolóngion, Methana, Pátrai (GGC), felt in Sicily (BI).

Aug.?   38·1°N, 28·2°E, R.26c, I_0 = VII (EGU).
Mugla, Koycegiz, Çine, Denizli.

1887

Aug. 13, 04^h, 45·7°N, 15·6°E, R.22b, $I_0 = $ VII (TD).
At Jastrebarsko a church fissured, at Krašić, Sv. Jana walls and chimneys collapsed (BA).

Oct. 3, 22^h53^m UT, $38\frac{1}{4}$°N, $22\frac{3}{4}$°E, $h = n$, R.26a, $I_0 = $ VIII−IX (GGC, MJD), $I_0 = $ = VIII, $r = $ 260 km (GG), $I_0 = $ VII−VIII (GGK), $I_0 = $ IX (CAM, MF), isos. map in (GGC), pp. 226, $M = $ 6·3 (GGE).
A disastrous shock in Korinthia, some houses collapsed, the others were partly uninhabitable, partly seriously damaged, at Xilókastron, Kiáton, Kokkonion, Nerantza, Tholeron, Diminion, Perakhóra, Velon and Khóstia, tsunami between Xilókastron and Sikiá (ANS, GGT), casualties, slumping and settling of the ground near Xilókastron and Kivéri; further information in (GGC).

Oct. 28,   an earthquake in Iceland, Eyrarbakki (63·9°N, 21·1°W), region of Flói, Kirkjubae (Rangarvalla Syssel), $I_0 = $ VII (MJD), $I_0 = $ IX (MF).

Nov. 9, 00^h30^m, 44·2°N, 12·0°E, R.19, $I_0 = $ VII? (MJD), isos. map in (BI).
Several chimneys fell at Forli and Rocca S. Gasciano, felt at Venezia, Verona, Parma and Siena (BI).

Nov. 14, 09^h, 43·9°N, 4·9°E, R.18a, $I_0 = $ VIII?, $r = $ 15 km (PRF, RO, MJD), not in (RF).
At Saint Saturnin the walls were fissured, felt at Avignon, Isle sur Sorgues, Cavaillon, Moriers (RO).

Nov. 29, 13^h30^m, 35·6°N, 0·3°E, R.15, $I_0 = $ IX−X (GA, RK).
Heavy damage at Kaláa where the mosque and 331 houses collapsed, 20 victims, the neighbouring villages Hillil, Debba and Thiouanet were also damaged VI−VII, rockslides, felt at Mascara, Rélizane and Oran (RK, HA).

Dec. 3,   39·5°N, 16·3°E (PET), R.20, $I_0 = $ VIII−IX, $I_0 = $ IX (CAM), $I_0 = $ IX−X (MR, MF), $I_0 = $ VIII (MJD), isos. map in (BI).
At Bisignano and Stazione Mongrassano many houses largely cracked and some collapsed, destructive (VIII) at Roggiano, S. Marco Argentaro, Staz. Lattarico, felt at Messina, Vietri, Stromboli (BI).

1888

Jan. 6, 23^h40^m, 36·5°N, 2·6°E, R.15, $I_0 = $ VII−VIII?, $I_0 = $ VIII (RK, GA), $I_0 = $ VII; Jan. 8 (MJD).
Much damage and panic at Mouzaïaville, La Chiffa and Oued Djer, between La Chiffa and El Affroun houses damaged, felt strongly at Blida and Alger (RK).

Febr. 25,   38·8°N, 15·2°E, R.20, $h = $ sup, $I_0 = $ VII (MR, MJD), $I_0 = $ IX (MF), volcanic.
At Stromboli many houses fissured, some old walls collapsed (BI).

March 30, 09^h UT, 45·5°N, 18·1°E, R.22b, $I_0 < $ VII?, $I_0 = $ VII (TD).
At Našice, Budimci, Podgorač, Kutjevo, Ruševo some walls fissured, etc., people could not stand (BA).

May ?,   38·4°N, 26·1°E, R.26c, $I_0 = $ VIII (EGU), not in (ONB).
Khíos, Çeşme, Urla, Karaburun.

May 4, 20^h05^m, 46·1°N, 2·8°E, R.11, $I_0 = $ VII, $r = $ 50 km (PEF).

May 20, 10^h30^m UT, 44·8°N, 17·2°E, R.22b, $I < $ VII?, $I_0 = $ VII, $r_{max} = $ 176 km (TD, JM), 45·0°N, 16·7°E (JM).

At Banja Luka a stone built house was fissured, VII at Kozarac, Bistrica, Omarska, VI(?) at Ch. Dubica, etc., felt at Narta, Tuzla, Sarajevo (BA).

July 8, $42 \cdot 7°$N, $13 \cdot 7°$E, R.19, $I_0 =$ VII (MR, MJD), $I_0 =$ IX (MF), local? $h =$ sup? At Teramo slight damage, destruction of some rural houses in the surroundings (BI).

Sept. 9, 15^h15^m, $38 \cdot 2°$N, $22 \cdot 1°$E, $h = n$, R.26a, $I_0 =$ X, $r = 150$ km (GG, CAM, MF), $I_0 =$ IX—X (GGK), $M = 5 \cdot 8$ (GGE), isos. map. in (GGC), p. 226.
A disastrous shock in Akhaïa, almost all houses were totaly destroyed at Kouloura, Valimitika, St. Konstantinos, St. Heleni and Aigos Athanasios, some houses collapsed at Aigion (1 victim) and Murla, serious damage (houses partly unin- habitable, many fissured) at Temeni, Selianitika, Dimitropulon, Diakopton, Neos-Erineos, Kamaraes, in many places fissures, subsidence of the ground, e. g. at the coast of Aíyion (3^m) and in the sea between Zákinthos and Kórinthos, chan- ges in the springs (GGC, GG), the disaster was limited on a 10 km wide band on both sides of Aíyion (GGK). During strong aftershocks on Sept. 10, 03^h, 04^h, 18^h, Sept. 11, night, Sept. 15/16, Sept. 17. 12^h, Sept. 23, 01^h, some other houses were damaged or destroyed in Aíyion (GGC).

Sept. 22, 13^h, $41 \cdot 3°$N, $43 \cdot 3°$E (BZ), $41 \cdot 1°$N, $42 \cdot 7°$E, R.33, $I_0 =$ VIII, $M = 6 \cdot 1$ (ONT, MJD). A strong shock at Keda damaged houses at Okam (victims) and was felt at Ardagan (BZ), disastrous, many victims in collapsed houses, many aftershocks; Sept. 23—24, VII—VIII, Sept. 23, 03^h30^m, many houses destroyed (MO).

Oct. ? $38 \cdot 2°$N, $28 \cdot 0°$E, R.26c, $I_0 =$ VII (EGU).
Izmir, Aydin, Ödernis.

March 6, 00^h32^m UT, $45 \cdot 8°$N, $16 \cdot 0°$E, R.22b, $I_0 = ?$, $I_0 =$ VII, $r = 16$ km (TD).
A heavy shaking at Zagreb (BA).

April ?, $39\frac{1}{2}°$N, $20\frac{1}{2}°$E, $h = n$, R.26a, $I_0 =$ VII? (GG, GGC), $I_0 =$ VI—VII (GGK), $M = 5 \cdot 2$ (GGE).
"Serious damage at Paraníthia and Margarítion" (GGC, GGK).

May 21, $35 \cdot 7°$N, $0 \cdot 8°$W, R.16, $I_0 =$ VII—VIII (RK, GA).
A violent shock at Oran (chimneys thrown down, walls fissured), felt at Mers el Kébir, Tamzoura, Sidi Chami and Tlélat (RK).

May 30, $49 \cdot 2°$N, $1 \cdot 7°$W (RF), R.11, $I_{max} =$ VII?, $r = 200$ km about (PEF, MJD).
Some chimneys were thrown down at Condé, Lisieux, Valognes (RF).

Aug. 25, 19^h13^m, $38\frac{1}{4}°$N, $22°$E, $h = i$, R.26a, $I_0 =$ VIII, $r = 210$ km (GG, MJD, GGC), $I_0 =$ VII—VIII (GGK), $M = 6 \cdot 3$ (GGE), isos. map in (GGC), p. 227.
Some houses collapsed at Pteri and Vella, houses fissured and seriously damaged at Agrinion VIII, Aitolikon VIII, Diakopton VIII, Klimenti, Matzani, Dokimion VII?, Spolaita VII?, Akrata, Pátrai VII—VIII, etc., cable breaks, mud and sand volcanoes (GGC), felt strongly (VII?) at Gjirokastër and Delvinë (MD), felt in the region of Otranto and SE Sicily (MCA, BI).

Sept. 13, $46 \cdot 4°$N, $13°$E, R.18b, $I_0 =$ VI?, $I_0 <$ VII? (MR), isos. map. in (BI).
A sequence of shocks in the region of Tolmezzo during June-November culminated by a shock on Sept. 13 which caused some slight damage (BI).

1889

Oct. 25, 23^h19^m, $39\frac{1}{4}°$N, 26°E, $h = n$, R.26c, $I_0 = X-XI$ (GGK), $I_0 = XI$, $r = 370$ km (GG), 39·3°N, 26·2°E, $I_0 = VII$ (MJD, ONB, EGU), $I_0 = IX$ (CAM), isos. map in (GGC), p. 228, $M = 6·9$ (GGE).

A disastrous shock in Lésvos, almost all houses destroyed at Eressós, Telonia, Chydira, Agra, Tzithra, Vatussa, Revma, serious damage at Mitilíni, Mesotopon, Sigrion, etc., 25 victims, changes in springs, large rockslides in the mountains (GGC). Strong at Izmir, Khíos (Sakiz) and Canakkale, slight at Mugla, Tekirdag and Istanbul (PL).

Aftershock on Nov. 21, 15^h50^m, at Vatussa the destruction was completed (GGC).

Dec. 8, 05^h15^m UT, epic. in the sea near Isole Tremiti (42·1°N, 15·5°E) and Lesina, R.19, $I_{max} = VII$ (MR), isos. map in (BI).

A largely felt earthquake along the whole Adriatic coast of Italy and Dalmacia, houses fissured at Apricena, Cagnano Varano, S. Marco in Lamis, Isole Tremiti, felt at Lecce, Roma, Bologna, Salo, Belluno, Pola (BI).

Dec. 25, 05^h23^m UT, 37·7°N, 15·1°E, $h = $ sup, R.20, $I_0 = VIII$ (MR), local.

Many houses rended uninhabitable at Zafferana and Acireale, felt at Catania (BI).

1890

Febr. 21, (38·7°N, 9·2°W), R.12, $I_0 = VII$? (FA), $I_0 = V$ (MU), $I_0 = VIII-IX$ (MF).

A violent local shock in Lisboa made important damage to churches at Batalha and Maceira (GS).

March 9, many houses destroyed and some others damaged at Kos (36·9°N, 27·3°E) $I_0 = VII$? (GGC), no other report, local? Not mentioned in (GGK).

May 20, 39·9°N, 38·8°E, R.33, $I_0 = IX-X$? (MJD), $I_0 = X$, $M = 7·3$ (ONB, CAM), $I_0 = IX$, 39·4°N, 39·8°E (ANV), Nov.?, 38·8°N, 38·3°E, $I_0 = VI$ (EGU).

No details available. Quite strong at Malatya (EGU).

May 21, $39\frac{1}{2}°$N, $20\frac{1}{2}°$E, $h = n$, R.26a, $I_0 = VII-VIII$ (GGK), $I_0 = VII$ (GG), $M = 5\frac{1}{2}$ (GGE).

Some walls and three houses collapsed and the others were seriously injured at Margarítion, felt at Préveza and Ioánnina (GGC, GGK).

May 26, 39·9°N, 38·8°E, R.33, $I_0 = VIII$?, $M = 6·1$ (ONB, MJD), $I_0 = X$ (CAM), not in (EGU). Kayi.

May 26, $38\frac{1}{2}°$N, $25\frac{1}{2}°$E, $h = n$, R.26c, $I_0 = VII-VIII$ (GGK), $I_0 = VII$, $r = 330$ km (GG).

One house collapsed in Psará, other houses were much fissured, felt in Khíos and Ródhos (GGC), cracks in the ground (GGK).

May 30, 18^h40^m UT, 43·3°N, 17·9°E, R.22a, $I_0 = VII-VIII$?, $I_0 = VII$, $r_{max} = 19$ km (TD), $h = $ sup.

VIII at Blagaj, VII at Mostar (roofs of some houses collapsed), VI at Nevesinje, etc. (BA).

June 7, in the villages Tadzhilar and Khadzhi-Amza south of Ikhtiman, Bulgaria, R.23, chimneys were thrown down (WS), $I_0 = VII$?

June 10, 15^h34^m, 47·6°N, 4·3°E, R.11, $I_0 = VII$, $r = 150$ km ca (PRF).

Sept. 13, 66·2°N, 12·5°E, R.3, $I_0 = VII$, $M = 4·6$, $r = 63$ km (BMF).

Helgeland, Svenningdal-Rödö (KF).

1890

Nov. 15, 17^h50^m, Inverness (57·5°N, 4·2°W), R.10a, $I_0 = $ VII (DBC, MJD).

Nov. 28, 00^h37^m, 48·3°N, 17·0°E, R.21a, $I_0 = $ VII, $P = 2750$ km^2 (RA, MJD, KC), isos. map in (RA).
At Stupava 8 chimneys fell and the walls were largely fissured, fissured walls or a few collapsed chimneys also at Vajnory, Malacky, Záhorská Bystrica (RA).

Dec. 14, 16^h30^m, $38\frac{3}{4}$°N, $26\frac{3}{4}$°E, $h = n$, R.26c, $I_0 = $ VIII, $r = 280$ km (GG), $I_0 = $ VI (EGU), $I_0 = $ VIII−IX (GGK), $M = 6·5$ (GGE).
35 houses collapsed at Ephesus, 150−200 houses damaged, strongly felt in Sámos, felt in Thíra (GGK), epic. Selçuk (EGU).

1891

Jan. 15, 04^h, 36·5°N, 1·8°E, R.15, $I_0 = $ IX−X, $I_0 = $ X, $r = 200$ km (RK, GA, CAM).
A destructive shock at Gouraya where 53 houses were thrown down, 36 victims, at Villebourg 22 houses were destroyed as well as many "gourbis in the dours", a crack 40 cm wide crossed the village, at Alger and El Affroun houses were fissured, at Blida some houses collapsed, at Mouzaïville an earthslide cut the road, felt at Mira, Perregaux, Boghari, Teniet el Haad, Tiaret, Saïda, Ain el Hadjar, Ain Bessem, Djelfa and Djelala (RK).

Apr. 3?, 39·1°N, 42·5°E, R 33, $I_0 = $ VII?, $M = 5·5$ (ONV), $I_0 = $ VI (EGU), $I_0 = $ VIII, 38·8°N, 42·5°E (ANV), $I_0 = $ IX (MJD).
Damage at Malazgirt, city walls collapsed (PL), 146 houses destroyed at Adilcevaz, 100 victims (LC).

May 9, 00^h15^m UT, $41\frac{1}{2}$°N, $13\frac{3}{4}$°E, R.19, $I < $ VII?, $I_0 = $ VII (MR), isos. map in (BI).
Slight damage at Sora and Isola del Liri, strongly felt at Solmona, Avezzano, Filettino, etc. (BI).

May 11, 18^h, $37\frac{1}{2}$°N, $24\frac{1}{2}$°E, $h = n$, R.26a, $I_0 = $ VII?, $I_0 = $ VI, $r = 250$ km, swarm (GG), $I_0 = $ V−VI (GGK), $I_0 = $ X (CAM), $M = 6·1$ (GGE).
Many houses, especially the old ones, were fissured at Kythnos during the strongest shock of a swarm, felt at Syra, Tínos, Andros, Khíos, Çeşme, Athínai, Aíyion (GGC).

June 7, 45·4°N, 11·2°E, R.19, $I_0 = $ VIII (MR, MJD), $I_0 = $ IX (CAM), $I_0 = $ IX−X (MF), isos. map in (BI).
Houses destroyed or uninhabitable at Badia Calavena, Tregnago, S. Andrea, Castagne, Cogolo, Marcenigo, Scorgnago and Trettene, etc., felt as far as Torino, Pisa, Arezzo, Trieste (BI).

June 27, 13^h45^m, 39°N, $20\frac{3}{4}$°E, $h = n$, R.26a, $I_0 = $ VIII?, $I_0 = $ VII (GG), $I_0 = $ IX−X (MF), $M = 5\frac{1}{2}$ (GGE).
Seriously damaged houses at Préveza, some houses collapsed, felt at Pálairos (Zaverda) (GGC).

Aug. 22, $04−05^h$, epic. in the sea? R.12, $I_{max} = $ VII (FA, GS), 42·0°N, 8·6°W, $I_{max} = $ V (MU).
A strong earthquake caused damage at Lisboa, felt at Oporto and Pontevedra (GS).

Sept. 18, 04^h, $39\frac{1}{4}$°N, 23°E, $h = n$, R.26a, $I_0 = $ VI−VII (GGK), $I_0 = $ VII (GG), $I_0 = $ IX (MF), $M = 5\frac{1}{2}$ (GGE).

One house collapsed at Lekhónia, felt at Vólos (GGC), two houses destroyed near Vólos (GGK), doubtful intensity.

Oct. 24—26, strong volcanic shocks at Pantelleria, (R.20), $I_0 \leqq$ VII?, accompanied by a submarine volcanic eruption (BI), $I_0 =$ VII (MJD).

1892

Jan. 5 or 6, $(45\frac{1}{2}°N, 10\frac{1}{2}°E)$, R.18b, $I_0 =$ VII (MR, MJD), isos. map in (BI).
At Campazzi a porch ruined, at Salo houses cracked and chimneys fell down, etc., felt at Pavia, Parma, Padova, Modena (BI).

Jan. 9, 06^h15^m, $39\frac{3}{4}°N$, $22\frac{1}{2}°E$, $h = n$, R.26a, $I_0 =$ VIII, $r = 130$ km (GG), $I_0 =$ VII to VIII (GGK), $I_0 =$ VII (GGC), $P = 50\,000$ km^2, $M = 5\cdot5$ (GGE).
At Lárisa several houses collapsed and the others were seriously injured (large fissures, walls or roofs thrown down), at Tyrnavos and Agya some walls were fissured and some roofs damaged, felt at Kazaklar, Skíathos, Skópelos and Thessaloníki (GGC).
An aftershock on Jan. 11, 07^h, damaged houses at Lárisa (GGC).

Jan. 22, 22^h25^m UT, $41\cdot7°N$, $12\cdot7°E$, R.19, $I_0 =$ VII?, isos. map in (BI), $h =$ sup? Jan. 29, $I_0 =$ IX (MF).
Damaging shock at Genzano, Civitalavinia and Velletri, felt as far as Spoleto, Aquila, Campobasso (BI).

March 5, 17^h26^m UT, epicentre? R.18a, $I_0 =$ VII? (MR), isos. map in (BI).
Houses fissured and some chimneys thrown down at Hône, Bard, Donnaz, Pont S. Martin and Lillianes, etc., felt at Torino, Baceno, Vercelli (BI).

March 7, 00^h UT, $38\frac{1}{2}°N$, $14\frac{1}{2}°E$, $h =$ sup, R.20, $I_0 =$ VIII?, $I_0 =$ IX (MR), volcanic? isos. map in (BI).
A very strong shock in the Lipari islands, felt in NE Sicily (BI, MR). An aftershock on March 16, 12^h38^m UT, $I_0 =$ VIII (MJD), $I_0 =$ IX—X (MF).

April—June, Peninsula Garganica $(41\cdot8°N, 15\cdot9°E)$, R.19, $I_0 =$ VII (MJD), isos. map in (BI).
Strong shocks on April 20, June 7, June 16, fissured houses at Mattinata, Vieste, Monte Saraceno and Isola Tremiti (BI).

June 22, 01^h35^m UT, $46\cdot7°N$, $18\cdot4°E$, R.21b, $I_0 =$ VII (CD), $I_0 =$ VIII, $P = 2700$ km^2, $r = 80$ km^2 (RA, RAB), isos. map in (RAB).
A few chimneys fell at Belecske, Miszla, Kisszékely, all houses damaged and chimneys thrown down at Pincehely (RAB).

June 23, 23^h20^m UT, $46\cdot2°N$, $12\cdot7°E$, R.18b, $I_0 =$ VII (MR, MJD), isos. map in (BI).
At Claut houses largely fissured, some chimneys fell down, etc., felt at Trieste, Rovinj, Follina (BI).

July 7, 00^h UT, $37\cdot7°N$, $15\cdot1°E$, $h =$ sup, R.20, $I_0 =$ VII (MR), isos. map in (BI).
At Zafferana some walls collapsed, the others fissured, strong at Nicolosi, Biancavilla and Linguaglossa (BI), 12 houses destroyed (LC).

Aug. 18, 00^h24^m, $(51\cdot7°N, 4\cdot9°W)$, R.10b, $I_0 =$ VII? (DCB).
Pembroke in S. Wales (DCB), 3 houses collapsed (LC).

Aug. 25, $45\cdot4°N$, $3\cdot4°E$, R.11, $I_0 =$ VII (RF).
Epicentre in the valley of Allier, at Vichy $(46\cdot2°N, 3\cdot4°E)$, Clermont $(45\cdot8°N,$

3·1°E), Brioude and Le Puy (45·0°N, 3·9°E) chimneys were thrown down and walls fissured (RF).

Oct. 14, 05^h UT, epic. unknown, Black Sea?, R.21d?, I_{max} = VIII?, $h = i$ (KGI), $I_{max} =$ = VII (MJD).

A largely felt earthquake in the eastern part of the Balkan peninsula, maximum intensity VIII reported from Ignatievo, district Varnensko, Bulgaria, felt over all Bulgaria, further in southern Rumania at Braşov, Sfîntu Gheorghe, Cohalm and also at Istanbul (FA, KGI).

Nov.15—16,   damaging shocks in the islands of Ponza (40·9°N, 12·9°E), W of Ischia (Italy), R.19, I_0 = VII?, I_0 = IX (MF), isos. map in (BI).

Nov. 21, 19^h UT, 43·5°N, 12·3°E, R.19, I_0 = VII (MJD), isos. map in (BI).

At Frazzano, Cafirenze, Citta di Castello a church was seriously fissured and many chimneys thrown down (BI).

Dec. 27, 18^h30^m, $37\frac{3}{4}$°N, 27°E, $h = n$, R.26c, I_0 = VII (GG), I_0 = VI—VII (GGK), $M = 5\frac{1}{2}$ (GGE).

A swarm in Sámos (GG), several poor houses collapsed in the eastern part of Sámos (GGK).

Dec. 1892 — Jan. 9, 1893, 44·6°N, 10·5°E, R.19, I_0 = VII? (MJD), isos. map in (BI).

Slight fissures in some houses and collapsed chimneys at Castel del Rio, at Monzuno some walls fissured (BI).

1893

Jan. 24, 23^h21^m UT, 40·6°N, 15·4°E, R.19, I_0 = VII (MR, MJD), isos. map in (BI).

Fissures of different size in the houses at Petina, Auletta, Pertosa, strong at Caggiano, etc., felt Andria, Canosa, Foggia, Napoli (BI).

Jan. 31, 03^h30^m, 37·7°N, 20·9°E, $h = n$, R.26a, I_0 = IX—X (MJD), I_0 = IX, $r = 230$ km (GG), I_0 = IX—XI (GGK), I_0 = X (CAM), I_0 = XI (MF), $M = 6·4$ (GGE).

A disastrous shock in the southern part of the island Zákinthos, mostly affected were Agalas, Kerí, Gaïtani, felt in Kérkira and as far as the eastern coast of Greece (GGK).

Febr. 1, 00^h15^m, 37·7°N, 20·9°E, $h = n$, R.26a, I_0 = VIII—IX (GGK), I_0 = IX (GG), $M = 6\frac{3}{4}$ (GGE).

Strong aftershock in Zákinthos ruined most houses at Kerí and Agalas which were damaged by the main shock, felt at Pátrai (GGK).

Febr. 9, 18^h, $40\frac{1}{2}$°N, $25\frac{1}{2}$°E, $h = n$, R.27, I_0 = IX, $r = 180$ km (GG, MJD), I_0 = VIII to IX (GGK), I_0 = X (CAM, MJD, MF), $M = 5·9$ (GGE), Jan. 28, I_0 = VII (EGU).

Extensive damage in Samothráki, where about 10% of houses collapsed and 50% was rended uninhabitable, in the capital of the island I = VII, tsunami 1 m at Agistron (GGT, ANS), rockslides (GG), felt along the coast of Thrakia and in S. Bulgaria with I_{max} = V (KGI), strong in Imroz and at Alexandroúpolis (Dedéagach).

March 11, 09^h25^m, 47·9°N, 23·1°E, $h = n$, R.21d, I_0 = VII, $r = 35$ km (PRA, RA, EU, FA).

I = VII at Halmeu, Dobolt and Turţ, felt as far as Khust, Bicsad, Negreşti, Seini, Somes, Beregovo, Kölcse, etc. (FA), walls fissured, chimneys thrown down, churches damaged (RA).

March 12, 16^h, $37\frac{3}{4}°$N, $27°$E, $h = n$, R.26c, $I_0 = $ VII−VIII, $I_0 = $ VII (GG), $I_0 = $ VI−VII (GGK), $M = 5\frac{1}{2}$ (GGE).
In Sámos some country houses collapsed and several houses were fissured (GGK).

March 31, $38\cdot3°$N, $38\cdot3°$E, R.35, $I_0 = $ IX, $M = 6\cdot7$ (ONB), $38\cdot4°$N, $38\cdot7°$E (EGU).
Severe damage at Malatya, 400 victims, damage and victims also at İzolu, epic. probably N of Malatya (PL).

April 8, 13^h UT, $44\cdot2°$N, $21\cdot2°$E, R.22b, $I_0 = $ VIII−IX, $I_0 = $ IX, $r_{max} = 445$ km (TD, CAM, MF), $44\cdot9°$N, $21\cdot4°$E, $h = n$ (PRA), $I_0' = $ X?, $P = 11\,000$ km^2 (RA), $I_0 = $ VIII (KGI).
A largely extended earthquake, felt over NE Yugoslavia, Hungary, SW Rumania, Bucureşti, Ploeşti, Olteniţa, W. Bulgaria ($I_{max} = $ VI−VII at Belogradchik) (FA, RA, KGI), two victims at Popović (MCA). At Svilajnac new springs originated, many houses rended uninhabitable, Sedlare, at Subotica the mill damaged, one victim, at Vel. Popovič many houses collapsed, no house undamaged, at Rudnik all tiles fell down, at Čuprija the railway station was seriously fissured, VII at Požarevac, 300 m long crack in the ground of an island in the Morava river, etc. (BA).

Aftershocks with $I_0 = $ VII or more:

April 8, 13^h50^m UT, $r_{max} = 303$ km (TD), 14^h09^m UT, $r_{max} = 85$ km (TD), $I_0 = $ VIII (MJD),
16^h41^m UT, April 9, 00^h50^m UT, April 10, 02^h29^m UT, 22^h57^m UT, April 21, 10^h12^m UT, April 23, 17^h36^m UT, April 29, 22^h45^m UT, $r_{max} = 51$ km, May 20, 20^h42^m UT, $I_0 = $ VIII, $I_0 = $ VII (MJD), May 31, 19^h13^m UT, July 27, 19^h54^m UT, Sept. 4, 13^h29^m UT, $43\cdot7°$N, $21\cdot4°$E, $r_{max} = 60$ km, Sept. 8, 09^h38^m UT, $43\cdot9°$N, $21\cdot1°$E, $I_0 = $ VIII, $r_{max} = 157$ km, $I_0 = $ IX (CAM), 1894, March 16, 02^h44^m UT, $44\cdot0°$N, $21\cdot2°$E (2 shocks), $r_{max} = 72$ km, Dec. 10, 14^h06^m UT, $44\cdot0°$N, $21\cdot2°$E (TD, MJD).

Apr. 17, $05^h37\cdot7°$N, $20\cdot9°$E, $h = n$, R.26a, $I_0 = $ X, $r = 230$ km (GG, CAM, MJD), $I_0 = $ IX−XI (GGK), $M = 6\cdot4$ (GGE).
Main disastrous shock at Gaïtani-Kerí, Zákinthos, casualties (GG), felt in Ipiros (MCA).
Both shocks (Jan. 31, Apr. 17), ruined totally 2000 houses and rended 1700 other houses uninhabitable, also the villages Gaïtani, Kerí, Agalas and Lithakiá were ruined (GGK).
Aftershock on Oct. 13, 05^h30^m, $I_0 = $ VII (GG), $I_0 = $ VI−VII (GGK), houses damaged in Zákinthos.

Apr. 22, 01^h16^m UT, $38\cdot0°$N, $14\cdot9°$E (PET), R.20, $I_0 = $ VII (MR, MJD), isos. map in (BI).
At Montalbano di Elicona (Sicily), S. Piero Patti, S. Barbara serious damage observed, felt in Lipari, at Messina, Militello (BI).

May 2, 15^h15^m UT, $41\cdot4°$N, $20\cdot8°$E, R.25, $I_0 = $ VIII? (TD), local?
A foreshock with $I_0 = $ VII at 07^h (TD), Gorizia, Ohrid.

May 23, 20^h04^m, $38\frac{1}{4}°$N, $23\frac{1}{2}°$E, $h = n$, R.26a, $I_0 = $ VIII−IX, $I_0 = $ VIII, $r = 220$ km (GG), $I_0 = $ X−XI (MF), $I_0 = $ VII−VIII (GGK), $M = 6\cdot1$ (GGE).
The main shock of a sequence of strong foreshocks on June 11, July 22, 1892, March 27, May 22, 1893, and of a very strong aftershock on May 31, 1893 (GG, threw down 100 houses at Thebes and made 800 houses uninhabitable (GGK))

1893

June 14, (40·2°N, 19·7°E), R.25, I_0 = X?, I_0 = XI (MD), I_0 = X—XI (GGK), I_0 = IX to X (MF), I_0 = IX (MA), M = $7\frac{1}{2}$ (GGE).
"A great catastrophe", highest intensities (I = XI?) at Himarë, Dhërmi, Kuç, Kudhesi, (I = X?) at Vlorë, Kanino, Narta and Smokhtinë (I = IX?), high tsunami in the bay of Vlorë (MD, MA, ANS), felt in Puglia (BI, MCA).

June 20, 21^h UT, 45·8°N, 16·0°E, R.22b, I < VII, I_0 = VII (TD).
At Zagreb a heavy shock, the tower of a church and the roof slightly damaged (BA).

Aug. 5, an earthquake in the valley of Mur, in Styria, I_0 = VIII (MJD), I_0 = IX (MF), questionable, not mentioned in (TT), i. e. I < VI.

Aug. 10, 09^h UT, ($41\frac{3}{4}$°N, 16°E), h = sup?, R.19, I_{max} = VIII (MJD), I_0 = IX (CAM), I_0 = IX—X (MR), I_0 = X (MF), isos. map in (BI).
Destructive at Mattinata where houses were partly destroyed, very strong at M. S. Angelo and Manfredonia (BI).

Aug. 17, 14^h35^m, 43·8°N, 26·6°E, h = i, R.21d, I_0 = VII?, I_0 = VIII, M = 6·6 (PRA, FA, EU).
An earthquake in the region of Vrincioaia, felt in Rumania, Bessarabia, N. Bulgaria and W. Ukraina, without serious damage (FA).

Sept. 10, 03^h40^m, 45·8°N, 26·6°E, h = i, R.21d, I_0 = VII? 46·4°N, 27·7°E, I_0 = VIII? (FA), 46·7°N, 28·1°E, I_0 = VI (ES).
Another earthquake with intermediate depth of focus, felt again over Rumania, Bulgaria (not mentioned in (KG)) and Ukraina, more severe than on Aug. 17 but also without serious damage (FA).

Oct. 11, 05^h25^m, 45·8°N, 16·0°E, R.22b, I_0 = VI—VII?, I_0 = VII, r_{max} = 97 km (TD).
At Zagreb a church fissured (BA).

Nov. 2, 17^h45^m, Carmarthen in Wales, 51·9°N, 4·3°W, R.10b, I_0 = VII (DCB), local?

Nov. 14, 08^h39^m, 38°N, $23\frac{1}{2}$°E, h = n, R.26a, I_0 = VII, r = 80 km (GG, MJD), I_0 = VI to VII (GGK).
At Salamís some old houses collapsed and the walls of the monastery St. Mina heavily fissured, felt as fas as Levádhia (GGK).

1894

Febr. 9, 12^h48^m UT, ($45\frac{1}{2}$°N, $11\frac{1}{4}$°E). R.18b, I_0 = VII? (MJD), isos. map in (BI).
At Bosco Chiesa, Velo Veronese, Adige, Chiampo, Badia Calavena, Tregnago walls fissured, felt at Belluno, Venezia, Trento etc. (BI).

March 1, 15^h25^m, 45·2°N, 27·2°E, R.21d, I_0 = VII (PRA, FA).
A strong shock in the district of Putna, I = VII at Panicu, Adjud and Focsani (walls fissured), felt in the districts of Iaşi, Rîmnicul Sărat, Ialomiţa and Tulcea (FA).

March 4, 06^h35^m UT, 45·8°N, 26·6°E, h = i, R.21d, I_0 = VII, M = 6·1 (PRA), I_0 = VIII (FH).
An earthquake which was felt over a large area, i. e. in Rumania and SW Ukraina, maximum intensity observed at Bereşti (FH).

March 25, 23^h UT, 41·8°N, 15·4°E, R.19, I_0 = VII? (MJD), r = 35 km, isos. map in (BI).
Fissures in the houses at Lesina, more serious in old houses (BI).

1894

April 20, 16^h52^m, 38·6°N, 23·2°E, $h = n$, R.26a, $I_0 = X$, $r_5 = 150$ km, $r = 310$ km (GG), $I_0 = X-XI$ (GGK), $M = 6·7$ (GGE).

A disastrous shock in the peninsula of Lokris, the localities Proskyma, Masi, Malesína, Martino suffered much, 200 m long fissures, many rockslides, felt as far as Spárti, Trípolis, Skyros, Lárisa, Ios and Thíra (Santorin) (GGK, GG).

April 27, 19^h42^m, 38·7°N, 23·1°E, $h = n$, R.26a, $I_0 = XI$, $r_5 = 150$ km, $r = 350$ km (GG), $I_0 = XI-XII$ (MF), $I_0 = X$ (CAM), $I_0 = X-XI$ (GGK), $M = 6·9$ (GGE).

Continuation of the activity in Lokris, a cumulative effect along a fault trace of 55 km, 1—4 m wide, 0·3—1·5 m jump, tension cracks up to 7 km, left hand strike-slip movement?, many landslides, changes in wells, tsunami of 3 m between Ata-lándi and Agiós Konstantinos (GGT, ANS, GG, RFC), main disaster between Agiós Konstantinos and Arkitsa up to Kiparíssia, both shocks destroyed 3783 houses at 69 localities, 255 persons were killed, by a 1—2 m subsidence in the basin of Atalándi the Nesos Patroklou (Gaidarion) peninsula became an island, 4250 aftershocks till May 6 (GGK).

May 28, 20^h14^m UT, 40·0°N, 16·0°E (PET), R.19, $I_0 = VII$ (MJD), $I_0 = VIII$ (MR), isos. map in (BI).

Houses fissured at Monte Pollino, Viggianello (5 houses destroyed), Rotonda, Castellucio, etc., felt at Taranto, Trivigno, Bisignano (BI, MJD).

July 10, 12^h33^m, 40·6°N, 28·7°E, $I_0 = IX$ (EGU), 40·8°N, 30·4°E, R.27, $I_0 = VIII$, $M = 6·1$ (ONB), $I_0 = X$ (CAM), $I_0 \geq IX$ (MJD).

At Istanbul a catholic church partly collapsed, the cupola of a mosque at Galata fissured, Pera, S. Stefano, gulf of Izmit, largest destructions in the islands of Khálki and Burgas Ada (Antigoni) (MJD, CAM), tsunami at Istanbul (ANS), changes in the efficiency and temperature of springs (DJ). Some weak houses collapsed at Istanbul, the roof of the Bazaar fell down, no damage, to the mosques, cracks on the shore, some damage in the Princes islands (Kizil Adalar), mainly in Heybeliada, felt at Izmit, Bursa, Bandirma, Tekirdag, epic. in the Marmara Sea (PL).

July 26, 00^h15^m, $37\frac{3}{4}$°N, $21\frac{3}{4}$°E, $h = n$, R.26a, $I_0 = VII-VIII$ (GGK), $I_0 = VIII$ (GG), $M = 6\frac{1}{4}$ (GGE).

Seriously damaged houses at Dukat, four of them collapsed, felt at Divri, Erithraí (Kriekouki), Pátrai and Zákinthos (GGK).

Aug. 8, 05^h16^m UT, 37·6°N, 15·1°E, $h = $ sup, R.20, $I_0 = IX$ (CAM), $I_0 = X$ (MR, MF), $I_0 = VIII$ (MJD), $r = 5$ km, isos. map in (BI).

Many houses ruined in the region of Fleri, Zerbate and Pisano (SE of Etna), 13 victims (MF, BI).

Aug. 31, 12^h, 45·8°N, 26·6°E, $h = i$?, R.21d, $I_0 = VIII-IX$?, $M = 6·9$ (PRA), 45·5°N, 27·7°E, $I_0 = VIII$ (FA), $I_0 = IX$ (CAM, KGI), $I_0 = IX-X$ (MF), Sept. 2 (MF, CAM, MJD).

In the districts of Putna, Ilfov, Tecuci, Covorlui and at' Bucureşti $I = VIII$ (FA), felt almost in the whole Rumania, in Bulgaria, $I_{max} = V$ at Silistra (KGI).

Nov. 16, 17^h48^m UT, 38·3°N, 15·8°E (PET), R.20, $I_0 = IX$ (CAM), $I_0 = X$ (MR, MF), $I_0 = VIII$ (MJD), isos. map in (BI).

A disastrous shock in the western part of the province of Reggio, particularly at Palmi and S. Cristina, 96—101 victims, destructive also at Seminara, S. Procopio,

Bagnara and Castellaci d'Oppido, etc., felt at Salerno, Cotronei, Palermo, for details see (BI).

Nov. 27, 05^h40^m UT, R.19, $I_0 =$ VII? (MJD), isos. map in (BI).
Slight damage at Brescia, Nave, Adro, Chiari, Provezze, Rodengo, etc., felt at Venezia, Oropa, Pavia, Treviso (BI).

Nov. 28, 16^h16^m UT, 46°N, 11·8°E, R.18b, $I_0 =$ VII? (MJD), isos. map in (BI).
Houses fissured and chimneys overthrown at Fonzaso (BI).

Dec. 19, 21^h35^m, 45·0°N, 21·7°E, R.22b, $I_0 =$ VII, $r = 105$ km (PRA, MJD, RA, FA).
Probable epicentre near Oraviţa in Banat, felt in Banat, SE Transylvania and Serbia (FA), houses collapsed or cracked, roofs thrown down (LC).

Dec. 27, 05^h55^m UT, 38·6°N, 14·6°E, R.20, $I_0 =$ VII−VIII (MR), $I_0 =$ VII (MJD).
In the island of Filicudi in the Lipari Islands some poor houses and walls collapsed, other houses were fissured, strong at Salina, felt at Messina and Catania (BI).

1895

Date? 38·6°N, 27·1°E, R.26c, $I_0 =$ IX? (EGU), Menemen; not in (ONB).

Febr. 27, 15^h37^m, R.18b, $I_0 =$ VII? (MJD), isos. map in (BI).
At Claut (Udine) several rural houses were fissured, felt at Treviso, Belluno, Avviano, V−VI, Cansiglio V, etc. (BI).

March 23, 12^h15^m, R.19, $I_0 =$ VII (MJD), isos. map in (BI).
At Comacchio walls of several houses were fissured, large pieces of ceilings fell, several roofs damaged at Ostellato poor houses were slightly fissured and some chimneys ruined, felt slightly at Venezia, Bologna, Ferrara, Ravenna, Pesaro, etc. (BI).

April 13, 15^h00^m, 37·2°N, 14·7°E, R.20, $I_0 =$ VII−VIII, $I_0 =$ VII (MR, MJD), $I_0 =$ IX (MF), isos. map in (BI), local?
At Vizzini two walls collapsed, in the church and in some houses the frames fell down, houses were slightly damaged, very strongly felt at Mineo, strongly felt at Caltagirone, etc. (BI).
A local but destructive shock at Mineo and Vizzini (Catania, Siracusa) (MF).

April 14, 22^h17^m UT, 46·1°N, 14·5°E, R.22b, $I_0 =$ VIII−IX, $M = 6·1$, $h = 16$ km, $r_2 = 350$ km, $r_4 = 250$ km, $r_5 = 180$ km, $r_6 = 110$ km, $r_7 = 52$ km, $r_8 = 18$ km, isos. map in (RV, BI), $I_0 =$ VII (MJD), $I_0 =$ IX (CAM, TD).
The most severe earthquake in Slovenija in the 19th century, 10% of all houses in Ljubljana were damaged, some public buildings were seriously injured, some houses were destroyed at Vodice, Domzale, Kosarje, Stranska Ves, cracks occurred in the ground, rockslides near Zagorje, 7 victims, the outer isoseismal crossed Wien, Split, Assisi, Firenze, Alessandria (RV, BI, HR). For further details see (HR, BI, BA). One aftershock with $I_0 =$ VII on April 14, 23^h01^m, $r_2 = 135$ km according to (RV), (TD) gives, however, six aftershocks with $I_0 =$ VII on April 14, two on April 15, one on April 20 (new damage).

May 1, Atlanta (?) (Greece), R. 26a, $I_0 =$ VII (MJD), not mentioned in (GG), doubtful.

May 14, 03^h or/and 05^h, $39\frac{1}{2}$°N, $20\frac{1}{2}$°E, $h = n$, R.25, $I_0 =$ IX−X (GGK), $I_0 =$ X (GG, CAM, MD), $I_0 =$ X−XI (CAM), $M = 7\frac{1}{2}$ (GGE).

Heavy damage at Margarítion, Filiátes, Paramithia and surroundings, casualties (GGK), "a great catastrophe"; damage in Ipiros and in the district of Ioánnina, Himarë (X), Durrës (VIII), Karbunarë (VIII) (MD), many victims, $h = $ sup? (MCA). Main shock (?) on May 13, 23^h, Himarë X, Durrës and Karbunarë VIII, great catastrophe in the district of Ioánnina (MD).

Aftershocks according to (MD):

May 15, $I_0 = $ X?, disastrous in S. Albania and N. Ipiros at Margarítion, Filiátes, Dragomi (MD, MA), $M = 6.4$ (GGE).

June 21, $I_0 = $ VIII—IX?, great damage at Elbasan (40.1°N, 20.1°E) and surrounding villages (MCA, MD, GGK), $M = 6\frac{1}{4}$ (GGE).

Aug. 6, $I_0 = $ IX—X?, destruction in the region of Shiak (MCA, MA) give July 6.

Sept. 5, $39\frac{3}{4}$°N, $20\frac{3}{4}$°E, R.25, $I_0 = $ VII—IX (GGK).

1896, Jan. 11, Ipiros, $I_0 = $ VIII (MA).

Febr. 9, $I_0 = $ VII?, Durrës.

Febr. 10, 01^h, $I_0 = $ VIII—IX (GGK), $I_0 = $ X?, Durrës, $I = $ VIII (MA, MCA), $M = 5\frac{1}{2}$ (GGE).

12^h, $I_0 = $ VII—VIII (GGK), $I_0 = $ VIII?, Korçe, $M = 6\frac{1}{4}$ (GGE).

15^h, $I_0 = $ VIII—IX (GGK), $I_0 = $ IX?, damage at Elbasan, $M = 6\frac{1}{4}$ (GGE).

18^h, $I_0 = $ VI—VII (GGK), $I_0 = $ VII?, strong at Korçë and surroundings.

Febr. 11, 19^h, $I_0 = $ VIII—IX (GGK), $I_0 = $ IX? A ruinous shock in the region of Ohrid (41.1°N, 20.8°E), extremely severe at Ohrid and Struga, Starove IX?, Korçë VI?, Ohrid lake, $I_0 = $ IX (MA, MCA), $M = 5\frac{1}{2}$ (GGK). 19^h30^m, $I_0 = $ VI—VII (GGK), $I_0 = $ VII?

21^h30^m, $I_0 = $ VII—VIII (GGK), $I_0 = $ VIII?, Elbasan, $M = 6\frac{1}{4}$ (GGE).

Febr. 14, $I_0 = $ VI—VII (GGK), 16^h, $I_0 = $ VII?, Korçë.

May 18, 19^h54^m, 43.9°N, 11.1°E, R.19, $I_0 = $ IX (CAM), $I_0 = $ IX—X (MF), $I_0 = $ X (MR), isos. map in (BI).

Houses ruined at S. Martino, Bossi, Appeggi, Gentilino, Villino, Spedaletto, felt at Parma, Ravenna, Arezzo, Livorno (BI), some destruction at Firenze, Pozzolatico, Bagno, Ripoli, etc.. (VIII R. F.), collapse of chimneys, all houses seriously damaged.

May 20, 15^h32^m UT, (42.7°N, 12.7°E), R.19, $I_0 = $ VII—VIII (MR), $I_0 = $ VII (MJD), $I_0 = $ IX (MF), isos. map in (BI).

Almost all houses damaged (fissured walls, some ceilings and many chimneys collapsed) at Spoleto, strongly felt at Giano and Terni, felt as far as Perugia and Norcia (BI, MF).

May 22,   "An earthquake damaged all houses around the village Sharodil", Caucasus, R.32 (BZ).

May 25, 11^h42^m UT, 45.0°N, 11.8°E, R.19, $I_0 = $ VI—VII?, $I_0 = $ VII (MJD), isos. map in (BI).

At Crespino, Papozze, Rovigo, Villonova Marchesana some walls fissured and some chimneys fell (BI).

June 10, $01^h 47^m$ UT, 45.9°N, 12.2°E, R.18b, $I_0 = $ VI—VII?, $I_0 = $ VII? (MJD), isos. map in (BI).

At Follina and Valmareno slight fissures in the walls, some chimneys thrown down (BI).

June 16, 10^h UT, $39\frac{1}{2}°$N, $20\frac{1}{2}°$E, R.25, $I_0 = $ X? (MD), $M = 6\frac{3}{4}$ (GGE), local.
A destructive shock at Durrës, V at Vlorë, aftershocks until Febr. 1896 (MD).

Aug. 9, 17^h36^m UT, Isole Tremiti, Chiesino (42·2°N, 15°E), R.19, $I_{max} = $ VII (MJD), isos. map in (BI), epic. in the sea?

Aug. 19, 37·8°N, 27·8°E, R.26c, $I_0 = $ VIII (EGU), $I_0 = $ VII−IX (GGK), $M = 6\frac{1}{4}$ (GGE).
A damaging shock between Aydin and Nasli (PL).

Sept. 4, 13^h30^m, 14^h49^m UT, 43·8°N, 11·2°E, $h = $ sup, R.19, $I_0 = $ VIII (MR), $I_0 = $ VII (MJD), $r = $ 15 km, isos. map in (BI).
At Querciolano houses largely fissured, many chimneys fell, some houses destroyed (BI).

Sept. 5, $39\frac{3}{4}°$N, $20\frac{3}{4}°$E, $h = n$, R.26a, $I_0 = $ VIII? (GG), $M = 6\frac{1}{4}$ (GGE).
Ioánnina was partly destroyed (MCA).

Oct. 25, 00^h24^m UT, 43·4°N, 11·1°E, R.19, $I_0 = $ VII (MJD), isos. map in (BI).
Several chimneys were thrown down and the church cracked at Poggibonsi (BI).

Nov. 1, epic. in the sea in front of Roma, R.19, $I_{max} = $ VII (MJD), $I_{max} = $ VIII (MR), isos. map in (BI).
Houses largely fissured in Roma, Malpasso, Decime, Fiumicino, Istia, Castel Porziano, felt at Spoleto, Anzio, Vico, Viterbo (BI).

1896

March 11, 23^h, Vrîncioaia, $h = n$?, R.21d, $I_0 = $?, $I_0 = $ VIII (FA), $I_0 = $ VII (MJD), not mentioned in (PRA, KR), region of Kishinev, $I_0 = $ VI (ES).
An earthquake felt almost over the whole historical Rumania, maximum intensity VIII at Avrameşti (FA), no information on damage.

March 15, 12^h, 41·8°N, 19·6°E, R.25, $I_0 = $ VIII? (MD, MA).
Damage in the estuary of Bojana, VIII? at Shengjin and Lesh (MD).

March 18, (40·1°N, 20·1°E), R.25, $I_0 = $?, $I_0 = $ IX, destruction at Gjirokastër (MCA, MA), not mentioned in (MD), $I_0 = $ X (MA), $I_0 = $ VII−IX (GGK), $M = 6\frac{1}{4}$ (GGE).

April 16, 39·3°N, 29·2°E, R.26c, $I_0 = $ VII?, Emet (EGU).

June 2−4, R.25, $I_0 = $?, $I_0 = $ VII at Korçe (MA), not in (MD).

June 26, 36·9°N, 28·1°E, R.26c, $I_0 = $ VII?, Marmaris, Kerme Bay (EGU).

June 29, 23^h, 34°N, 33°E, $h = i$?, R.31, $I_{max} = $ VIII, $P > 260\,000$ km^2 (ANC), $I_0 = $ X (CAM), $I_0 = $ IX (MJD).
At Akrotiri a number of local houses were destroyed, fissures opened in the ground and slides developed along the coast, VII at Limassol and Cape Gata, VI−VII throughout the island, III−IV in Lebanon, Syria and Israel (ANC), IV−V Jerusalem, V Haifa, Tiberias, Tabgha, Safad (KAD).

July 9, Caspian Sea, $I_{max} = $ VII? (MJD), not mentioned in (BZ, PV), doubtful.

July 21, R.25, $I_0 = $?, $I_0 = $ VIII at Paramithia (39·6°N, 20·5°E, R.25) (MA), not in (MD), "a very strong earthquake" (MCA).

1896

July 27, 04^h46^m UT, 45·2°N, 20·9°E, R.21c, $I_0 =$ VII, $r = 16$ km (TD), local?
At Ilandža clocks stopped, walls fissured, tiles were thrown down (BA).

Aug. 4, 18^h52^m UT, 41·5°N, 20·5°E, R.25, $I_0 =$ VII, $r_{max} = 110$ km (TD).
Three shocks, the first one on 18^h47^m felt at Durrës and Shkodër, the second one at 19^h52^m was stronger and was felt at Debar, the third one at 20^h17^m caused damage at Kruje (MCA, BA).

Aug. 26,   Sept. 10. About 20% houses were destroyed at Kot between Hekla and the sea, in the districts Rangarvalla Syssel and Arnes Syssel, fissures up to 15 km long, 4 victims, $I_0 =$ X (MF), $I_0 =$ IX−X (MJD), Sept. 6, $I_0 =$ VIII (MJD), R.2.

Sept. 2, 09^h10^m, 50·3°N, 2·9°E, R.9, $I_0 =$?, $I_0 =$ VII, $r = 150$ km (PRF).
Strong shocks largely felt, particularly in Hainaut, Brabant and some localities of Vlaanderen (Flandres), the centre of the shaken area near Arras and Douai (LAB), no information on damage.

Sept. 28,   (41·1°N, 20·9°E), R.25, $I_0 =$?, $I_0 =$ IX (MD).
A destructive shock in the region of the Lake Ohrid and at Struga in Yugoslavia, IX? at Durrës, Pogradec, (MD).

Oct. 16, 06^h15^m UT, $43\frac{3}{4}$°N, 8°E, R. 18a, $I_{max} =$ VII (MJD), not in (RO), isos. map in (BI), epic. in the sea.
Intensity VII at S. Remo and Oneglia, felt at Genova, Asti, Nice (BI).

Oct. 27, $36\frac{1}{2}$°N, 28°E, $h = n$, R.26c, $I_0 =$ VII−VIII (GGK), $I_0 =$ VI (EGU), $I_0 =$ VIII, $r = 240$ km (GG), $M = 6·2$ (GGE).
Partly destructive effects in Ródhos; Budrun and Marmaris in Asia Minor were slightly damaged, strongly felt at Kenydieghiz, Mugla and Elmali, slight at Aydin and Izmir (SAM).

Nov. 13, 04^h56^m, $39\frac{3}{4}$°N, $20\frac{3}{4}$°E, R.25, $I_0 =$ VII (GG), $I_0 =$ VI−VII (GGK), $M = 5\frac{1}{2}$ (GGE).
Ioánnina VII, Paramythiá VI, felt at Korçe (MD, MCA).

Nov. 13, 14, Aydin, Bergama, (39·1°N, 27·2°E), R.26c, $I_0 =$ VII? (MJD), not in (ONB, EGU).

Dec. 17, 05^h30^m, Hereford (52·1°N, 2·7°W), England, R.10b, $I_0 =$ VII (MJD).

Dec. 20,   R.25, $I_0 =$?, $I_0 =$ IX at Ioánnina (MA), not mentioned in (MD), "an earthquake recorded at Ioánnina" (MCA). Dec. 29, $I_0 =$ VII−VIII (GGK).

Dec. 29, 05^h, 37°N, $22\frac{1}{4}$°E, $h = n$, R.26a, $I_0 =$ VII−VIII (GGK), $I_0 =$ VII (MJD, GG).
Some houses were fissured and some houses collapsed at Kalámai and Ioánnina, strongly at Kiparissía and Spárti (GGM, MCA).

1897

Jan. 7, 02^h17^m UT, Jan. 19, 19^h25^m UT, R.19, $I_0 =$ VII? (MJD).
Two strong shocks of a sequence in Umbria fissured houses and threw down some chimneys at Spello, and Mucciafero (BI).

Jan. 17,   40·0°N, 20·1°E, R.25, $I_0 =$?, $I_0 =$ IX (MD, CAM, MA), $I_0 =$ VIII (MJD), $I_0 =$ IX−X (MA).
Delvinë almost completely destroyed (MD, MCA).

Jan. 22, 20^h UT, 46·2°N, 14·5°E, R.22b, $I_0 =$?, $I_0 =$ VII, $r_{max} = 23$ km (TD).
VII at Mengeš and Dob (BA), no description of damage available.

Febr. 3, 04^h UT, 42,6°N, 21·9°E, R.24, $I_0 = $ VII, $r_{max} = 27$ km (TD).

At Vranje all houses were damaged (BA).

Febr. 11, 23^h33^m UT, epic.?, R.20, 25, 26a, $I_{max} = $ VII (MD), $I_{max} = $ VI−VII (CGK, BI), isos. map in (BI); two shocks?

An earthquake felt largely from Malta to E of Sicily and over Calabria to Albania (Gjirokastër, Ioánnina), highest intensity in Sicily was observed at Pachino, Noto and Floridia (BI, MD, MCA), the epicentre was probably in the Ionian Sea even in the case that the reports correspond to two shocks (one in Albania).

Febr. 15, R.22b, $I_0 = $?, $I_0 = $ IX (TD, CAM), $I_0 = $ VIII (MJD), $I_0 = $ IX−X (MF).

Vranja, Meštanica, Ristovac (MF), not mentioned by (TD), a wrong date? July 15?

March 5, 19^h, 40·5°N, 19·5°E, R.25, $I_0 = $ VII? (MD), $I_0 = $ VI−VII (GGK).

Slight damage at Vlorë and surroundings (MD, MCA).

Apr. 2, 14^h30^m, 20^h UT, 46·2°N, 14·6°E, R.22b, $I_0 = $?, $I_0 = $ VII (TD), very local.

Only VII at Dob, no other reports (BA).

Apr. 4, 22^h43^m UT, 46·2°N, 14·5°E, R.22b, $I_0 = $?, $I_0 = $ VII (TD), very local.

VII at Vodice, no other reports (BA).

May 3, North coast of Iceland, R.2, $I_0 = $ VII (MJD), not in (MF).

May 15, 13^h45^m UT, epic. in the sea?, R.19, 20?, $I_{max} = $ VII? (MJD), isos. map in (BI).

Heavily felt from Palermo (some fissured houses) to Monte S. Guiliano and Ustica, felt strongly at Gaglieri (BI).

May 28, 22^h35^mUT, $37\frac{1}{2}$°N, $20\frac{1}{2}$°E, $h = i$, R.26a, $I_{max} = $ VI−VII (GGK), $I_0 = $ VI, $r = 800$ km, $I_0 = $ VII (MJD), $M = 7·6$ (GGE).

Ionian Sea, at Trípolis walls were fissured, felt in Pelopónnisos, in Kithíra, Zákinthos, Mesolóngion, Atalándi, Mólos (GGK), Sicily VI, S. Italy, Albania Vlorë VII, Malta, Bitolj, Cetinje (MD, MCA), for macroseismic effects in Italy see isos. map in (BI).

June 30, 14^h48^m UT, $39\frac{1}{2}$°N, 21°E, $h = n$, R.25, $I_0 = $ VII (GG), $M = 5\frac{1}{2}$ (GGE).

Prámanda (GG), at Ioánnina damage and panic, slightly felt in Puglia (MCA, BI).

July 15, 05^h53^m UT, 46·1°N, 14·5°E, R.22b, $I_0 = $ VII−VIII, $h = 6$ km, $M = 4·9$, $r_7 = $ $= 11$ km, $r_6 = 25$ km, $I = $ IX at Ljubljana, $r_5 = 40$ km, $r_4 = 69$ km, $r_5 = 110$ km, isos. map. in (RV), $I_0 = $ IX, $r = 112$ km (TD), $I_0 = $ VII (MJD),

VIII at Ježica, Vodice, Rožnik, VII at Kamnik, Tunice, Škofja Loka, etc. (BA). Felt in Italy at Udine, Pozzuolo, Zappolo, Sacile, Livenza, Marano (BI).

July 27, 09^h02^m UT, 43·6°N, 10·6°E, R.19, $I_0 = $ VII? (MJD).

At Ponsacco several chimneys thrown down, 5−6 houses damaged (BI).

Aug. 29, Morava valley, R.22b, $I_0 = $ VII (MJD), not in (TD), questionable.

Sept. 21, 12^h56^m UT, 43·7°N, 13·2°E, R.19, $I_0 = $ VII−VIII, $I_0 = $ VIII (MR), $I_0 = $ VII (MJD), isos. map in (BI).

At Senigallia many chimneys collapsed, all houses fissured or cracked, very strong at Fano and Mondolfo, largely felt as fas as Roma, Livorno, Bolzano, etc. (BI).

Nov. 2, an earthquake in the Levkás, felt also in Ipiros (MCA), R.26a, not in (GG), i. e. $I_0 < $ VI?, $I_0 = $ IX (MF), $I_0 = $ VII (MJD).

1897

Dec. ?   40·1°N, 28·0°E, R.27, $I_0 = $ VIII? (EGU), Balikesir region; identical with Febr. 28, 1898?

Dec. 18, 07^h24^m, 43·5°N, 12·2°E, R.19, $I_0 = $ VIII (MR), $I_0 = $ IX (MF), $I_0 = $ VII (MJD), isos. map in (BI).
At Citta di Castello, Aggiglioni, Montemaggiore, Pietralunga, etc., poorly constructed houses were destroyed, the others largely fissured, many chimneys thrown down, felt as far as Cesena, Bologna, Firenze, Pienza, Spoleto, Macerata (BI, MR, MJD, MF).

Dec. 31, 17^h30^m UT, 43·3°N, 17·9°E, R.22a, $I_0 = $ VII?, $r_{max} = $ 70 km (TD).
At Blagaj some walls collapsed, stones fell down from the hills, etc. (BA).

1898

Jan. 16, 12^h10^m UT, (44·7°N, 11·8°E), R.19, $I_0 = $ VII?, isos. map in (BI).
A damaging shock at Portomaggiore, Argenta, VI at Molinella, felt at Bologna and Verona (BI).

Febr. 11, 19^h, $40\frac{3}{4}$°N, $20\frac{3}{4}$°E, R.25, $I_0 = $?, $I_0 = $ VII−VIII (GGK), $I_0 = $ VIII (MD), $M = 5\frac{1}{2}$ (GGE).
VIII? at Korçë and Pogradec (Albania) (MD).

Febr. 20, 04^h56^m UT, 46·1°N, 13·4°E, R.18b, $I_0 = $ VII (MR).
Almost destructive at Cividale del Friuli, slight fissures at Udine, Vernasco, Stregna, etc., felt at Pontebba, Treviso, Venezia and in Austria at Packenstein, Celje, Schalkendorf (BI).

Febr. 28,   39·6°N, 27·9°E, R.26c, $I_0 = $ VIII (ONB, MJD), $I_0 = $ IX (CAM).
Balikesir (CAM, MJD).

Febr. 22,   (46·8°N, 6·6°E), R.18a, $I_0 = $ VII, $I_0 = $ VIII (MFS).
Chimneys fell at Grandson and many walls fissured, ceilings and walls damaged at Yverdon, Longueville, St. Croix, Provence, Épendes (MFS).

March 4, 21^h04^m UT, (44·7°N, 10·3°E), R.19, $I_0 = $ VII (MJD), $I_0 = $ VIII (MR), isos. map. in (BI). Torrechiara (valley of Parma, Emilia), Reggio (MR, MJD).
Maximum intensity VII observed at Felino Calestano and Langhirano, felt as far as Livorno, Ferrara, Padova, Belluno, Trento, Cressa, Genova (BI).

April 17, 23^h55^m, 46·2°N, 14·5°E, R.22b, $I_0 = $ VII, $r_{max} = $ 95 km (TD).

May 6,   (46·6°N, 7·6°E), R.18a, $I_0 = $ VII, $I_0 = $ VIII (MFS).
At Frutigen the walls of the church fissured, six chimneys fell, fall of plaster at Spiez, rockslides in the Kandertal and Kiental, felt almost over all Switzerland (MFS).

May 14, 04^h45^m UT, 37·6°N, 14·8°E, $h = $ sup?, R.20, $I_0 = $ VII−VIII, $I_0 = $ VII (MJD), $I_0 = $ VIII−IX (MR), volcanic?
Serious damage et St. Maria di Licodia near Etna (MR, BI).

June 2, 21^h42^m, $37\frac{1}{2}$°N, $22\frac{1}{2}$°E, $h = n$, R.26a, $I_0 = $ VII−VIII (GGK), $I_0 = $ VIII, $r_s = $ 120 km, $r = $ 320 km (GG), $I_0 = $ VIII−IX (MF), $I_0 = $ VII (MJD), isos. map in (MCA), $H = 22^h08^m$ UT (BI), $M = 6·4$ (GGE).
Disastrous shock near Trípolis (Arcadia), felt at Megalópolis, Pasia (Pássion?), Messíni, IV−V (GGM, GGK), felt in S. Italy and in Sicily (MCA, BI).

1898

June 7, 05^h, (47°N, $9\frac{1}{2}$°E), R.18b, $I_0 =$ VII−VIII? (SW), not in (MFS), doubtful.
Epic. Vaduz (WS).

June 27, 23^h38^m, 42·4°N, 13·0°E, R.19, $I_0 =$ IX, $I_0 =$ IX−X (MR), $I_0 =$ X−XI (MF),
$I_0 =$ VII (MJD), $r =$ 100 km ca, isos. map in (BI).
A complete destruction of Coppaelli Basso, heavy damage at Rieti and Citta-
ducale, $I =$ VIII at S. Rufina, etc., felt as far as Mondolfo, Pergola, Foiano,
Velletri, Fine, Scanno, Atri, Sinigallia (BI).

July 2, 04^h20^m UT, 43·6°N, 16·7°E, R.22a, $I_0 =$ IX, $I_0 =$ X (MF, CAM), $I_0 =$ XI,
$r =$ 307 km (TD), $I_0 =$ VII (MJD), $I_0 =$ VIII−IX (MC).
At Trilj all houses almost completely destroyed, victims, rockslides, at Vojnič the
church was damaged, the school building collapsed, at Košuta all houses des-
troyed, cracks on the road (3 victims), many roofs destroyed and the church
damaged at Vedrine, at Gardun serious damage, the school building collapsed,
at Brnaze many houses collapsed or damaged, cracks on the road, etc., felt at
Sarajevo, Zenica, Krupa, Ljubinje, Zadar, etc. (BA).
Aftershocks with $I_0 \geq$ VII according to (TD):
July 2, 04^h25^m UT, 43·5°N, 16·2°E, $I_0 =$ VII, $r_{max} =$ 23 km.
July 23, 10^h28^m UT, 43·6°N, 16·7°E, $I_0 =$ VIII, $r_{max} =$ 12 km.
July 28, 10^h45^m UT, 43·7°N, 16·6°E, $I_0 =$ VII, $r_{max} =$ 13 km.
Aug. 12, 06^h20^m UT, 43·7°N, 16·6°E, $I_0 =$ VII.
Oct. 10, 17^h46^m UT, 43·6°N, 16·7°E, $I_0 =$ VII, $r_{max} =$ 24 km.
Oct. 18, 20^h05^m UT, 43·7°N, 16·6°E, $I_0 =$ VII, $r_{max} =$ 18 km.
Dec. 9, 00^h30^m UT, 03^h15^m, 43·6°N, 16·5°E, $I_0 =$ VII, $r_{max} =$ 13 km.
Dec. 11, 05^h15^m, 43·7°N, 16·6°E, $I_0 =$ VII, $r_{max} =$ 220 km?

July 31, 05^h40^m, $20\frac{3}{4}$°N, $20\frac{3}{4}$°E, $h = n$, R.25, $I_0 =$ IX−X (GGK), $I_0 =$ IX (MF), $I_0 =$ X
(MD), $M =$ 6·2 (GGK).
Ioánnina X?, Préveza VIII, Pátrai VI, Vlorë (MD), Ioánnina was destroyed, very
slihgtly felt in the region of Otranto (MCA).

Aug. 10,   (38·2°N, 15·5°E), R.20, $I_0 =$ VII (MJD), Aug. 12, 12^h53^m UT (AC).
Rometta VII−VIII, S. Lucia VII, Messina V−VI, etc. (AC).

Aug. 25, 16^h37^m UT, 42·9°N, 13·1°E, R.19, $I_0 =$ VII? (MR, MJD).
Visso VII, Preci VI−VII (AC).

Nov. 8, 10^h36^m UT, 44·2°N, 15·2°E, R.22a, $I_0 =$ VI−VII, $I_0 =$ VII, $r_{max} =$ 345 km? (TD).
At Nin panic, objects moved, in some houses mortar fell down and walls fissured,
etc. (BA).

Nov. 9, 18^h15^m, $37\frac{1}{4}$°N, $21\frac{3}{4}$°E, $h = n$, R.26a, $I_0 =$ VII−VIII (GGK), $I_0 =$ VII, $r_5 =$
$=$ 70 km, $r =$ 200 km (GG, MJD), $I_0 =$ X−XI (MF), $r =$ 180 km (GGM), $M =$
$=$ 6·0 (GGE).
An aftershock at Kyparissía (GG), 6 small houses collapsed at Kyparissía, slight
damage (some small fissures in walls) at Ligudista, Messini and Androusa, felt
over Pelopónnisos (GGM), felt at Taranto (MCA).

Dec. 3, 05^h50^m, $37\frac{3}{4}$°N, 21°E, $h = n$, R.26a, $I_0 =$ VII−VIII, $I_0 =$ VII, $r =$ 90 km (GG),
$I_0 =$ VI−VII (GGK).
At Zákinthos some poor houses and a part of the theater collapsed, small
fissures at Amilias (GGK), rockslides also in the island Voidi, changes in wells,
tsunami (GG), light tsunami (ANS).

1899

Date ?   A destructive shock at Montemurlo (43·9°N, 11·0°E), R.19, $I_0 =$?, $I_0 =$ IX (MF), doubtful.

Jan. 22, 07^h49^m, $37\frac{1}{4}$°N, $21\frac{3}{4}$°E, $h = n$, R.26a, $I_0 =$ VIII−IX (GGK), $I_0 =$ IX, $r_5 = 100$ km, $r = 200$ km (GG), $M = 6·7$ (GGE), $I_0 =$ X (CAM, MJD, MD), isos. map. in (MCA). Serious damage at Kyparissía and surroundings, main shock, casualties (GG), Gjirokastër IV (MD), serious damage also in Messinía, in total 245 houses destroyed, more than 275 houses uninhabitable, more details in (GGM); tsunami at Messíni, Kyparríssía, Marathos (ANS).

Febr. 14, 16^h58^m, 48·1°N, 7·6°E, R.8, $I_0 =$ VII (SW, RSF), $r = 100$ km (PRF). A damaging shock in Kaiserstuhl (Germany), at Sasbach VII (chimneys thrown down, tiles fell down, rocks loosened in quarries), at Achkarren fissures in walls, at Balzenheim chimneys fell, etc., VI−VII at Bischoffingen, Burkheim, Jehtingen Kiechlinsbergen, Königschaffhausen, Kühnham (Kuhnheim), Lieselheim, Oberrothweil (SW).

Febr. 27,   Few houses damaged in the Reykjanes peninsula, R.2, $I_0 =$ VII (MJD).

Apr. 15, 04^h32^m, $37\frac{3}{4}$°N, $21\frac{1}{4}$°E, $h = n$, R.26a, $I_0 =$ VII, $r_5 = 70$ km, $r = 100$ km (GG), $I_0 =$ VI−VII (GGK). Serious damage at Katákolon, Pírgos, houses slightly fissured at Amaliás, Zákinthos IV−V (GGK).

May 15, 10^h57^m UT, 43·6°N, 16·7°E, R.22a, $I_0 =$ VIII, $r_{max} = 164$ km (TD). At Turjaci one third of all houses collapsed, new cracks on the roads, people injured, largely felt, etc. (BA).

June 26, 20^h04^m UT, 43·2°N, 16·6°E, R.22a, $I_0 =$ VII?, $r_{max} = 165$ km (TD). At Stari Grad in the island of Hvar strong shocks, tsunami observed on a ship (BA). Aftershock of the earthquake of June 2, 1898.

June 26, 23^h18^m UT, 43·9°N, 11°E, R.19, $I_0 =$ VII. Montemurlo VII−VIII, Cantagalo VII, Firenze V−VI, Vernio VI−VII etc. (AC).

July 2, 04^h48^m UT, 37·7°N, 15·2°E, $h =$ sup, R.20, $I_0 =$ VII? (MR), local? Slight damage at Malati, Guardia, Carico, Stazzo, Mangano, not observed in Catania (AC).

July 19,   41·7°N, 12·6°E, R.19, $I_0 =$ VII (MJD), $I_0 =$ VIII (MR), $I_0 =$ IX (MF). Large fissures in the walls, some chimneys and houses damaged at Frascati, Marino, Grottaferrata, Rufinella, Mondragone, Rocca di Papa VII, Roma VI−VII, etc. (MR, MF, AC), a ruinous shock (MF).

Aug. 13, 21^h, 38·7°N, 9·2°W, R.12, $I_0 =$ VII (FH, MJD), $I_0 =$ VI (MU). Panic at Lisboa and Sintra, felt at Coimbra, Evora (GS).

Sept. 18, 05^h15^m UT, 46·1°N, 14·4°E, R.22b, $I_0 =$ VII?, $I_0 =$ VII, $r_{max} = 84$ km (TD). At Treska near Ljubljana bricks thrown down from roofs and chimneys, etc. (BA).

Sept. 20,   37·8°N, 28·1°E, R.26c, $I_0 =$ X (CAM, MJD, EGU), $I_0 =$ IX, $M = 6·7$ (ONB), $I_0 =$ VIII−X (GGK), $M = 6\frac{3}{4}$ (GGE). Aydin, Nazilli, Ortaköy, Denizli (CAM), large cracks, VI at Mugla and Izmir (GGK). Great damage along the river Büyük Menderez, at Aydin 400 m long crack (ESE− WNW), earthslides, the villages between Aydin and Nazilli heavily damaged, some victims, E—W cracks of total length of 50 km, subsidence 1·5 m, heavy damage at Buldan, damage at Denizli and Tire, slight damage at Uşak, felt over the whole W. Anatolia (PL).

Dec. 31, 10^h50^m, 41·6°N, 43·5°E, R.32, $I_0 =$ VIII—IX, $I_0 =$ IX (MJD), $M =$ 5·6 (AV), isos. map in (BZ, TAD).

A destructive earthquake of Akhalkalaki, felt over the major part of Zakavkazie, 664 houses injured, 247 victims, maximum intensities $I =$ VIII at Azavret (a part of the village seriously suffered, changes in springs), Bezhano (seriously damaged), Malyj Samsar, Mereniya, $I =$ VII at Abazbekskaya, Agara, Alastani, Akhalkalaki Bakuriani (chimneys fell, walls fissured), Balkho (an extensive destruction), Bolshoi Samsar, Borzhomi, Godolar (almost all houses damaged), Gom, Gumurdo, Ikhtila, Kizil-Kilissa, Kilda, Kochio, Lamaturtskh, Moliti, Ordzha, Toria, Tyrkna, Khumrisi, Tsikhisdziri, Tskhra-Tskaro, etc., see (BZ).

1900

Jan. 8, 14^h45^m UT, 43·7°N, 16·7°E, R.22a, $I_0 =$ VII, $r =$ 4 km (TD), local.
At Glavice all objects were swinging (BA).

Jan. 27, 07^h20^m UT, 43·6°N, 16·7°E, R.22a, $I_0 =$?, $I_0 =$ VII, $r =$ 6 km (TD), local.
At Turjaci two small landslides (BA), no damage reported.

Jan. 28, 09^h35^m UT, 45·5°N, 16·9°E, R.22b, $I \leq$ VII, $I_0 =$ VII (TD).
At Medžurič, Banova Jaruga some tills thrown down from one roof (BA).

Jan. 29, 02^h15^m, 46·0°N, 21·2°E, R.21c, $I_0 =$ VIII?, $P =$ 6500 km^2 (FA).
At Vinga intensity VIII, felt at Variaş, Ghertiam, Periam, Timişoara, Fibiş, Arad (FH).

Jan. 31, 11^h, epic.?, R.21d, $I_0 =$ VII?, $I_0 =$ VIII (FA), local?
Panic and walls fissured at Avramesti (Tutova) (FA).

March 10, 02^h05^m UT, 44·0°N, 17·2°E, R.22a, $I_0 =$ VI?, $I_0 =$ VII, $r_{max} =$ 83 km (TD).
Orebič, Stari Grad, strongly shaken, etc. (BA).

March 20, 05^h44^m UT, 45·5°N, 16·9°E, R.22b, $I_0 =$ VI—VII, $I_0 =$ VII (TD).
At Medžurič some tiles thrown down, walls slightly fissured, etc. (BA).

May 19, 42·6°N, 12·6°E, R.19, $I_0 =$ VII (SGA).

Sept. 20, 37·8°N, 29·1°$\frac{1}{3}$, R.26c, $I_0 =$ VIII, $M =$ 6·1 (ONB).
Some damage at Isparta, Denizli, Sarayköy and Aydin.

Dec. 15, 17^h10^m UT, 44·1°N, 15·4°E, R.22a, $I_0 =$ VI?, $I_0 =$ VII, $r_{max} =$ 37 km (TD).
Strong at Zemunik, some walls fissured, etc. (BA).

2.41. EARTHQUAKE CATALOGUE OF THE EUROPEAN AND MEDITERRANEAN AREA, UNIDENTIFIED NAMES OF LOCALITIES

Abazbekskaya (Caucasus) 1899, Abdine (Egypt)1847, Abeto (Italy) 1859, Abruzzo Ult. (Italy) 1840, Acilatena (Italy) 1818, Adzhidara (Caucasus) 1872, Agalas (Greece) 1893, Agistron (Greece) 1893, Aggiglioni (Italy) 1897, Agya (Greece) 1892, Aiwali (Greece, Turkey) 1858, Alarnatajo (Spain) 1884, Alazata (Greece?) 1883, Alberton (England) 1884, Albuquerra (Spain) 1884, Alcancin (Spain) 1886, Alemtejo (Portugal) 1883, Alfenz (Austria) 1876, 1885, Allan Goulandes (France) 1873, Alquiedao (Portugal) 1858, Amilias (Greece) 1898, Amtagalo (Italy-Toscana) 1899, Anaziri (Greece) 1846, Appeggi (Italy) 1895, Argigliano (Italy) 1837, Arkitsa (Greece) 1894, Arpatchai (Armenia) 1868, Atlanta (Greece?) 1895, Avramești (Rumania) 1896, Azavret (Caucasus) 1899, Azeitao (Portugal) 1858,

Bab-el-Charie (Egypt) 1847, Babulinska (USSR, Poland) 1879, Bacas (Rumania) 1838, Baglio (Italy) 1865, Bakon (Yugosl.) 1870, Balaj (Albania) 1851, Baliba (Yugosl.) 1861, Balinka (Hungary) 1810, Balkho (Caucasus) 1899, Balzenheim (Germany) 1899, Bares (Yugosl.) 1883, Baretta (Italy) 1866, Barkuks (Egypt) 1847, Baošič (Yugosl.) 1853, Bastida (Greece?) 1863, Bauco (Italy) 1876, Belecske (Hungary) 1892, Belyam (Caucasus) 1828, Belyasuvar (East Caucasus) 1867, 1868, Belyj Klynch (Caucasus) 1867, Benja (Albania) 1851, 1865, Besztereczen (Hungary) 1829, Bezovand (Caucasus) 1869, Bikaa (Libanon?) 1802, Bizovac (Yugosl.) 1883, Bicsad (Rumania) 1893, Bischffingen (Germany) 1899, Blagaj (Yugoslavia) 1897, Bolshoy Samsar (Caucasus) 1899, Borchaloia (Georgia) 1827, Boseo Chiesa (Italy) 1894, Bosjakovina (Yugosl.) 1880, Bossi (Italy) 1896, Bolshaya Kendura (Georgia) 1844, Bozhi-Vody (USSR-Kuban) 1870, Bozhie (Caucasus) 1860, Bozunavac (Yugosl.) 1883, Bratto (Italy) 1834, Branze (Yugosl.) 1898, Budrun (Turkey) 1869, 1896, Buerdos (Spain) 1854, Bugag (USSSR-Kuban) 1830, Buchati (Albania) 1855,

Cafirenze (Italy) 1892, Cagheri (Italy) 1897, Cajeta (Italy-Calabria) 1830, Campeggio (Italy) 1881, Capo del Colle (Italy) 1859, Caprio (Italy) 1833, Cartignano (Turkey) 1877, Casio (Italy) 1869, Casone (Italy) 1881, Cassone (Italy) 1866, Cassotis (Greece ?) 1870, Castellaci d'Oppido (Italy) 1894, Casamenello (Italy) 1828, Castelnuovo di Mis. (Italy) 1846, Catagne (Italy) 1891, Cauceres (Portugal) 1858, Cfiri (Albania) 1851, 1865, Cellara (Italy) 1870, Chaili (Caucasus) 1869, 1872, Charagan (Caucasus) 1872, Chekond-Kassaba (Turkey) 1862, Chiavrie (Italy, Piemonte) 1886, Chobani (Caucasus) 1869, Churkhur-Yurt (Caucasus) 1872, Chrysso (Italy) 1879—80, Cohalm (Rumania?) 1892, Coppaelli Basso (Italy) 1898, Covorlui (Rumania) 1894, Crevassa (Turkey) 1884,

Dak (France) 1858, Dakhorsh (USSR-Kuban) 1870, Darachiachga (Georgia) 1827, Dashanly (Iran) 1878, Delimeri (Greece ?) 1846, Pellene (Greece) 1876, Demotiku (Bulgaria) 1859, Dengeleg (Hungary) 1829, 1834, Dhermi (Albania) 1859, 1866 (3×), 1869, Dimilia (Greece) 1863, Djelala (Algeria) 1891, Djesr (Syria) 1822, Dob (Yugosl.) 1897 (2×), Dobolt (Rumania) 1893, Dogheria (Italy) 1870, Dolgobychavo (Ukraina) 1875, Domici Soprano (Italy) 1854, Distelrath (Germany) 1878, Dragalevski (Bulgaria) 1858, Droh (Algeria) 1869, Drenkova (Yugosl.) 1879, Dusham (Albania) 1855, Dzhabani (Caucasus) 1872, Dzhafarkhan (Caucasus) 1872, Dzhangigan (Armenia) 1869, Dzhengi (Caucasus) 1869, 1872,

Ecamis (Turkey) 1834, Ehrenberg (Yugosl.) 1876, Ejenel (Tunisia) 1887, El Hebbab (Algeria) 1869, El Jish (Izrael) 1836, Engerea (Greece ?) 1867, Engestingen (Germany) 1828, Enkluvi (Greece?) 1885, Ergheni (Albania) 1858, Er Reina (Izrael) 1836, Erul (Rumania) 1829, Esbekije (Egypt) 1847, Eyafjördr (Iceland) 1837,

Féhérvácsurgó (Hungary) 1810, Felmo Calestano (Italy) 1898, Felsund (Scandinavie) 1866, Ferti (Italy) 1847, Faggio (Italy) 1865, Foglieri (Greece ?) 1845, Fostat (Egypt) 1847, Fusch-Bardhë (Albania) 1858, 1858, 1860, 1866,

Galignano (Italy) 1869, Gallichio (Italy) 1857, Galospetri (Hungary) 1834, Gamminella

(Italy) 1828, Gadita (Lebanon, Israel) 1836, Gardun (Yugoslavia) 1898, Garizogli (Greece) 1846, Gelovinsk (USSR-Kuban) 1870, Genezzano (Italy) 1844, Gentilino (Italy) 1896, Gerrace (Italy) 1886, S. Giovanni di Moriana (Italy) 1839, Gomance (Yugosl.) 1870, Gom (Caucasus) 1899, Gordolar (Caucasus) 1899, Goriachie Vody (Caucasus) 1830, Ghertiam (Rumania) 1900, Grandina (Portugal) 1858, Grassnitz (Alps) 1885, Grumevano (Italy) 1883, Guert or Gnert (Caucasus) 1864, Guicciardini (Italy) 1812, Guinady (Italy) 1834, Guia (Italy) 1860, Gumurdo (Caucasus) 1899, Gurta (Algeria) 1869, Gzig (Albania) 1854, Gzrig (Greece ?) 1823,

Haad (Algeria) 1867, Heleddi (Turkey) 1880, Heurnoes (Norway, Sweden) 1819, Hexamilia (Greece) 1858, Hobzelfingen (Germany) 1828, Hohenwang (the Alps) 1885, Horosco (Turkey) 1880, Hospitakia (Greece ?) 1846, Hrkanovci (Yugoslavia) 1883, Hochstetter (Yugoslavia) 1880,

Iegordzhevan (Caucasus) 1828, Ikhtila (Caucasus) 1899, Ichsi (Italy, Adriat. coast) 1887, Infikheran (Caucasus) 1872, Innocenti (Italy) 1812, Ingar (Caucasus) 1828, Izetzino (Greece ?) 1897, Iskinit (Greece) 1858,

Janitsanika (Italy?) 1885, Jaroso (Spain) 1863,

Kala-Zeiva (Caucasus) 1828, Kalacha (Armenia) 1871, Kalifa (Egypt) 1847, Kadykovka (Crimea) 1875, Kalamaki (Greece) 1858, Kalodjik (Turkey) 1884, Kaplony (Hungary) 1824, Kara-Asanlu (Armenia) 1840, Kara-Shiran (Iran) 1878, Karami (Crimea) 1875, Karabarm (Greece ?) 1845, Karakoyulu (USSR) 1883, Karma (Albania) 1855, Karnugi (Norway) 1819, Karpan (Turkey) 1855, Kastrovala (Greece) 1864, Kalakhen (Caucasus) 1872, Kazaklar (Greece) 1892, Keldukverf (Iceland) 1885, Kenydieghiz (Turkey) 1896, Kerkis (Turkey) 1846, Kevrag (Caucasus) 1841, Keyvandy (Caucasus) 1828, Khadra (Greece ?) 1863, Khadzhi-Amza (Bulgaria) 1890, Khumrisi (Caucasus) 1899, Kicherskoe (Causcasus, Georgiewsk) 1830, Kilda (Caucasus) 1899, Kilva (Caucasus) 1872, Kirphis (Greece ?), Kiskombo (Hungary) 1878, Kismer-Kaissua (Egypt) 1847, Kizil-Kilissa (Caucasus) 1899, Kohlstetten (Germany) 1828, Kochio (Caucasus) 1899, Kolodtsy (Caucasus) 1860, Kolohalfön (Swed., Norway) 1819, Koman (Albania) 1855, Kondurič (Yugosl.) 1884, Korentzik (Greece ?) 1883, Kordelio (Turkey) 1880, Korgo (Turkey ?) 1856, Kosarje (Yugosl.) 1895, Kosmaci (Albania) 1855, Kosterna (Yugosl.) 1838, Kostambul (Turkey) 1884, Kot (Iceland) 1869, Kourtkousi (Greece) 1845, Kovak (Greece ?) 1870, Krute (Albania) 1851, 1865, Kurdskhami (USSR) 1842, Kurgansk (USSR) 1879, Kumetsi (Turkey) 1835, Kush-Engidzha (Caucasus) 1872, Kynchetan (Caucasus) 1828, Kythnos (Greece) 1883, Kayi (Caucasus) 1890,

La Caminate (Italy) 1870, Lack (Yugosl.) 1840, Lacro (Italy) 1828, La Fajola (Italy), Landeyiar (Iceland) 1828, Langenwald (Alps) 1885, Lamaturtskh (Caucasus) 1899, Lappano (Turkey) 1855, Leiria (Albania) 1858, Liedolo (Italy) 1836, Lieselheim (Germany) 1899, Limpidi (Italy) 1886, Linen (Italy) 1885, Loi (Italy) 1885, Lokris (Province-Greece) 1894 (2×), Longenjon (Belg.) 1828, Lubya (Lebanon, Izrael) 1836, Lusttal (Yugoslavia) 1840, Lulut (Caucasus) 1828,

Magherus (Rumania) 1862, Malati (Italy) 1899, Malesine (Italy) 1877, Mandzofer (Turkey) 1835, Mannington (England) 1884, Mannried (Alps) 1885, Marrabotto (Italy) 1869, Mare (Rumania, Yugosl.) 1887, Maricaj (Albania) 1851, 1833, 1858, 1860, 1862, 1866), Mashagi (USSR) 1842, Mataguši (Yugosl.) 1855, Matrasy (Caucasus) 1872, Medžurič (Yugoslavia) 1900 (2×), Megdi Malaya (Caucasus) 1828, Malye Dzhamzhili (Armenia) 1869, Mechnnele (Alger) 1869, Meisari (Caucasus) 1872, Maldovassi (Turkey), Mependiti (Rumania) 1837, Merenyia (Caucasus) 1899, Meshkidzhik (Iran) 1878, Metsér (Hungary) 1810, Meyné (France) 1873, Mikale (Turkey) 1846, Mira (Algeria) 1891, Misala (Hungary) 1892, Moldava (Yugoslavia) 1879, 1880 (2×), Moldovani (Turkey) 1875, Molgan (Caucasus) 1872, Moliti (Caucasus) 1899, Molla-Gaspar (Armenia) 1871, Mouhalitch (Turkey) 1855, Monochilo (Sicily) 1879, Montebaranzone (Italy) 1811, Montesidaio (Italy) 1846, Monte Brandone (Italy), 1882, Mucciafero (Italy) 1897, Mugdy (Caucasus), Mukharavani (Caucasus) 1897,

Nardoran (USSR) 1842, Nea Kaimeni (Turkey) 1866, Niar (Caucasus) 1894, Nipéros (Greece) 1869, Nouskhar (Caucasus) 1864, Novokracine (Yugosl.) 1870, Nugdy Caucasus) 1828, Nurat (Caucasus) 1828,

Oliveira-de-Arenas (Portugal) 1858, S. Onofrio (Italy) 1828, Osman Zera or Zeza (Albania) 1851, 1865, Otabad (France) 1958, Ouled Boussaf (Algeria) 1886, Ouled Mériern (Algeria) 1886,

Packenstein (Italy) 1897, Palendzhukana (Mts., Turkey, Greece) 1867 (2×), Parchinis (Armenia) 1871, Peligna (Italy) 1840, Penczen (Hungary) 1863, Peperiksa (Greece ?) 1846, Pera (Turkey ?) 1894, Perighiali (Greece) 1858, Perivari (Turkey) 1884, Pharmezi (Greece ?) 1846, Phoutzala (Greece) 1846, Piavicina (Yugosl., Korčula) 1881, Pietrafessa (Itally) 1826, Pir-Abul-Konsul (Caucasus) 1828, Pnos (Italy) 1873, Pocretta (Italy) 1864, Pontepigadia (Albania, Greece) 1858, Poppiano (or Toppiano) (Italy) 1812, Pravi (Greece) 1829, Presolanana (Italy) 1882, Prochmookap (USSR-Kuban) 1870, Promysla (Caucasus) 1860, Provecze (Italy) 1894, Puerno (Italy) 1857, Pühlheim (Germany-Bav.) 1847, Pyrna (Greece ?) 1870,

Querciolano (Italy) 1895,

Rac (France) 1873, Racz-Kanisza (Hungary) 1838, Raedestos (Greece) 1887?, Rann (Alps) 1885, Reda (Caucasus) 1888, Regnano (Italy) 1837, Rocca Respampani (Italy) 1881, Rodinella (Italy) 1865, Rodolfi (Italy) 1812, Ronna (Rumania, Ukraina) 1830, Rovella (Italy-Calabria) 1870, Rožnik (Yugosl.) 1897, Rivière-Thilon (France) 1877, Rufinella (Italy) 1899,

Safdere (Greece) 1883, Sagir (Turkey) 1886, Sala (Greece) 1821, Sarsura (Caucasus) 1828, Salka (Schabka) (USSR-Bessarab.) 1878, Salut (Turkey) 1862, Salyany (Caucasus) 1828, Sardanus (Caucasus) 1828, Sardob (Caucasus) 1860, Santa Magdalena (Italy, Yugoslavia) 1856, Scanello (Italy) 1881, Scarconacci (Italy) 1865, Scorgnano (Italy) 1891, Shandrukovskaya (Caucasus) 1830, Sharodil (Caucasus) 1895, Sharopan (Caucasus) 1867, Sharur (Armenia) 1840, Shiak (Albania) 1895, Shishdanak (Caucasus) 1828, Shumen (Bulgaria ?) 1875, Serandumo (Greece) 1883, Selec (Ukraina) 1875, Serrafontana (Italy) 1828, Servia (Italy) 1875, Shoragial (Gruzie) 1827, Schalkendorf (Yugoslavia) 1898, Schwanzenberg (Yugosl.) 1870, Sibistra (USSR) 1864, Sierra Tejla (Spain) 1885, Sinja Gorica (Yugosl.) 1882, Skalnica (Yugosl.) 1870, Smokhtine (Albania) 1833, 1851, 1958, 1865, 1866 (3×), So'eure (Switzerland) 1853, Solmona (Italy) 1891, Solygia (Greece) 1873, Somaki (Armenia) 1880, Sottano (Italy) 1854, St. Lattario (Italy) 1887, Stranska Ves (Yugosl.) 1895, Sugatag (Rumania, Ukraina), Sulikadon (Greece) 1840, Sundi (Caucasus) 1869, 1872, Surd (Hungary), Syra (Greece ?) 1883,

Tadzhilar (Bulgaria) 1890, Tamzoura (Algeria) 1889, Tarkie Kolodtsy (Caucasus) 1868, Taraktash (Caucasus) 1869, Teldon (England) 1884, Tensaut (Algeria) 1860, Tepeidjik (Turkey) 1855, Terrapilata (Italy) 1823, Tesong (Azerbaidzhan) 1857, Thionanet (Alger) 1887, Thinia (Greece ?) 1867, Thomar (Portugal) 1858, Tchouda (Algeria) 1869, Tirli (Italy) 1874, Tirzhil (Caucasus) 1828, Tlébat (Alger) 1889, Todiano (Italy) 1859, Tole (Italy) 1864, Tomaso (Turkey) 1880, Teria (Caucasus) 1899, Torrechiara (Italy) 1898, Trebbiano Nizza (Italy) 1828, Trente-Nova (Greece) 1840, Tressen (Yugosl.) 1871, Trettene (Italy) 1891, Tunice (Yugosl.) 1897, Turami (or Turani) 1833, 1851, 1858, (2×) 1859, 1860, 1862, 1866 (2×), Turi (Yugosl.) 1855, Tyrkna (Caucasus) 1899, Tyrnavos (Greece) 1892,

Uffugo (Italy) 1885, Uglian Caldo (Italy) 1837, Ura Besirit (Albania) 1860,

Vagliagli (Italy) 1871, Vanka (USSR-Caucasus) 1867, Varensko (Bulgaria) 1892, Varignana (Italy) 1877, Vascari (or Vasciari) (Albania) 1833, 1851, 1858, 1860, 1862, 1866 (2×), Vayes (Italy, Piemonte) 1886, Vedrine (Yugoslavia) 1898, Vegliaturo (Italy, Calabr.) 1870, Veenburg (the Netherlands) 1850, Veis-Aga (Greece) 1846, Vella (Greece) 1889, Venta de los Cazadores (Spain) 1863, Villa di Monto (Italy) 1828, Vinneda (Spain) 1884, Vintmille (France) 1854, fort Vnezapnaya (Caucasus) 1830, Voidi (Greece) 1898, Vralo (Yugosl.) 1876, Vrabec (Yugosl.) 1880,

Wanbach (the Netherlands) 1873, Welkeri (Turkey) 1835,

Yaka (Turkey) 1875, Yakov-monastery (Armenia) 1840, Yarli-Dag (Turkey) 1859,

Zarbitza (Greece ?) 1867, Zarnava (Caucasus) 1828, Zaidakhani (Caucasus) 1828, Zegna (Croatia) 1843, Zemienik (Yugosl.) 1900, Zerbate (Italy) 1894, Zergiran (Caucasus) 1828, Zhulat (Albania) 1858, 1860, 1866, Zigrova (Caucasus) 1828, Zocca (Spain), 1864, Zukany (Yugosl.) 1883, Zurnabad (Caucasus) 1867, 1868.

Europe and Mediterranean

2.42. REFERENCES TO THE CATALOGUE OF SHOCKS 1801—1900

(arranged according to the alphabetical order of the abbreviations)

AC G. AGAMENNOME, A. CANCANI: Notizie sui terremoti osservati in Italia durante l'anno 1898, 1899, 1900. R. Ufficio Centrale di Met. e. Geod., Roma.

AI I. ATANASIU: Cutremurele de pâmînt din Romînia, Editura Academici Republicii Populare Romîne, Bucarest, 1961, 164 pp.

ANA N. N. AMBRASEYS: On the seismicity of south-west Asia. Data from a XV century Arabic manuscript. Revue pour l'étude des calamités, 1962, No 37, 3—15.

ANC N. N. AMBRASEYS: The seismic history of Cyprus. Revue de l'Union Int. de Secours, Mars 1965, No. 3, 25—48.

ANS N. N. AMBRASEYS: Data for the investigations of the seismic sea-waves in the Eastern Mediterranean. Bull. Seism. Soc. Am. 52, (1962), 895—913.

ANT N. N. AMBRASEYS: The seismicity of Tunis. Ann. di Geof. XV (1962), No 2—3, 233—244.

ANV N. N. AMBRASEYS: Seismic activity in the Varto region, Table III, mimeographed, 1966.

ANW N. N. AMBRASEYS: A note on the chronology of Willis'list of earthquakes in Palestine and Syria. Bull. Seism. Soc. Am. 52 (1962), 77—80.

ASMZ Annalen der Schweizerischen Meteorologischen Zentralanstalt, Zürich.

AV I. V. AIVAZOV: Zavisimost mezhdu ballnostyu, intensivnostyu í glubinoy ochaga dlya kavkazskikh zemletraseniy, Soobsh. Ak. Nauk Gruzinskoy SSR, 1961, T. XXVI, Nr 2, 149—152.

BA British Association for the Advancement of Science, Seismological Committee, Seismic Activity, 1899—1903.

BCK E. BISTRICSÁNY, D. CSOMOR and Z. KISS: Earthquake zones in Hungary. Annales Universitatis Sci. Budapestinensis de Rolando Eötvös Nominatae, Sectio Geologica 4 (1960), 35—38.

BFS M. BÅTH: An earthquake catalogue for Fennoscandia for the years 1891—1950. Sverige Geologiska Undersökning, Ser. C., No 545, Årsbok 50 (1956), No 1, Stockholm 1956.

BI M. BARATTA: I terremoti d'Italia. Torino 1901, 950 pp.

BIN M. BARATTA: Notize sui terremoti avvenuti in Italia, Boll. Soc. Seism. Ital., I (1895).

BMF M. BÅTH: Earthquake catalogue for Fennoscandia for the interval 1800—1890, unpublished report, prepared for the project of seismicity of Europe, Uppsala, August 1961, 4 pp.

BMP I. BÓBR-MODRAKOVA: Catalogue des tremblements de terre polonais. Manuscript, 1959, 63 pp.

BS V. I. Bune, A. A. SORSKY: Seismotectonic principles of distinguishing zones of probable origin of strong earthquake foci on the example of the Caucasus. Proceedings, 9th Assembly of the E. S. C., Copenhagen, 1966, Akademisk Vorlag, Koebenhavn 1967.

BRS J. BONELLI RUBIO: Sismos de grado IX en adelante occuridos antes des ano 1800, Sismos de gradio VII en adelante ocurridos entre les anos 1800 a 1900. Manuscript, 1961, 4.

BZ   E. I. Bjus: Seysmicheskiye usloviya Zakavkazya, Chast I, II, III. Chronologia zemlyetryaseniy v Kavkazii. ANGrSSR, Tbilisi 1948.

CA   A. Cavasino: I terremoti d'Italia nel trentacinquennio 1899—1933, Appendice al Vol. IV, Ser. III, delle Memoire del R. Ufficio Centrale di Meteorologia e Geofisica, Roma 1935, 7—266.

CAM   A. Cavasino: Note sur catalogo dei terremoti distruttivi dal 1501 al 1929 nel bacino del Mediterraneo, 29—36.
A. Cavasino: Catalogo dei terremoti avertiti nel bacino del Mediterraneo del 1501 al 1929, 37—60.
R. Ac. Nat. del Lincei, Publ. della Com. It. per 10 studio delle grandi calamita, Vol. II, Mem. Sc. e Techn., Roma 1931.

CD   D. Csomor: Das zur Konstruktion der europäischen Seismizitätskarte benötigtes Datenmaterial für das Gebiet von Ungarn. Manuscript, 1963, 5 pp.

CHP   P. Choffat: Les tremblements de terre 1903 en Portugal. Lisbonne 1904.

DCB   Ch. Davison: A history of British earthquakes. Cambridge Univ. Press, 1924, 409.

DCE   Ch. Davison: Great earthquakes. London 1936.

DJ   J. Dück: Die Erdbeben von Konstantinopol. Erdbebenwarte, III. Jg., 1904, 177—196.

EGU   K. Ergin, U. Güclü, Z. Uz: Catalog of earthquakes for Turkey and surrounding area, Techn. Univ. Istanbul, 1967.

ES   S. V. Evseev: Dopitaniya pro seysmichnost Ukrainskoy SSR. Ak. Nauk USSR, Geol. zhurnal, XIV (1954), vyp. 4, 57—68.

EU   S. V. Evseev: Zemletraseniya Ukrainy. Izd. Ak. Nauk USSR, Kiev, 73 pp.

FA   A. Florinesco: Catalogue des tremblements de terre ressentis sur le territoire de la RPR, Académie de la RPR. Bucarest 1958, 167 pp.

FC   C. W. C. Fuchs: Statistik der Erdbeben von 1895—1885. Sitzungsberichte d. M. mathem.-naturw. Ges., XCII. Bd. I, Abt. 215—625, Wien 1886,

FH   H. A. Ferreira: Earthquake catalogue of Portugal for the periods 1701—1800 ($I_0 < IX$; $I_0 \geq IX$), 1801—1900 ($I_0 < VII$; $I_0 \geq VII$), 1901—1960 ($I_0 < VI$). Servico Meteorologico National, No. G 11 (June 61), Manuscr., 6 pp.

GA   A. Grandjean: Epicentres des séismes algériens, mimeographed, 1959, 11 pp.

GC   G. P. Gorshkov, V. P. Spetsivtseva, V. V. Popov: Katalog zemletryaseniy na territorii SSSR. AN SSSR, Trudy Seysm. Inst., Moskva—Leningrad 1941, No 95.

GGB   E. I. Grigorova, B. M. Grigorov: Epicentre i sesmichnite linii N. R. Blgariya. Blgarskaya Akademiya na Naukite, Sofia 1964, 83 pp.

GG   A. G. Galanopoulos: A catalogue of shocks with $I_0 \geq VI$ or $M \geq 5$ for the years 1801—1958. Seism. Lab., Athens Univ., Athens 1960, 119 pp.

GGC   A. G. Galanopoulos: Katalog der Erdbeben in Griechenland für die Zeit von 1879 bis 1892. Annales Géologiques des Pays Hélléniques, Athens 1953, 111—229.

GGD   A. G. Galanopoulos: A catalogue of shocks with $I_0 \geq VII$ for the years prior to 1800. Seism. Lab., Athens Univ., Athens 1961, 19 pp.

GGE   A. G. Galanopoulos: Evidence for the seat of the strain-producing forces. Ann. di Geof., XVIII (1965), No 4, 399—409.

GGK   A. G. Galanopoulos: Earthquake catalogue of Greece. Manuscript, 1966, 181.

GH   A. G. Galanopoulos: Die Seismizität der Insel Chios. Gerl. Beitr. z. Geophys. 63 (1954), Heft 4, 253—264.

GR   B. Gutenberg, C. F. Richter: Seismicity of the earth and associated phenomena. Princeton Univ. Press 1949.

GS   J. Galbis Rodriguez: Catalogo sísmico de la zona comprendida entre les meridianos 5°E y 20°W de Greenwich y los paralelos 45°E y 25°N. Instituto geográfico, catastral y de estadística, I. part, Madrid 1932, 807 pp., II. part, Madrid 1940, 279 pp.

HA   A. Hée: Catalogue des séismes algériens de 1850 à 1911. Annales de l'Inst. du Phys. du Globe de Strasbourg, Nouvelle Série, Tome VI, Géophysique, 1950, 41—49.

HK   K. S. A. v. Hoff: Geschichte der durch Überlieferung nachgewiesenen natürlichen Veränderungen der Erdoberfläche, IV Teil: Chronik der Erdbeben und Vulcan-Ausbrüche mit vorausgehender Abhandlung über die Natur dieser Erscheinungen. Gotha 1840.

HR   R. Hoernes: Das Erdbeben von Laibach und seine Ursachen. Vortrag in der Versammlung des naturw. Vereines für Steiermark am 20. April 1895, Graz 1895, 61.

JM   M. Janković, Z. Mičević: List of earthquakes with $I_0 \geqq$ VI in Bosnia and Herzegovina in 1801—1900. Manuscript, 1966, 1.

KAD   D. H. Kallner-Amiran: A revised earthquake-catalogue of Palestine. Israel Exploration Journal, 1950—51, No. 4, 223—236.

KC   V. Kárník, E. Michal, A. Molnár: Erdbebenkatalog der Tschechoslowakei. Travaux de l'Inst. Géophys. de l'Ac. Tchécosl. Sc., Prague (1957), No 69, 411—598.

KF   C. F. Kolderup: Die norwegischen Erdbebenuntersuchungen (Norges Jordskjaelv). Bergens Museums Aarbok, 1913, Nr. 8, 152 pp.

KGI   K. Kirov, E. Grigorova, N. Ilev: Beitrag zur Untersuchung der Seismizität Bulgariens. Izvestiya na Geofiz. Inst., T. I., Sofia 1960, 137—183.

KR   V. Kárník, L. Ruprechtová: Seismicity of the Carpathian region. Travaux de l'Inst. Géophys. de l'Ac. Tchécosl. Sc., (1963), No 182, 143—187.

KRS   R. Kjellén: Förteckning på jordskalf i Sverige t. o. m. 1906. Göteborgs Högskolas Årsskrift, Band XV, 1909, Göteborg.

LAB   A. Lancaster: Les tremblements de terre en Belgique. Bruxelles 1901, 37 pp.

LC   Lersch: Erdbeben-Chronik für die Zeit von 2362 v. Chr. bis 1897. Manuscript, archives of the Zentralinstitut für Physik der Erde, Jena.

LI   I. Lehmann: Danish earthquakes, abstract of I. Lehmann: Danske Jordskaelv. Meddelelser fra Dansk Geologisk Forening, Bd. 13, Kobenhavn, 1956, 4 pp.

MC   C. Morelli: La seismicità a Trieste. Tectonica Italiana, N. S., Anno IV (1949), No 5, 8 pp.

MCA   C. Morelli: Carta sismica del Albania. Reale Academia d'Italia, Commissione Italiana di Studio per i problemi del soccorso alle popolazioni, Vol. X, Firenze 1942, 121 pp.

MCB   C. Morelli: I terremoti in Albania. Reale Accademia d'Italia, Bolletino della Commissione Italiana di studio per i problemi del soccoreo alle popolazioni, Fasc. II, Roma, 1942, 25 pp.

MD   D. J. Mihailović: Catalogue des tremblements de terre épiro-albanais. Archive Séismologique de l'Inst. Séismologique de Beograd, Zagreb 1951, 73 pp.

ME   A. E. Mourant: A Catalogue of earthquakes felt on the Channel Islands. Reprinted from the Transactions of La Société Guernésiaise for 1936, Vol. XII (1937), 523 to 540.

MF   F. Montandon: Les tremblements de terre destructeurs en Europe. Genève 1953, 195 pp.

MFS   F. Montandon: Les séismes de forte intensité en Suisse. Revue pour l'étude des calamités, Bull. de l'Union Internat. de Secours, Fasc. 18—19 (1942), 20—21 (1943), 105 pp.

MJA   D. J. Mihailović: Mouvements séismiques épiro-albanais. Série B, Monographies Fasc. No 1, UGGI Comité national du Royaume des Serbes, Croates et Slovènes, et travause scientifiques, Sect. de Séismologie, Beograd 1927, 78 pp.

MJD   J. Milne: Catalogue of destructive earthquakes, AD 7 to AD 1899. Porthmouth 1911, 92 pp.

MMC   R. Mallet and J. W. Mallet: The earthquake catalogue of the British Association with the discussion, curves and maps, etc.. Trans. of the British Assoc. for the Advanc. of Sc., 1852 to 1858, London 1858, 326 pp.

MO   I. Mushketov, A. Orlov: Katalog zemletryaseniy Rossiyskoy Imperii. Zapiski Imperatorskogo Russkogo Geograficheskogo Obshchestva, Tom XXVI, Sankt-Peterburg 1893.

MR   R. Malaroda, C. Raimondi: Linee di dislocazione e sismicità in Italia. Bolletino di Geodesia a Scienze Affini, Anno XVI, No. 3, 1957, 273—323.

MSK   R. N. Morozova, N. V. Shebalin: O zemletryseniakh Kryma 1800—1967 gg. Geofizicheskiy Sbornik, AN USSR, vyp. 26 (1968), 13—41.

MSM   S. V. Medvedev: Opyt novogo rayonirovaniya Moldavskoy SSR po zonam seysmicheskoy aktivnosti. Trudy Geof. Inst. No 5 (132), 40—48.

MU   J. M. Munuera: Datos básicos para un estudio de sismicidad en el área de la Peninsula Ibérica (Seismic Data). Memorias del Inst. Geográphico y Catastral, XXXII, cuaderno I, Madrid 1963, 93 pp.

NL   R. L. Nedeljković: Carte séismologique de Yugoslavie. Travaux de l'Inst. Séismologique de Beograd, Beograd 1950, 110 pp.

OB   E. B. Osman: Relève des secousses sismiques en Tunisie depuis 1892 jusqu'à 1955. Manuscript, 1959, 21.

OE   E. Odone: Sulla carta della frequenza dei terremoti disastrosi nel bacino del Mediterraneo. Microfilm.

ONB   N. Öcal: Kurze Liste der Erdbeben in der Türkei bis 1800 ($I_0 \geqq$ IX). Manuscript, 1961, 2.

PAA   A. Perrey: Note sur les tremblements de terre en Algérie et dans l'Afrique septentrionale. Mémoires de l'Acad. de Sci., Dijon 1845—46, 299—323.

PAB   A. Perrey: Mémoire sur les tremblements de terre dans le Bassin du Rhin. Mém. cour. et mem. des sav. étr. Ac. R. de Belgique, Bruxelles 1847, T. XIX, 4—113.

PAI   A. Perrey: Mémoire sur les tremblements de terre de la péninsule Italique présenté à la séance du 8 janv. 1847. Ac. R. de Belgique, Mém. T. XXII, 145 pp.

PAF   A. Perrey: Mémoire sur les tremblements de terre ressentis en France, en Belgique et en Hollande depuis le quatrième siècle de l'Ere Chrétienne jusqu'à nos jours, (1843 inclusive). Mém. cour. et mém. des sav. étr., Ac. Sc. de Belgique, Bruxelles 1844, T. XVIII, 1—110.

PAP   A. Perrey: Mémoire sur les tremblements de terre ressentis dans la Péninsule Turco-Hellénique et en Syrie, Présenté à la séance du 1er juillet 1848. Mém. Ac. R. de Belgique, T. XXIII., 75 pp.

PAT   A. Perrey: Notes sur les tremblements de terre en 1854—1871, avec suppléments pour les années antérieurs. Bulletin de l'Ac. Royale des Sciences, des Lettres et des Beaux-Arts de Belgique, 1855—1872.

PEF   E. Peterschmitt: Catalogue des principaux séismes ressentis en France de 1843 à 1918. Bureau central séismologique francais, manuscript 1959, 2 pp.

PET   E. Peterschmitt: Quelques données nouvelles sur les séismes profonds de la mer Tyrrhénienne. Ann. di Geof., Vol. IX (1956), No 3, 305—334.

PK   J. Plessard, B. Kogoj: Catalogue des séismes ressentis au Liban. Annales-Mémoires de l'Observatoire de Ksara, Tome IV, Cahier 1, 12.

PL   N. Pinar, E. Lahn: Türkiye depremleri izahli katalogu. T. C. Bayindirlik Bakanligi, Yapi ve Imar Isleri Reisligi Yayinlarindan, Seri 6, Sayi 36, Ankara 1952, 153 pp.

PRA   G. Petrescu, C. Radu: Seismicitatea teritoriului R. P. Romîne in perioada anterioara anului 1900. Acad. Republ. Populare Romîne, Centr. de Cercetari Geofizice, Probleme de geof., Vol. II, 1963, 79—85.

G. Petrescu et C. Radu: La séismicité du territoire de la RPR pendant la période antérieure à l'année 1900. Manuscript, Bucarest 1961, 5 pp.

PV   V. V. Popov: Katalog zemletryaseniy Soyuza SSR, vyp. 1, Krym (1908—1936 gg.) AN SSR, Trudy Seism. Inst. No 89, Moskva—Leningrad 1940, 24 pp.

RA   A. Réthly: A Kárpátmedencék földrengései (455—1918). Akadémiai Kiadó, Budapest 1952, 510 pp.

A. Réthly: Erdbeben im Karpatenbecken von 455 bis 1918 (a table compiled using data in the book). Manuscript, 1961, 23 pp.

RAB   A. Réthly: Erdbeben in der Umgebung des Balatonsees. Resul. der wiss. Erfor. des Balatonsees, 1. Band, 1. Teil, IV. Sektion, Wien 1912, 47 pp.

RE   A. Rey Pastor: Sismicidad de la regiones litorales espanolas del Mediterraneo. I. region geografica Catalana. Association pour l'Étude Géologique de la Mediterranée Occidentale, 3 (1935), 19 pp.

RF   J. P. Rothé: Note sur la séismicité de la France métropolitaine. Manuscript, 1962, 14, additional information in the letter of July 16, 1968.

RFC   C. F. Richter: Elementary seismology. S. Francisco 1958, 768 pp.

RG   G. Roux: Notes sur les tremblements de terre ressentis au Maroc avant 1933. Mémoires de la Société des sci. naturelles du Maroc, No. XXXIX, 31 juillet, 1934, 42—72.

RH   H. Renquist: Erdbeben in Finland (Findlands Jordskalv). Helsinkfors 1930, 113 pp.

RI   A. Rey Pastor: Traits sismiques de la péninsule Ibérique. Direction Génerale de l'Inst. Géograf. et Catastral d'Espagne, Madrid.

RK   J. P. Rothé: Les séismes de Kerrata et la séismicité de l'Algérie. Bulletin du Service de la carte géologique d'Algérie, 4e Série, No. 3, 1950, 3—40.

RO   J. P. Rothé: La séismicité des Alpes. l'Inst. de Phys. du Globe de Strasbourg, Tome III, Géophysique, 1938, 105, Mende, Clermont-Ferrand 1941.

J. P. Rothé: La séismicité des Alpes occidentales (compl.). Annales de l'Inst. de Physique du Globe, Nouvelle Série, Tome IV, Géophysique, 89—105, Mende 1948.

ROE   J.P.O'Reilly: Alphabetical Catalogue of the Earthquake recorded ... Trans. Royal Irish. Acad. (1880—86), 489—708.

ROI   Ottende Raelke: Erdbeben in Island in den Jahren 1013—1908. Det Kongelige Danske Videnskabernes Selskabs Skifer, Naturvidenskabelig og mathematisk afdeling, Kobenhavn 1925.

RSF   J. P. Rothé, G. Schneider: Catalogue des tremblements de terre du Fossé Rhénan (1021—1965). Institut de Physique du Globe de Strasbourg, Landeserdbebendienst Baden-Württemberg Stuttgart, Stuttgart 1968, 91 pp.

RU   F. H. Van Rummelen: Overzicht van de tusschen 600 en 1940 in Zuid-Limburg en omgeving waargenomen aardbevingen.... Mededeelingen Behoorende Bij Het Jaarverslag 1942—43, Geologisch Bureau Heerlen 1945, 139 pp.

RV   Vladimír Ribarič: Studia seizmičnosti ozemlja SR Slovenije s posebnim ozirom na dinamične vplive potresov na gradbene objekte. Katalog potresov v Slovenii do leta 1914 z dodatim posisom močnejših potresov do leta 1960. Astronomsko Geof. Observatorij Univerze v Ljubljani, Ljubljana 1963, 82 pp.

RVP   P. v. Radics: Das Erdbeben in Österreich-Ungarn am 14. Jänner 1810. Erdbebenwarte, VI. Jg., 1906—1907, 116—121.

SAB   A. Staehelin: Das Erdbeben von Basel. Basler Jahrbuch 1956, 11—44.

SAE   A. Sieberg: Erdbebengeographie. Handbuch der Geophysik, 1932, Band IV, Abschnitt IV, 688—1006.

SAG   A. Sieberg: Beiträge zum Erdbebenkatalog Deutschlands und angrenzender Gebiete für die Jahre 58 bis 1799, Mitteilungen des Deutschen Reichs-Erdbebendienstes, Berlin 1940, Heft 2, 112 pp.

SAM   A. Sieberg: Untersuchungen über Erdbeben und Bruchschollenbau im Östl. Mittelmeergebiet. Denkschriften der med.-naturw. Ges. zu Jena, 18. Band, 2. Lief., Jena 1932, 161—273.

SB   B. Simon: Die Erdbebentätigkeit des Ungarischen Beckens. Veröff. d. Reinchsanst. f. Erdbebenf. in Jena, 1941, Heft 40, 80—84.

SF   J. F. Schmitt: Studien der Erdbeben, Leipzig 1875, 324 pp.

SGA   G. A. Shenkareva: Materialy po seismogeografii Apeninskogo poluostrova. Manuscript, 43 pp, 1964.

SNP   N. Shalem: Seismicity in Palestine and neighbouring areas (macroseismical investigation). 15 tables, manuscript, 79, 1960.

SG   G. Schneider: Erdbeben und Tektonik in SW Deutschland. Tectonophysics, 5 (1968), No 6, 459.

SW   W. Sponheuer: Erdbebenkatalog Deutschlands und der angrenzenden Gebiete für die Jahre 1800 bis 1899. Mitteil. d. Deutschen Erdbebendienstes, Berlin 1952, Heft 3, 195 pp.

TAD   A. D. Tskhakaya: Seysmichnost Dzhavakhetskogo nagorya i prilegaushchikh rayonov. Trudy Inst. Geofiziki AN GrSSR, T. XVI, 1957.

TBI   E. Tillotson: British earthquakes and the structure of the British Isles. Manuscript 1952, 62 pp.

TD   D. Trajić: Les tremblements de terre pour les années 361—1800 dans la Yougoslavie (table, $I_0 \geqq$ IX). Manuscript 1961, 3 pp.

TEA   E. Tryggvason: Jardskjálftar á Islandi og nyrzta hluta Atlantshafsins, Sérp. úr Náttúrufraedingnum, 25 (1955), 194—197.

TEN   E. Tams: Die seismischen Verhältnisse des europäischen Nordmeers. Zentralblatt f. Min. etc., 13 (1922), 385—397.

TL   L. Torfs: Fastes des calamités publiques survenues dans les Pays-Bas et particulièrement en Belgique, dépuis les temps les plus reculés jusqu'à nos jours. Paris 1962, 132—204.

TM   M. Taşdemiroğlu: General Explanation on North Anatolian Fault Zone. T. C. Imar ve iskân bakanliği afet işleri genel müdürlüğü, Ankara 1969, mimeographed.

TR   Th. Thorkelsson: Earthquakes in Reykjavik. Edited by the Municipality of Reykjavik, 1935, 8 pp.

TT   M. Toperczer, E. Trapp: Ein Beitrag zur Erdbebengeographie Österreichs nebst Erdbebenkatalog 1904—1948 z. Chronik der Starkbeben. Mitteilungen der Erdbeben-Kommission. Neue Folge, Wien 1950. Nr. 65, 59 pp.

TTP   E. Tryggvason, S. Thoroddsen, S. Pórarinsson: Greinardgerd Járdskjálftanefndar um Jardskjálftahaettu á Islandi. Sérp. úr Timariti Verkfraedingafélags Islands 43 (1958), 6, 1—19.

TU   E. Trapp: Die Erdbeben Österreichs 1949—1960, Ergänzung u. Fortführung des österreichischen Erdbebenkatalogs. Mitteilungen d. Erdbeben-Kommision, Neue Folge, Wien 1961, Nr. 67, 23 pp.

WS   S. Watzof: Tremblements de terre en Bulgarie au XIXe siècle. Centralna Meteorologicheskaya stancija 96, Sofia 1902.

3. Some characteristics of seismic activity

The final objective of any investigation into the seismicity of a region is to find symptoms of an approaching shock, its location, magnitude and time of origin. So far, only average numbers on seismic risk can be estimated because there is no complete physical understanding of focal processes, therefore seismologists combine different empirical relations, simple statistical models or some regularities in the spatial distribution of foci for solving some partial tasks. Among the most frequently used characteristics are the magnitude-frequency (or energy-frequency) relation with its parameters (slope, M_{max}, M_{min}), strain (or energy) release with time and space and fault plane solutions. Statistical methods are frequently applied because they can supply us with numerical characteristics of seismic activity, e. g. the earthquake phenomena are correlated between different regions or with other geophysical parameters (Earth tides), the sequences of shocks are analyzed using the theory of time series, the theory of largest values, Poisson or other distributions, etc.

In the following paragraphs we will discuss the magnitude-frequency relations, the application of the theory of largest values and the position of seismogenetic lines as inferred from seismic maps.

3.1. DISCUSSION OF THE MAGNITUDE-FREQUENCY RELATION

The frequency distribution of earthquakes as a function of their magnitude is of primary importance for the seismicity investigation of a region. We have applied the relation in the form [1]

$$\log N = a - bM, \tag{1}$$

where a and b are constants.

The exponential distribution function $N(M) = 10^{a-bM}$ belongs to the family od similar relations of the type $f(x) = e^{-cx}$ which are common for different phenomena observed in natural sciences. The $N(M)$ graphs relating to the combined seismic regions are presented in Part I. In this volume the magnitude-frequency graphs for

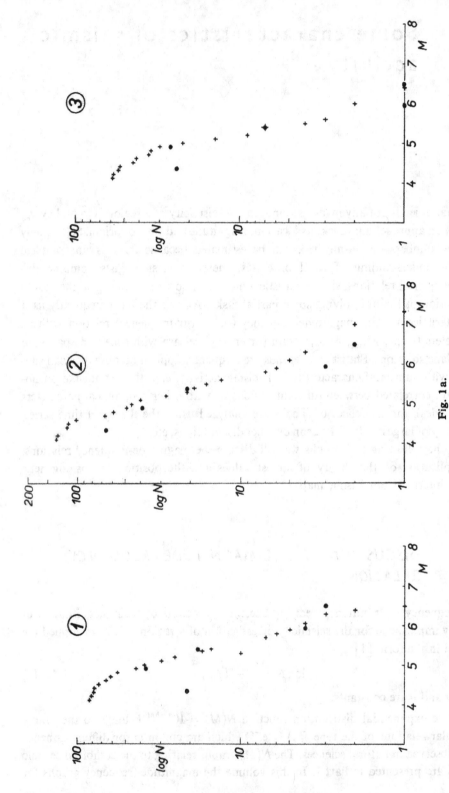

Fig. 1a.

Fig. 1. Magnitude-frequency relations for individual seismic regions (for the delineation of regions see Part I, Fig. 1); crosses correspond to cumulative frequencies, circles to frequencies within $\Delta M = 0.5$; period 1901—1955.

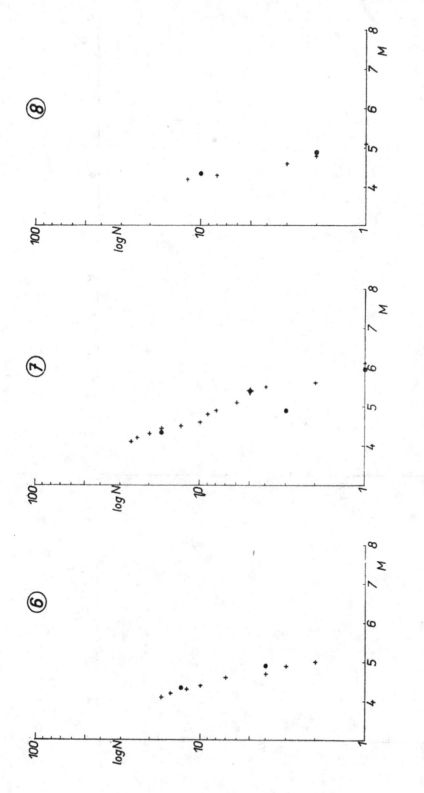

Fig. 1 b.

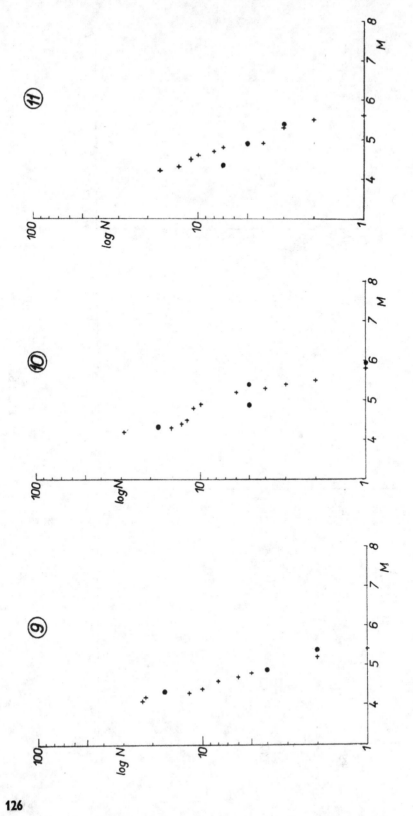

Fig. 1c.

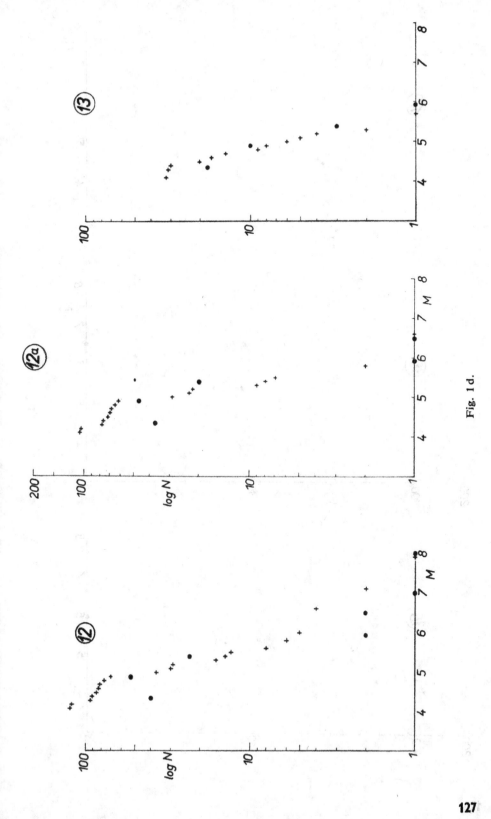

Fig. 1 d.

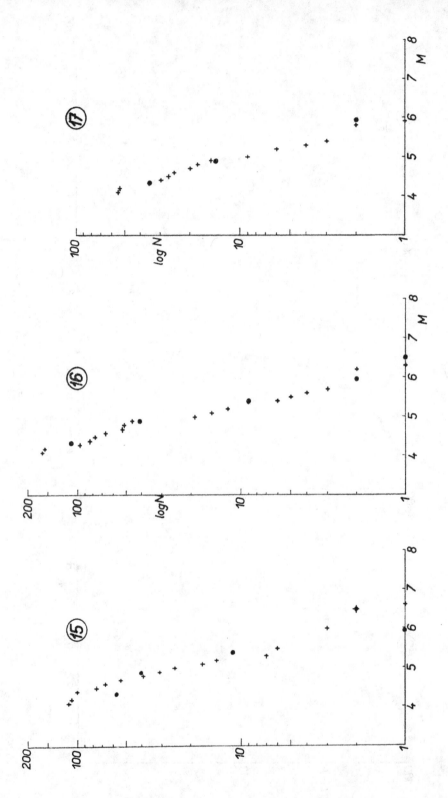

Fig. 1e.

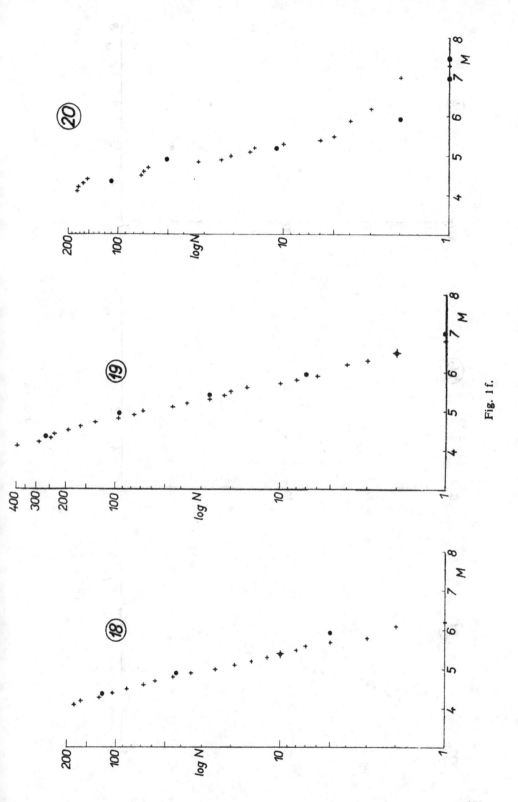

Fig. 1 f.

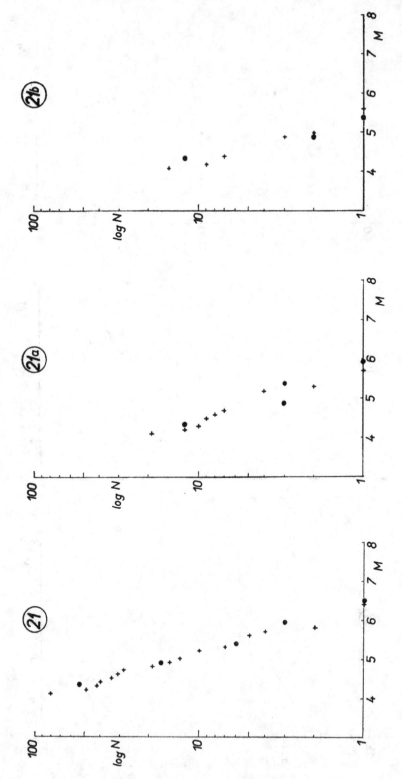

Fig. 1g.

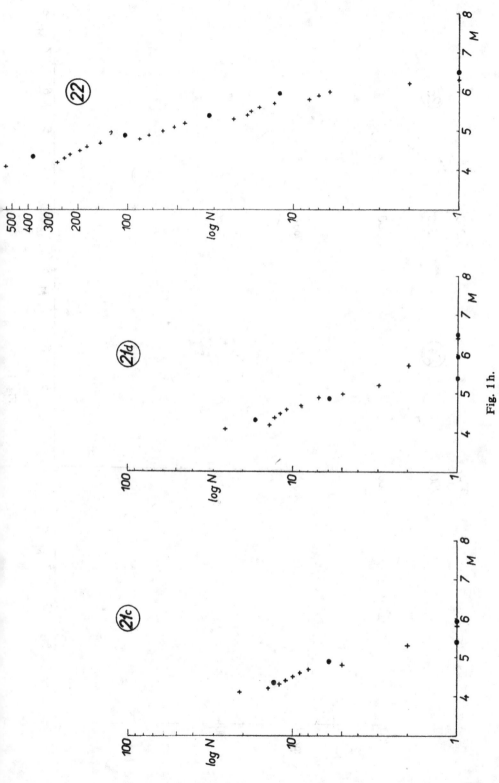

Fig. 1h.

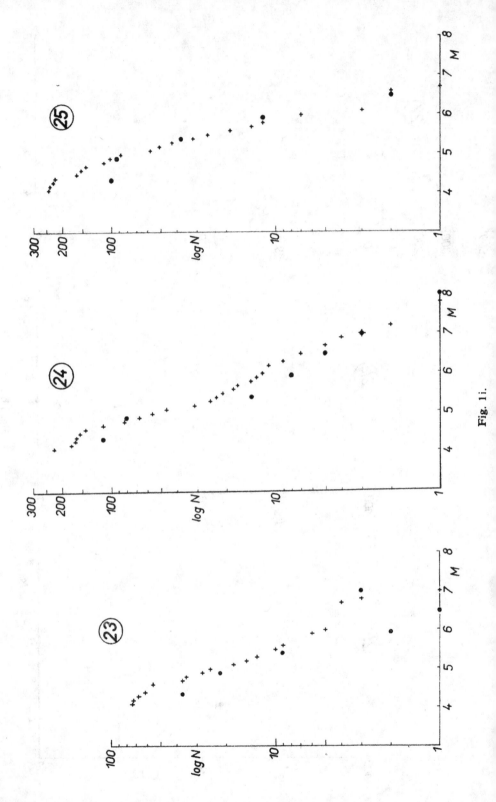

Fig. 1i.

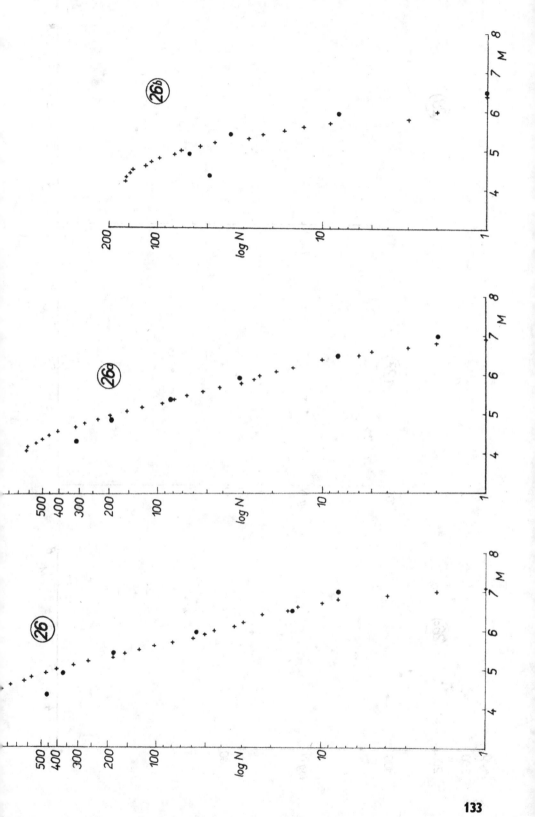

133

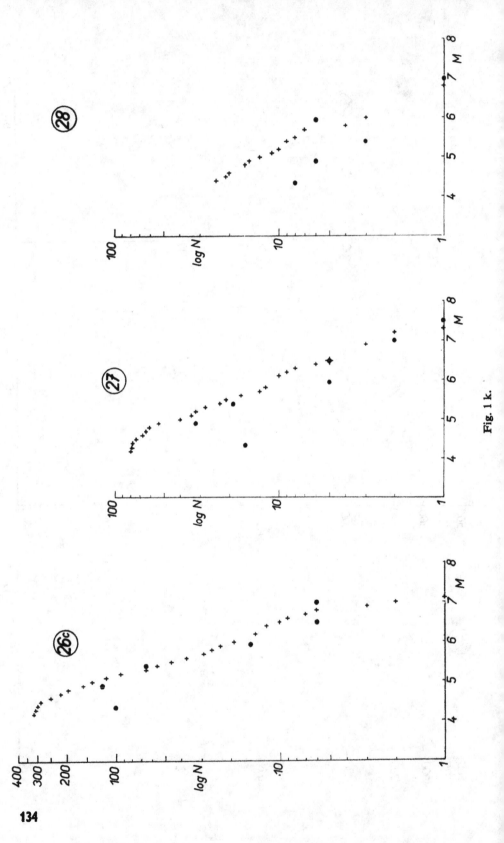

Fig. 1 k.

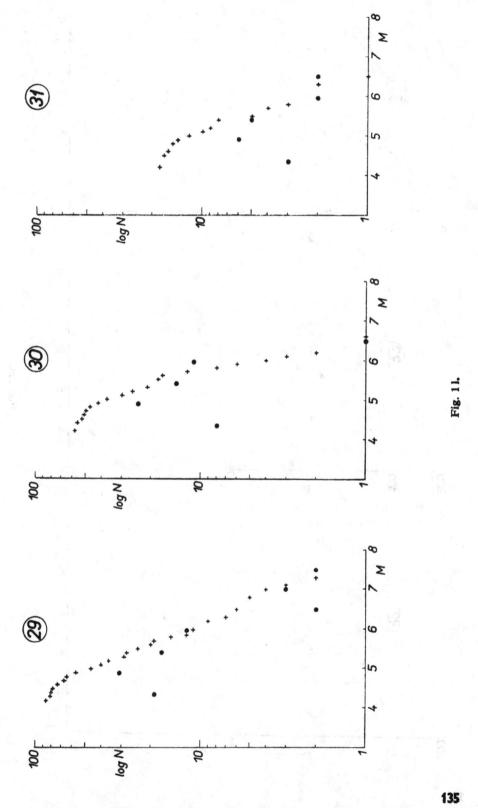

Fig. 11.

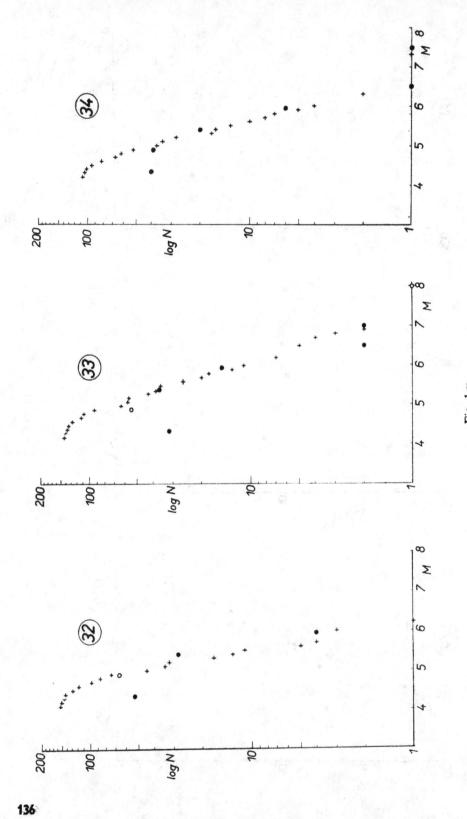

Fig. 1 m.

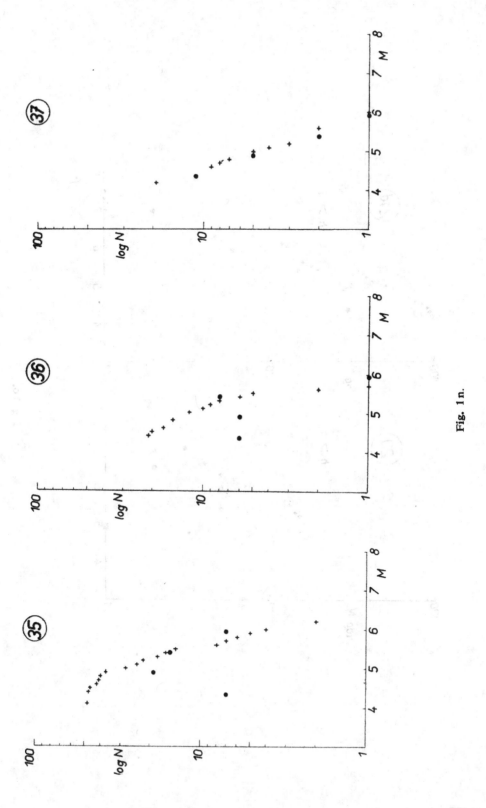

Fig. 1 n.

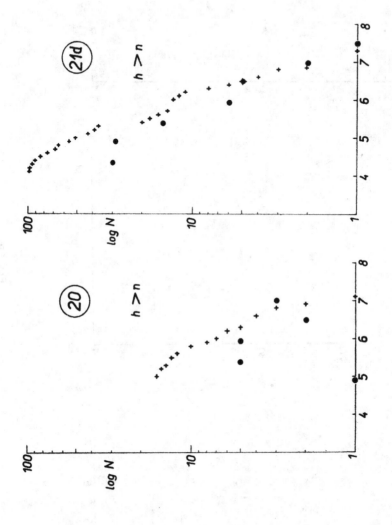

Fig. 1 o.

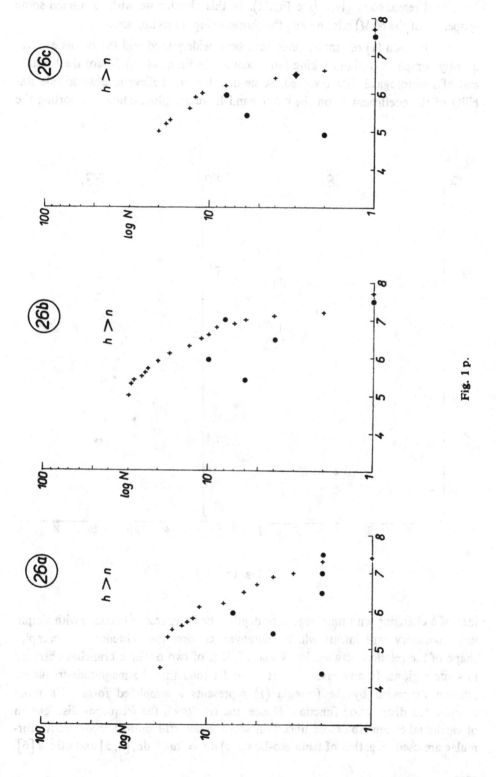

Fig. 1 p.

individual regions are given (see Fig. 1). In this chapter we wish to discuss some properties of the $N(M)$ relation and the shortcomings in its application.

The relation (1) or similar ones have been widely used and the results are frequently compared without taking into account the limits of validity of the formula and of seismological data involved. Some investigators believe in a worldwide stability of the coefficient b, on the other hand there are observations supporting the

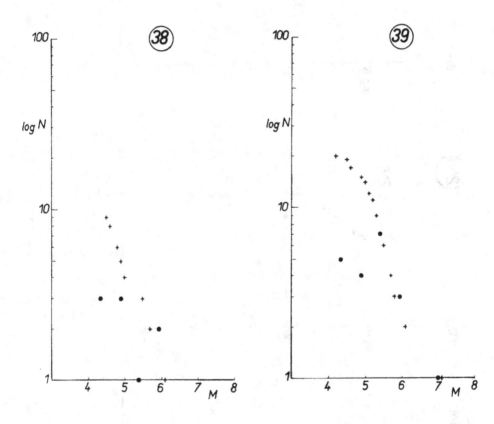

Fig. 1 r.

idea of b changing with time, region or depth. The latter results coincide with simplified laboratory experiments which, moreover, confirm the existence of a complex shape of the relation between $\log N$ and M, i. e. of two or three branches observed in some regions. In any case, it must be understood that the magnitude-frequency relation expressed by the formula (1) represents a simplified form of a more complicated distribution function. Hence, the results on the frequency distribution of earthquakes must be more uncertain when other still more approximative formulae are used, e. g. that of Ishimoto-Iida $n(a)\, \mathrm{d}a = ka^{-M}\, \mathrm{d}a$, [4, 5] and others [6].

They are usually derived under certain simplifying assumptions, not always valid, e. g. a homogeneous distribution of earthquake foci, independence of absorption on frequency, etc.

The magnitude (or energy)-frequency relation need a more physical interpretation and a more detailed analysis. Only then can it be of more use for the estimation of earthquake risk.

3.11. DELINEATION OF FOCAL REGIONS AND RELATED PROBLEMS

The valid of the results depends on our ability to delineate natural seismogenetic units within the crust or upper mantle where the earthquake generation is governed by the same system of forces and where the building material and structure are of the same type. As yet there exists no definite idea of how to define a focal or seismic zone. Ch. Tsuboi [7, 8] estimated by the correlation of seismic activity between neighbouring narrow belts that $160-200$ km is the most appropriate width of a typical zone in Japan. In delineating seismic regions for statistical studies in the European area (see Part 1, Fig. 1) the boundaries were drawn in such a way that they involve one or more clusters of epicentres and cross the regions of minimum density of epicentres but do not intersect the main tectonic provinces. In this way regions of very differing size (see Table 2) originated. Some of them are probably not well defined, e. g. region No. 24 (Macedonia, N. Greece, W. Bulgaria), some regions on the margin of the area are not closed, e. g. No 36 (Syria), No. 38 (Egypt). Some regions are too large and had to be subdivided into smaller zones. This step, however, was made also for geological reasons, e. g. in Nos. 10, 18, 21, 22 and 26, respectively. On the other hand, we cannot arbitrarily shrink the size of a zone because of the necessity of having sufficiently numerous data for a reliable statistical treatment (about 100 earthquakes). Because of inaccuracies in location the epicentres are somewhere irregularly dispersed and no distinct pattern is seen.

We must inevitably expect that some magnitude-frequency relations represent an average distribution function as a result of the superposition of several $N(M)$-functions corresponding to individual focal zones involved in the region. This superposition may distort the final result; supposing e. g. a set of $N(M)$ relations of the same slope but different level we obtain a stepwise shape of the function which with the usual presence of a scatter of values, may lead to a resulting approximative line of a greater slope than the individual partial lines have. Different combinations are possible when the slopes of contributing relations are different; we can obtain also a non-linear form of the graph in this way. For example the convex shape

Table 2

Region	$\bar{h}$	P (10^6 km^2)	E (erg/1 year/1 km^2)
1	n	0·078	$2·6 \times 10^{15}$
2	(3—5)	0·158	$9·6 \times 10^{15}$
3	(50)	0·500—1·040	$7·5—\ 16 \times 10^{13}$
6	(5—10)	0·030—0·053	$4·5—\ 8·0 \times 10^{14}$
7	(5—20)	0·030—0·049	$5·5—\ 8·9 \times 10^{14}$
8	n	0·030—0·045	$2·2—\ 3·2 \times 10^{13}$
9	n	0·020—0·041	$8·8—\ 18 \times 10^{13}$
7, 8, 9	n	0·085—0·135	$2·3—\ 3·7 \times 10^{14}$
10	(15—20)	0·100—0·152	$8·7—\ 1·3 \times 10^{14}$
11	n	0·050—0·102	$7·1—15·0 \times 10^{13}$
12	n	0·390—0·783	$1·1—\ 2·3 \times 10^{16}$
13		0·065—0·259	$3·7—15·0 \times 10^{15}$
14	n	0·070—0·216	$9·3—28·0 \times 10^{12}$
15	5	0·192	$9·8 \times 10^{14}$
14—15	5	0·260—0·408	$2·5—\ 4·0 \times 10^{15}$
16	5; 15	0·180—0·270	$3·1—\ 4·6 \times 10^{14}$
17	5; 20	0·075—0·150	$2·1—\ 3·0 \times 10^{14}$
18	1—30	0·185—0·278	$3·0—\ 4·5 \times 10^{14}$
19	3—15	0·120—0·164	$2·4—\ 3·3 \times 10^{15}$
20	10—30	0·105—0·155	$9·1—13·0 \times 10^{15}$
20	100—450	0·080	$2·4 \times 10^{16}$
21	10—30	0·150—0·245	$2·9—\ 4·8 \times 10^{14}$
21	100	0·035	$1·2 \times 10^{17}$
22	5—15	0·150—0·225	$9·1—13·0 \times 10^{14}$
23	7; 40	0·065—0·094	$9·5—14·0 \times 10^{15}$
24	10	0·098	$7·6 \times 10^{16}$
25	5	0·080—0·111	$3·6—\ 4·9 \times 10^{15}$
26	n	0·520—0·676	$5·9—\ 7·7 \times 10^{15}$
26	150	0·180—0·335	$2·6—\ 4·8 \times 10^{17}$
26a	n	0·246	$6·0 \times 10^{15}$
26a	100	0·050	
26b	n	0·205	$6·9 \times 10^{14}$
26b	150	0·205	
26c	n	0·225	$1·1 \times 10^{16}$
26c	150	0·800	
27	n	0·080—0·102	$2·3—\ 2·9 \times 10^{16}$
28	n	0·041	$5·6 \times 10^{15}$
29	n	0·090—0·119	$2·8—\ 3·7 \times 10^{16}$
30	n	0·115—0·229	$9·5—19·0 \times 10^{14}$
31	n	0·105—0·135	$8·6—11·0 \times 10^{14}$
32	5—80	0·220—0·295	$3·0—\ 4·1 \times 10^{14}$
33	n	0·123	$9·9 \times 10^{16}$
34	n	0·210—0·274	$4·1—\ 5·3 \times 10^{15}$
35	n	0·150—0·201	$4·2—\ 5·7 \times 10^{14}$

Table 2 (*continued*)

Region	$\bar{h}$	P $(10^6$ km$^2)$	E (erg/1year/1 km^2)
36	n	0·125—0·168	1·2— 1·5 × 10^{14}
37	5—10	0·090—0·123	1·4— 1·9 × 10^{14}
38	n	0·150—0·297	7·8—15·0 × 10^{13}
39	n	0·105—0·140	3·9— 5·3 × 10^{15}

$\bar{h}$ — average depth of foci
P — approximate surface of the zone containing earthquake foci
E — average seismic energy release based on seismic activity from 1901—1955

or a broken line with the first part of a steeper slope, sometimes observed, may be the result of the superposition of two zones with quite a different type of activity. Such a figure was observed for regions 2, 12, 20 and 24 (Fig. 1). There is good reason to assume the existence of at least two separate focal zones or "layers" in all of them, the first one corresponding to very shallow shocks and being responsible for the steep branch of the $N(M)$-graph in the lower magnitude range, the other corresponding to foci of normal depth or slightly deeper.

A more accurate delineation of focal zones is desirable, particularly as far as the depth boundaries are concerned.

3.12.   INFLUENCE OF THE METHOD OF DATA TREATMENT

The relation between the number of earthquakes N and their magnitude M is usually approximated by a simple formula (1), i. e. by a single straight line in the coordinates $(\log N, M)$. The range of linearity (M_1, M_2) should be indicated because, for the weakest and the strongest shocks, the $(\log N, M)$ distribution usually deviates from linearity. It is also desirable to give the lowest and the highest observed magnitudes M_{min} and M_{max}. The level and the slope of relation (1) are defined by two parameters, a and b, which can be determined by several methods. The treatment usually begins with grouping N according to the magnitude classes. Taking into account that $\pm 0·3$ is the accuracy of the magnitude determination, one half of a magnitude unit is the most convenient class interval, although class intervals ranging from 0·1 to 1·0 are used. The values $\log N$ representing individual magnitude classes are usually plotted against values corresponding to the centres of the classes. Some authors prefer to plot cumulative numbers $N_c(M)$ to avoid errors introduced by unequal

Table 3a

$$\log N = a - bM, \quad 1901-1955$$

Region	$\sum N$	a_1	b_1	a_2	b_2	b_3	ΔM	Remark
1	83	5·15 ± 0·73	0·73 ± 0·13	5·28	0·76	0·78	4·4−6·5	
2	133	4·69 ± 0·45	0·66 ± 0·07	4·63	0·64	0·70	4·3−7·4	
3	59	5·96 ± 1·32	0·95 ± 0·23	6·82	1·10	1·04	4·3−6·4	
6	17					0·96	4·1−5·0	
7	26					0·75	4·1−6·1	
8	12					1·29	4·2−5·1	
9	23					1·18	4·1−5·4	
10	29					0·82	4·2−5·8	
11	17					0·72	4·2−5·6	
12	122	4·19 ± 1·01	0·57 ± 0·16	5·97	0·87		4·2−7·9	two branches
13	32					1·04	4·3−6·0	
14, 15	118	5·50 ± 1·15	0·84 ± 0·21	4·87	0·70	1·05	4·3−6·6	
16	162	6·43 ± 0·41	1·00 ± 0·07	6·24	0·96	1·05	4·3−6·3	
17	55	5·04 ± 0·36	0·81 ± 0·07	5·08	0·81	0·80	4·2−5·9	
18	180	6·00 ± 0·48	0·90 ± 0·09	6·08	0·92	1·02	4·3−6·1	
19	398	6·59 ± 0·20	0·95 ± 0·04	6·49	0·93	0·98	4·3−6·8	
20	176	4·89 ± 0·69	0·69 ± 0·12	5·72	0·84		4·3−7·3	two branches
20							5·5−7·0	$h = i$
21	81	5·11 ± 0·20	0·79 ± 0·04	5·41	0·85	0·80	4·3−6·4	
21d	98						5·0−7·3	$h = i$
22	523	7·60 ± 0·62	1·14 ± 0·11	6·96	1·01	0·96	4·3−6·3	
23	74	3·88 ± 0·84	0·55 ± 0·14	4·11	0·58	0·70	4·4−7·0	
24	223	4·44 ± 0·35	0·57 ± 0·06	4·91	0·65	0·62	4·3−7·8	
25	244	6·88 ± 0·32	0·98 ± 0·08	6·02	0·83	0·96	4·8−6·7	
26	1105	6·57 ± 0·36	0·81 ± 0·07	6·63	0·82	0·87	4·8−7·0	
26a	634	5·99 ± 0·32	0·77 ± 0·06	5·21	0·62	0·95	4·8−7·0	
26b	157	6·70 ± 0·84	0·98 ± 0·14	6·09	0·87	1·16	4·9−5·9	
26c	314	5·88 ± 0·45	0·77 ± 0·08	5·95	0·78	0·83	4·9−7·0	
27	80	4·28 ± 0·34	0·57 ± 0·06	4·39	0·58	0·58	4·9−7·3	
28	24				?	?		two branches
29	86	3·87 ± 0·60	0·49 ± 0·09	3·97	0·51	0·54	4·9−7·3	
30	58	5·43 ± 1·54	0·80 ± 0·27	4·00	0·053	0·91	4·9−6·6	
31	18					0·67	4·8−6·5	
32	151	7·60 ± 1·24	1·17 ± 0·23	6·59	0·97	1·32	4·9−6·2	
33	144	4·72 ± 0·27	0·60 ± 0·04	4·75	0·60	0·76	5·0−8·0	
34	107	5·16 ± 0·82	0·72 ± 0·10	4·82	0·66	0·85	5·0−7·3	
35	48					0·80	5·2−6·2	
36	21					?	5·3−6·1	
37	19					0·71	4·3−6·0	
38	9					0·59	5·0−6·1	
39	20					0·68	5·4−7·1	

Notes: $\sum N$ — total number of shocks with $M \geqq 4\cdot1$ b_3 — eye-fitting
a_1, b_1 — least squares ΔM — magnitude range
a_2, b_2 — maximum likelihood

For some regions the total number of shocks is too small and only b_3 can be roughly estimated.

classes, or to read the numbers of the earthquakes with magnitudes equal and larger than a given M. The relation between $N_c(M)$ and $N(M)$ can be defined by

$$N_c(M_k) = \sum_{i=k}^{s} N(M_i)\, dM_i \quad k = 1, 2, \ldots, s \qquad (2a)$$

where s is the number of different values of M arranged in ascending series M_1, $M_2, \ldots, M_s$, dM being the class interval. Instead of the sum, the integral definition is used by some investigators, e. g., in the form

$$N_c(M) = \int_{M}^{+\infty} 10^{a-bM}\, dM = 10^{a-bM}/b \ln 10 , \qquad (2b)$$

i. e. $\log N_c = a - bM - \log(b \ln 10) = a' - bM$.
It follows that the cumulative graph differs only by an additive constant from that of the "normal" distribution. However, it should be emphasized that this conclusion is valid exactly only if the $N(M)$ function is continuous. This condition is not satisfied in the observed $N(M)$ distributions and, consequently, $b_c \geq b$, as can be demonstrated on practical examples (see Table 3b). The examples in Table 3b also show the influence of the methods of computation of parameters a and b. The methods are:

a) The *eye-fitting* method, used frequently, however, involving personal judgement and being undefinable.

Table 3b

Region: W. Greece, 1901—1955

	Least squares		Generalized least squares		Max. likelihood		Page	Utsu
	a	b	a	b	a	b	b	b
centres of classes, $\Delta M = 0.5$;	6.03 ± 0.35	0.79 ± 0.06	5.29 ± 0.39	0.65 ± 0.08	5.34	0.66	0.60	0.64
weighted mean, $\Delta M = 0.5$	6.17 ± 0.37	0.82 ± 0.07	5.51 ± 0.41	0.69 ± 0.09	5.55	0.70	0.66	0.69
$\Delta M = 0.1$	4.50 ± 0.32	0.67 ± 0.06	4.22 ± 0.40	0.57 ± 0.08	4.04	0.54	0.56	0.60
Cumulative frequency								
centres of classes, $\Delta M = 0.5$	6.75 ± 0.40	0.89 ± 0.07	5.86 ± 0.31	0.72 ± 0.07	5.92	0.73	0.66	0.69
$\Delta M = 0.1$	6.74 ± 0.16	0.92 ± 0.03	5.88 ± 0.15	0.75 ± 0.03	5.91	0.75	0.75	0.77

b) The *least squares method*, based on the assumption that $\log N_i$ is a continuous random quantity, obeying the Gaussian distribution. It follows also from the definition that $a = f(b)$. Deviating isolated points may seriously distort the result. A large number $(n > 100)$ of points is needed.

c) The *generalized least squares* method introduces weights p_i for individual points. Other basic conditions and the computation procedure are the same as sub b).

d) The *maximum likelihood method* is based on the assumption that a random quantity N_i displays a Poisson distribution. If the logarithmic probability distribution and the principle of maximum truth are considered we get these relations for a and b:

$$a = \log \sum_{i=1}^{s} N_i - \log \sum_{i=1}^{s} 10^{-bM_i},$$

$$\sum_{i=1}^{s} M_i \, 10^{-bM_i} \left(\sum_{i=1}^{s} 10^{-bM_i} \right)^{-1} - \sum_{i=1}^{s} N_i M_i \left(\sum_{i=1}^{s} N_i \right)^{-1} = 0. \tag{2c}$$

Again $a = f(b)$.

The least squares method and the maximum likelihood method are both derived from the principle of maximum truth. The former requires a continuous Gauss distribution of the random quantity $\log N_i$, the latter requires a discrete Poission distribution of the random quantity N_i.

e) *The Page formula*

$$b = \left[\left(\sum_{i=1}^{s} N_i M_i \right) \left(\sum_{i=1}^{A} N_i \right)^{-1} - \left(M_{min} - M_{max} \cdot 10^{-b(M_{max} - M_{min})} \right) \right.$$

$$\left. \left(1 - 10^{-b(M_{max} - M_{min})} \right)^{-1} \right]^{-1} \cdot \log e, \tag{2d}$$

and the *Utsu* formula

$$b = \left[\left(\sum_{i=1}^{s} N_i M_i \right) \left(\sum_{i=1}^{s} N_i \right)^{-1} - M_{min} \right]^{-1} \log e. \tag{2e}$$

The former is derived using the principle of maximum truth under the assumption that the observations are independent and that the probability that $N(M_i)$ earthquakes will occur with a magnitude between M_i and $M_i + dM_i$ is $P = N(M_i) \cdot$ $\left[N_c(M_{min}) - N_c(M_{max}) \right]^{-1}$. This relation holds if $N(M)$ is a continuous function; the square brackets represent the sum of all shocks with magnitudes between M_{max} and M_{min}.

In the Utsu formula M_{min} is equal to the centre of the lowest class minus half the class interval, i. e. $- \Delta M/2$. Both formulae are similar and we can write $b(\text{Page}) = b(\text{Utsu}) + K$, where K is a correction which cannot be neglected for low values of b and $M_{max} - M_{min} = 2$ to 3, e. g., $K = 0.22$ for $M_{max} - M_{min} = 2$, $b = 0.5$ and $\bar{M} - M_{min} = 0.5$. It is also important that M_{min} corresponds to the lowest threshold of the homogeneous data and not of all data available. Utsu observed agreement with other method if $b \Delta M < 1/4$, i. e. for magnitude classes smaller than about 0.3, in other words if the class distribution $N(M_i)$ is closer to a continuous distribution for which the theoretical assumptions hold.

3.13. LINEARITY OF THE RELATION BETWEEN log N AND M

In the majority of cases the distribution of representative values of $\log N$ plotted for equal magnitude classes fits a straight line fairly well. A change to a steeper slope within the highest magnitude classes was mentioned already by Gutenberg and Richter [23]. Some recent studies give evidence of distributions which are better approximated by two or more linear branches or by a curve [11, 12]. These observations are supported by the laboratory experiments of Mogi [13] who relates the position of discontinuous points of the broken line to the dimension of the units of regular structure (blocks, cracks) in the sample of material tested. When the material is heterogeneous the $N(M)$-function is always linear and satisfies the basic simple relation (1). The experiments with the fracturing of different materials yield an acceptable explanation of the observed shape of the magnitude-frequency relations assuming that we do not combine the activities of two or three separate focal zones differing in structure and stress distribution pattern as mentioned in para 3.11.

3.14. REGIONAL, DEPTH AND TIME VARIATIONS OF THE COEFFICIENT b

Some seismologists believe in a constant b as a worldwide regularity in earthquake occurrence, the others support the opinion that b is variable. A comparison of $N(M)$-relations determined uniformly for individual regions of the European area shows systematic deviations from a "standard" value of b. For shallow shocks b ranging from 0.5 to 1.2 has been found. The regions with identical values of b are not distributed randomly, they can be grouped into larger provinces which are characterized by a common tectonic development (see Table 3a). Seismic regions in the Western Mediterranean, i. e. the regions in the belt from NW Africa and SE Spain to S. Italy and Albania, show elevated values of $b = 1.0 \pm$; in these regions very shallow foci are predominant. The lowest values of $b = 0.6\pm$ are concentrated along the North Anatolian fault system, in Bulgaria and Macedonia.

Intermediate values of $b = 0.8-0.9$ are typical for the Alps, the Balkans and the remaining part of Europe. This result seems to be realistic because a small scatter in most magnitude-frequency graphs (see Fig. 1) does not give very much freedom in drawing the lines. The accuracy can be estimated by comparing the standard deviations of the coefficients a and b (see Table 3a); the probable error of different methods is apparent from the same table where the results corresponding to different methods are confronted.

It may be of interest to bring into our discussion also results on *intermediate-depth shocks*. Before comparing both sets we must bear in mind that the magnitudes of shallow shocks are based on surface waves (M_{LH}) and those of shocks deeper than normal on body waves $(M_B = m)$, respectively. It is well known that the two magnitude scales are not identical except $M = m = 6\frac{3}{4}$, that means also that the "units" of the two scales are not the same. We can convert M to m or vice versa using the known formulae, e. g. that of Båth [15] $m = 0.56M + 2.9$ or that determined for Europe [13] $M - m = 0.53(M - 6.8)$. When accepting M as a common basis we get $b = 0.2-0.5$ for intermediate-depth shocks and we can observe a *tendency of b to decrease with depth*. This result is in good agreement with the laboratory experiments of Mogi [13] and Vinogradov [14], assuming that the degree of homogeneity or the stress rate increase with depth. It is noteworthy that in agreement with

Table 4

Region	22			
Period Activity b	1901—1911 increase 0·77	1912—1920 decrease 0·94	1921—1930 increase 0·93	1931—1955 decrease 1·20
$\bar{b}$	0·96			

Region	26a		
Period Activity b	1901—1922 * increase 0·83	1923—1942 decrease 1·03	1943—1955 * increase 0·78
$\bar{b}$	0·95		

the above mentioned experiments, the seismic regions with highest activity are marked by the lowest values of b. In the European area similar results on the variations of b were recently found by a group of Greek seismologists [17].

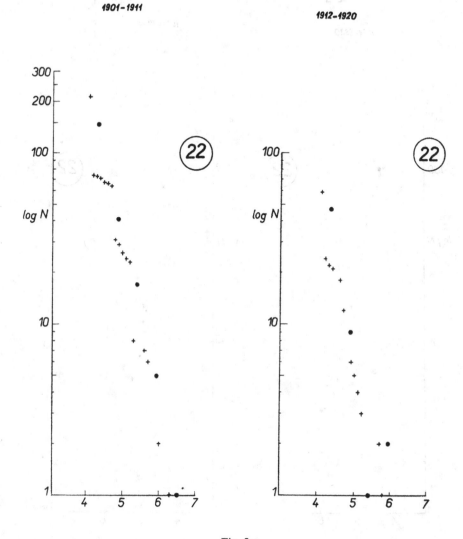

Fig. 2a.

Fig. 2. Magnitude-frequency relations for periods of increased and decreased activity within the regions No. 22 and No. 26a.

Furthermore, we must expect that b is not stable *with time* because of the changing level of seismic activity, changing stress rate or changing average focal depth. This assumption was tested in the example of regions Nos. 22 (Yugoslavia) and 26a (W. Greece), respectively (Fig. 2a, 2b, 2c). The total period of observation was divided into natural intervals of a relative activity or quiescence using the corresponding strain-release curves. The values of b obtained for those intervals can be seen in Table 4; they demonstrate an increase of b in the quiet interval and a decrease in the active one. Similar time variations were observed by Vinogradov during a sequence of rock bursts [6] and, recently, by Ikegami in Japan [26]. Also we checked

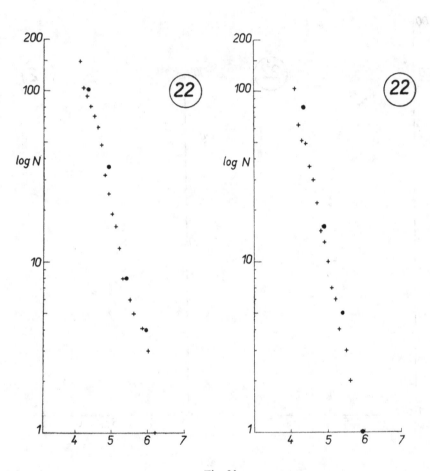

1931–1955

1921–1930

Fig. 2b.

a possible *influence of aftershocks* on the slope of the magnitude-frequency graph. The omission of aftershockss resulted in the decrease of b by amounts smaller than 0·1, i. e. by a difference which falls within the range of standard deviations or differences caused by various working procedures. A more substantial influence cannot, however, be excluded, when strong shallow foci earthquakes are followed by numerous aftershocks and the period of investigation coincides with such a sequence.

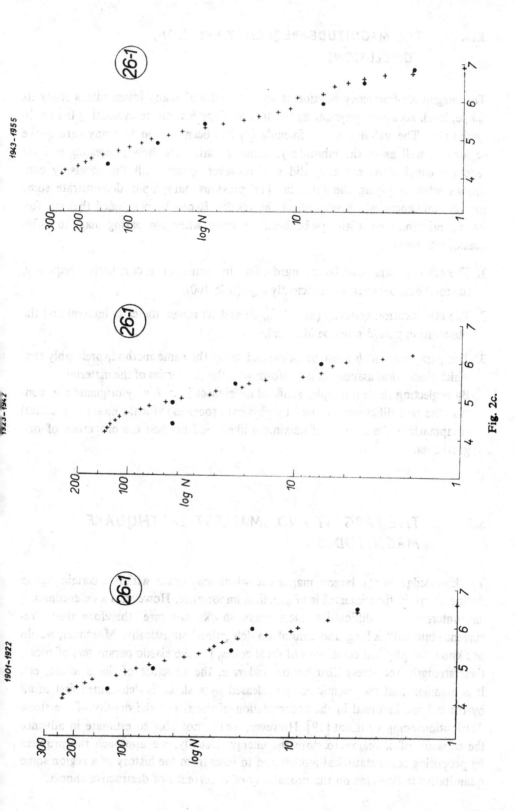

Fig. 2c.

3.15. THE MAGNITUDE-FREQUENCY RELATION; CONCLUSIONS

The magnitude-frequency relation is widely used and many investigators study its shape, level, accuracy, physical significance and applications in estimating the earthquake risk. The validity of the formula (1) has been proved in many earthquake regions as well as in the laboratory. Some investigators drew interesting and far reaching conclusions but they did not, however, observe all the necessary conditions when applying the formula. The previous paragraphs demonstrate some possible influences which may affect the results. It can be concluded that the following principles must always be borne in mind when comparing magnitude-frequency relations:

1. The original data must be arranged either in normal or in cumulative frequency, the total number must be sufficiently high $(n \geq 100)$.

2. The classification quantity (m, M, I_0, E) and its range, the class interval and the observation period must be identical.

3. The parameters a, b must be calculated using the same method, preferably that which theoretical assumptions conform with the properties of the material.
   By neglecting these principles artificial differences in a, b may originate and confuse the real differences caused by physical processes in focal zones. For practical computations the method of maximum likelihood fits best the properties of original data.

3.2. THE LARGEST AND SMALLEST EARTHQUAKE MAGNITUDES

The knowledge of the largest magnitude which may occur within a certain region during a certain time interval is of practical importance. However, its determination encounters serious difficulties. Destructive shocks are rare, therefore their "recurrence interval" is long and cannot be determined statistically. Moreover, we do not know the physical conditions of focal zones, i. e. the elastic parameters of rocks, their strength, the stress distribution and rate, the structure of the medium, etc. It is assumed that the seismic energy released by a shock is determined first of all by the volume involved in the accumulation of the stress, the density of the stress distribution being constant [19]. However, we are not able to estimate in advance the capacity of a region to store the energy. Usually, we approach the problem by proposing some statistical models and to infer from the history of a region some quantitative indications on the probability of occurrence of destructive shocks.

We may also use a simple empirical deduction when, exceptionally, the magnitude-frequency graph shows a threshold value in the range of large magnitudes. In this way the possible upper magnitude limit M_{max} can be estimated for the following regions $(h = n)$:

No. 6, 8:	$M = 5\frac{1}{4}$	No. 13–16:	$M = 6\frac{3}{4}$
No. 11:	$M = 5\frac{3}{4}$	No. 25:	$M = 6\frac{3}{4}?$
No. 7, 10:	$M = 6?$	No. 19:	$M = 7?$
No. 17, 35–38:	$M = 6\frac{1}{4}$,	No. 26a,b:	$M = 7\frac{1}{4}$
No. 3. 21, 30, 32:	$M = 6\frac{1}{2}?$	No. 2, 26c, 27, 29:	$M = 7\frac{1}{2}$
No. 18, 22:	$M = 6\frac{1}{2}$.		

M_{max} observed during 1901–1955 within areas 0·5° long. × 0·5° lat. are written in the map of the energy release (see enclosure) and these values may also serve for the estimation of the possible threshold magnitudes.

The statistical solution of the problem becomes more difficult due to lack of data when a rather limited small zone is considered.

3.21.   APPLICATION OF GUMBEL'S THEORY OF LARGEST VALUES

Nordquist was the first [46] to apply the Gumbel theory of largest values [45] to the earthquake statistics. He concluded that "the observed distribution of the largest earthquakes is in good agreement with the theory of largest values ..."; this conclusion related to world shocks, to southern California and to a small region selected in southern California. Recently, Gayskiy and Katok [50] in the Soviet Union and Epstein and Lomnitz in the USA [47] applied and developed the Gumbel theory for the example of Central Asia and of California, respectively. The latter authors proved that the relation for the expected number of earthquakes in a given year with magnitudes exceeding a definite value was of the same form as the magnitude-frequency relation.

J. V. Riznichenko relates the shock with maximum possible energy E_{max} to the level of earthquake activity defined by the ordinate of the energy-frequency plot [194]. The principle of his procedure is based on empirical formulae derived for a specific region. We can also find the maximum shock by "reasonably" extrapolating the magnitude-frequency curve converted to 10, 50 or 100 years, towards highest magnitudes. This procedure is simple but involves personal judgement. We can find different shapes of the $N(M)$-curve, a regular decrease towards large magnitudes, a sudden drop of the number of shocks, largely scattered values towards large magnitudes, etc. These different situations indicate the degree of statistical instability of the set of observations.

The details of the Gumbel theory of largest values and its further theoretical development by Epstein and Lomnitz are given in their papers [45, 47]. It seems, however, useful to review the main ideas.

Gumbel applied his theory to the forecasting of floods which represent a phenomenon similar to disastrous earthquakes. According to Gumbel's flood analysis we can consider the earthquake magnitude x as a random variable distributed with cumulative distribution function of the type

$$F(x) = 1 - e^{-x}, \quad x \geq 0. \tag{3}$$

It follows from Gumbel's probability theory that the largest annual magnitude y fits the cumulative distribution function

$$G(y) = \exp\left(-e^{-y}\right), \quad \text{where } y = c(x - u), c \text{ and } u \text{ are parameters}. \tag{4}$$

The probability of the most probable largest magnitude $\tilde{x}_n$ according to [45] is

$$F(\tilde{x}_n) = \exp\left(-1/n\right) \tag{5}$$

and the probability of the most probable smallest magnitude $\tilde{x}_1$ follows from the equation

$$n = \frac{1}{F(\tilde{x}_1)} + \frac{1}{F(\tilde{x}_1) \log F(\tilde{x}_1)} - \frac{1}{\log F(\tilde{x}_1)}. \tag{6}$$

The table of values $F(\tilde{x}_n)$ and $F(\tilde{x}_1)$ is given in [45] p. 74.

The probability $F(\hat{x}_j)$ of the j-th maximum magnitude $\hat{x}$ must lie between $F(\tilde{x}_n)$ and $F(\tilde{x}_1)$ and the corresponding formula is

$$F(\hat{x}_j) = F(\tilde{x}_1) + \frac{j-1}{n-1}\left[F(\tilde{x}_n) - F(\tilde{x}_1)\right]. \tag{7}$$

The return period (occurrence interval) of earthquakes having a magnitude equal to or greater than $\hat{x}_j$ is

$$T(\hat{x}_j) = \frac{1}{1 - F(\hat{x}_j)}. \tag{8}$$

Epstein and Lomnitz [47] added to Gumbel's original assumption a second one that the number of earthquakes in a year is a Poisson random variable with a mean α. They also give a slightly different distribution function for the magnitude

$$F(x) = 1 - e^{-\beta x}, \quad x \geq 0, \tag{9}$$

where β is a parameter.

154

From the two basic assumptions it follows that the cumulative distribution function

$$G(y) = P(Y \leq y) = \sum_{k=0}^{\infty} \frac{e^{-\alpha}\alpha^k}{k!} [F(y)]^k =$$

$$= \exp\{-\alpha[1 - F(y)]\} = \exp(-\alpha e^{-\beta y}), \quad y \geq 0. \tag{10}$$

The expected number of earthquakes in a given interval with magnitudes exceeding y

$$N_y = \alpha e^{-\beta y}, \quad \text{i. e. } \log N_y = \log[-\log G(y)] = \log \alpha - \beta y \tag{11}$$

which can be compared with the same form of the empirical formula connecting magnitude and cumulative numbers of earthquakes, valid for one year

$$\log_{10} N_c = A - bM. \tag{12}$$

Thus we can compare the coefficients in the formulae (11) and (12) using conversion relations

$$A = \log \alpha/\log 10, \quad b = \beta/\log 10. \tag{13}$$

In the case of the most frequently used magnitude-frequency formula, expressing the relation between M and the number of earthquakes N corresponding to a certain magnitude class which took place within a certain area,

$$\log_{10} N = a - bM, \tag{1}$$

valid for n-years, the conversion relations are

$$a = \frac{\log \alpha}{\log 10} + \log_{10} n + \log_{10} \frac{b}{q}, \quad q = \log_{10} e. \tag{14}$$

Epstein and Lomnitz derived several useful relations [47], e. g. for the modal annual maximum magnitude, for the value of the maximum magnitude which is exceeded with probability P in a D year period and the earthquake risk, i. e. the probability of occurrence of an earthquake of magnitude y or more in a D year period. The theoretical treatment by Smirnov and Dunin-Barkovskiy [38] is similar to that of Lomnitz and Epstein, except that they applied the method to flood analysis. Recently, Gayskiy and Katok employed the Gumbel theory for the determination of the number of shocks in different energy classes for the Pamir-Hindukush region [40].

For the practical treatment of data the largest annual magnitudes $y_1, y_2, ..., y_n$ must be arranged in the order of increasing magnitude. Then the representative values of y_j grouped into classes are associated with the probability $F(\hat{y}_j)$ calculated according to (7) or simply the individual values of y_j are associated with probabilities

$$G(y_j) = \frac{j}{n+1} \tag{15}$$

(n — number of intervals, e. g. years).

The first procedure was used by Nordquist [46], the second by Epstein, Lomnitz [47] and Smirnov, Dunin-Barkovskiy [48].

The couples y_j and $F(\hat{y}_j)$ or y_j and $G(y_j)$, respectively, are plotted on extremal probability paper. According to the theory, the points should fit a straight line if the distribution function (3) is authentic.

The parameters log α and β in the formula (11) can be calculated by the least-square method. The resulting values log α and β can be checked by comparing them with the parameters A or a and b of the corresponding magnitude-frequency relation (12, 1). The return periods may be read on the upper edge of the probability paper. The straight line can be extrapolated and the largest magnitude can be read for an arbitrary probability, e. g. $P = 0.99$.

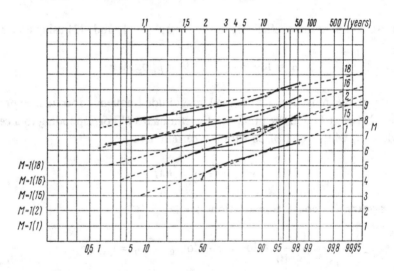

Fig. 3a.

Fig. 3. Largest magnitudes observed in the European seismic regions plotted on the probability paper. The zero of the M-scale is shifted for each region.

The coordinated research of the seismicity of the European area [1,30] has resulted in uniform catalogues and maps. The advantage of catalogued seismological information is its homogeneity as to time interval, area and classification. Among other problems also that of the largest possible earthquakes appeared. The period of 55 years (1901—1955) for which the data on earthquakes of $M > 4.5$ are available might be long enough for the full manifestation of earthquake forces in most regions. The Gumbel theory is suitable also for testing this assumption. For the experiment only regions, from the total number of 39 earthquake regions (see the map), with

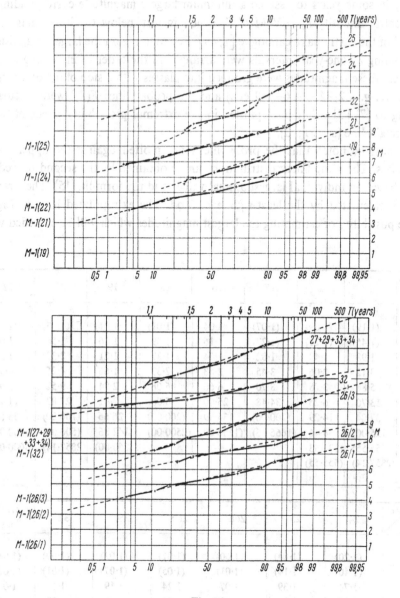

Fig. 3 b.

highest number of observations were selected, i. e. Nos. 1 (Jan Mayen), 2 (Iceland), 15 (Algeria), 16 (Spain), 18 (the Alps), 19 (the Appennines), 21 (the Carpathians), 22 (Yugoslavia), 24 (Macedonia, N. Greece), 25 (Albania), 26a (Greece), 26b (Crete), 26c (W. Turkey), 32 (the Caucasus), 27, 29, 33, 34 (The North Anatolian fault). Only shallow-focus earthquakes are involved in the calculation. The one-year interval was used as a most appropriate unit, shorter intervals are not suitable because then the number of "empty" intervals increases. Even for one-year intervals it is necessary in some years to assume a minimum largest magnitude corresponding to the epicentral intensity $I_0 = $ V M. C. S. which is just below the lower magnitude threshold of the earthquake catalogue [1, 49]. For each region the largest magnitudes corresponding to one-year intervals were arranged in the order of increasing magnitude. Then, these magnitudes were grouped into classes with a step of 0·4 of a magnitude unit $\left(\text{e. g. } 4\cdot 2-4\cdot 5, 4\cdot 6-4\cdot 9 \text{ etc.} \right)$ and the probabilities $F(\hat{x}_j)$ were calculated according to (7) [51]. The data plotted in the extremal probability paper fit fairly well a straight line (see Fig. 3).

It must be noted that the probabilities are plotted against the upper value of each class. The same linear distribution can be obtained by the second procedure [47, 48] based on individual largest annual values and the formula (15). The straight lines were calculated by the least-square method. The extrapolated lines in Fig. 3 yield the possibility of estimating the largest magnitudes which will be exceeded with

Table 5

Region \ M	1	2	15	16	18	19	21	22
3·5	(1·19)	(1·09)	(1·07)	1·04	(1·01)	(1·01)	(1·23)	(1·01)
4·0	(1·45)	(1·21)	(1·27)	1·16	1·09	1·06	1·56	1·07
4·5	1·98	1·49	1·82	1·54	1·37	1·21	2·33	1·28
5·0	3·12	2·06	3·45	2·63	2·33	1·66	4·25	1·92
5·5	5·88	3·28	8·70	6·66	6·06	2·94	9·52	3·92
6·0	13·70	6·25	34·48	25·00	25·00	7·41	27·82	11·76
6·5	37·04	14·29	(200·00)	136·99	(200·00)	26·66	(100·00)	55·55
7·0	(125·00)	38·46	(1666·66)	(1250·00)	(2500·00)	142·86	(500·00)	(400·00)
7·5	(555·55)	(133·33)	—	—	—	(1250·00)	(2857·14)	(5000·00)
8·0	(2857·14)	(555·55)	—	—	—	—	—	—

Region \ M	24	25	26a	26b	26c	32	27,29, 33,34
3·5	(1·20)	(1·04)	(1·00)	(1·01)	(1·02)	(1·00)	(1·01)
4·0	(1·40)	(1·14)	(1·01)	(1·05)	(1·07)	(1·01)	(1·02)
4·5	1·74	1·39	1·07	1·24	1·19	1·13	1·08
5·0	2·38	2·00	1·28	1·92	1·49	1·82	1·20
5·5	3·57	3·57	2·13	4·35	2·15	5·55	1·50
6·0	6·06	8·00	4·00	16·66	3·77	42·55	2·17
6·5	11·14	25·00	12·50	111·11	8·33	(833·33)	3·77
7·0	24·39	100·00	55·49	(1333·33)	22·22	—	7·69
7·5	58·82	(555·55)	(454·54)	—	(76·92)	—	21·74
8·0	(166·66)	(4000·00)	(5000·00)	—	(357·14)	—	66·66

Note: The values of return periods extrapolated over the largest observed magnitude are given in parentheses.

a given probability. Using the formula (8) the return periods T [year] of earthquakes of magnitudes equal to or greater than a given threshold value can be simply determined (see Table 5).

We have chosen the probability $F(M_n) = P(M \leq M_n) = 0.99$, i. e. the occurrence of an earthquake of $M > M_n$ has the probability 1% and we may regard M_n as a largest "possible" earthquake. This probability corresponds to a return period of 100 years which is probably sufficient for the manifestation of seismogenetic forces in most of the investigated regions. The results are given in the first line of Table 6 and can be confronted with the largest magnitudes, actually observed during 1901 to 1955. The differences are relatively small, the significant ones are only in regions Nos. 1, 22, 26c, indicating the possibility of occurrence of still stronger shocks. The same regions do not fulfil the condition $T(M_{\max}, \text{observed}) <$

Table 6

Region	1	2	15	16	18	19
$M_{\max}$, $P = 1\%$	6·9	7·4	6·7	6·4	6·4	6·9
$M_{\max}$, observed	6·5	7·3	6·6	6·3	6·2	6·8
δM	0·4	0·1	0·1	0·1	0·2	0·1
$T(M = 6·5)$	37	14	62	137	200	27
$TM_{\max}$, observed	37	77	80	62	56	71

Region	21	22	24	25	26a	26b
$M_{\max}$, $P = 1\%$	6·5	6·7	7·8	7·0	7·1	6·5
$M_{\max}$, observed	6·4	6·3	7·8	6·7	6·9	6·4
δM	0·1	0·4	0·0	0·3	0·2	0·1
$T(M = 6·5)$	100	56	11	25	12	111
$TM_{\max}$, observed	74	29	105	42	38	71

Region	26c	32	27, 29, 33, 34
$M_{\max}$, $P = 1\%$	7·6	6·2	8·2
$M_{\max}$, observed	7·1	6·2	8·0
δM	0·5	0·0	0·2
$T(M = 6·5)$	8	833	4
$TM_{\max}$, observed	29	125	67

$< n$ (for $T(M_{max}$, observed) see last line of Table 6). According to the theory [45] the inequality $T(M_{max}) > n$ means that if the value M_{max} was observed during the total period a larger value cannot be expected in the future within a comparable set and the same period, and vice versa for $T(M_{max}) < n$. The return periods of the largest observed magnitudes smaller than 55 years are revealed in regions Nos. 1, 22, 25, 26a, 26c. Return periods for different magnitudes and regions are shown in Table 5. They could be used as a quantitative measure of seismicity. However, this idea is not new, it was used in principle already by Riznichenko who maps the value of 10^{δ} of the formula $\log N = \delta - \gamma \log E$ [16].

According to the original conditions and formulae (12−15) the parameters $\log \alpha$ and β can be checked in an independent way by the parameters A, a and b. This comparison was made for the regions given in Table 7.

Table 7

Region	1	2	15
a; b	5·28; 0·76	4·63; 0·64	5·62; 0·86
a_1; b_1	4·61; 0·76	4·87; 0·67	5·47; 0·84

Region	16	18	19
a; b	6·24; 0·96	6·08; 0·92	6·49; 0·93
a_1; b_1	5·02; 0·76	5·89; 0·92	5·54; 0·80

Region	21	22	24
a; b	5·41; 0·85	6·96; 1·01	4·91; 0·65
a_1; b_1	4·63; 0·71	5·33; 0·77	4·33; 0·57

Region	25	32
a; b	6·02; 0·83	6·59; 0·97
a_1; b_1	5·46; 0·78	6·56; 1·02

Note: a_1 and b_1 were calculated using formulae (11) and (13).

Gumbel's statistical model reveals the possibility of a quantitative estimation of the probability of occurrence of large earthquakes. The return periods may also be used as a quantitative measure of seismicity. It is advantageous that this method requires only a knowledge of the largest magnitudes in selected intervals.

3.22.   THE MINIMUM EARTHQUAKE

Previously the magnitude-frequency relation was believed to continue without limitation towards the smallest magnitudes. This idea was supported by the observation of minimum shocks with $M = -1$ or smaller in zones of shallow activity. Recently, the investigation of the existence of weak shocks below a depth of about $h = 100$ km demonstrated that there are also threshold values in the "minimum" magnitude range. The value of M_{min} may be in relation with the strength of the building material and the stress rate and thus could represent another characteristic of the earthquake generating processes.

Increased sensitivity of seismographs in some regions of intermediate-depth or deep shocks did not result in the recording of very weak shocks in a number corresponding to the extrapolated magnitude-frequency graph. Either a decreased number was observed [11, 17] or simply no shocks were observed below a certain threshold magnitude. So far, there are only two examples of the latter case. No shocks below $m = 2\frac{1}{2}$ were recorded in the Vrancea region (E. Carpathians, Rumania) during a two year period of operation of sensitive seismographs [20]; in this region the main activity is concentrated between $h = 100$ km and $h = 150$ km, respectively. The second experience was described by Suyehiro who did not record in S. America, in a region of deep shocks, earthquakes with magnitudes smaller than $m = 3\frac{3}{4}$ within a three month operation of very sensitive instruments [18].

These two examples represent only an indication which should be confirmed or disproved in the future. The value of M_{min} could become an additional characteristic depending on the mechanical properties of the material or the stress rate. For this kind of investigation a more precise determination of the depth of focus is necessary, particularly in the range of shallow depths.

3.3.   SEASONAL AND DAILY VARIATIONS

It is of some interest to demonstrate briefly statistical data about average time variations of the number of shocks. The catalogue 1901—1955 was transferred to the IBM punched cards, i. e. 5100 events with complete earthquake parameters; other catalogued shocks were deleted. This step enabled to make various selections very quickly. No evident seasonal variation was found as seen from the monthly distribution of events given in the following table

Month	Jan.	Feb.	March	Apr.	May	June	July	Aug.	Sept.	Oct.	Nov.	Dec.
n	416	393	456	457	455	406	429	573	419	422	324	350

The differing value for August is influenced by many aftershocks of a sequence of strong shocks in Greece in 1953. Otherwise we can observe only some slight indication of a relative decrease in the number of shocks in winter. A more pronounced tendency can be found in the distribution of times of origin during 24 hours as seen from the table:

Hour	0—1	1—2	2—3	3—4	4—5	5—6	6—7	7—8	8—9	9—10	10—11	11—12
n	252	237	230	249	217	187	209	109	190	211	213	166

Hour	12—13	13—14	14—15	15—16	16—17	17—18	18—19	19—20	20—21	21—22	22—23	23—24
n	186	190	178	178	185	191	183	211	250	230	216	232

The minimum of shocks occurs around noon UT, which is difficult to explain although the tendency is evident. The day-time minimum and night time maximum remain stable also when the times of origin are corrected for the difference in longitudes of epicentres, i. e. converted to the "local time", see the following table:

Hour	0—1	1—2	2—3	3—4	4—5	5—6	6—7	7—8	8—9	9—10	10—11	11—12
n	233	230	205	257	226	220	196	180	205	202	215	185

Hour	12—13	13—14	14—15	15—16	16—17	17—18	18—19	19—20	20—21	21—22	22—23	23—24
n	164	195	206	169	184	186	161	201	199	240	240	201

It is also noteworthy that of all catalogued shocks only 212 have a focus deeper than normal, two of them being deep (i. e. $h > 300$ km). Because of the incompleteness of these shocks for lower magnitude classes (about $M < 6$) it is appropriate to compare the proportion in the range $m = M > 6\frac{1}{4}$. We obtain 88 shallow shocks within $M = 6\frac{1}{4}-8$ in comparison with 40 shocks of $h > n$ in the same range.

It is questionable whether these results can be checked using the catalogue $1801-1900$ because a large number of events from this period is not documented by a complete set of basic parameters and the whole set is not homogeneous.

3.4. ENERGY AND STRAIN RELEASE

The total seismic energy released in focal zones represents an objective measure of seismic activity. Table 2 gives the values of ΣE calculated using the formulae of Gutenberg

$$\log E = 11 \cdot 8 + 1 \cdot 5M \ \left(h = n, \text{ shallow focus earthquakes}\right) \tag{16}$$

$$\log E = 5 \cdot 8 + 2 \cdot 4m \ \left(h = i, d, \text{ intermediate and deep focus earthquakes}\right) \tag{17}$$

and reduced to a unit time of 1 year and a unit area of 1 km² (only the surface covered by epicentres in each region is considered). The sums are in fact determined by the two uppermost magnitude classes, weak shocks do not influence the total energy balance. The unified values of energy release are two orders higher in the provinces belonging to the Alpine folding system in comparison with the Caledonian and Variscian ones. It is suprising that slightly larger amounts of energy are released in the upper mantle than in the crust, the maximum being at about $h = 100-150$ km. This difference is still greater when individual zones of crustal and upper mantle activities are considered. It is, however, necessary to note that there are some uncertainties in the calculation of energy for shocks deeper than normal and in the mutual connection of the m- and M-scales, respectively. Nevertheless, in European regions with foci below the crust (Aegean region, E. Carpathians), the earthquake generating stress is greater in the upper mantle than in the crust. This phenomenon is most evident in the Vrancea zone (E. Carpathians). The geographical distribution of sums of seismic energy can be well surveyed in the seismic energy map attached to this volume and discussed in chapter 4.3. We need a more accurate determination of focal depths for a detailed study of energy release with depth.

The deformation process in focal regions is demonstrated in a simplified form by the *strain release curves* constructed by summing up the square roots of seismic energy released by individual shocks in relation with time. Benioff's assumption on the proportionality between strain and the quantity $E^{1/2}$ simplifies the

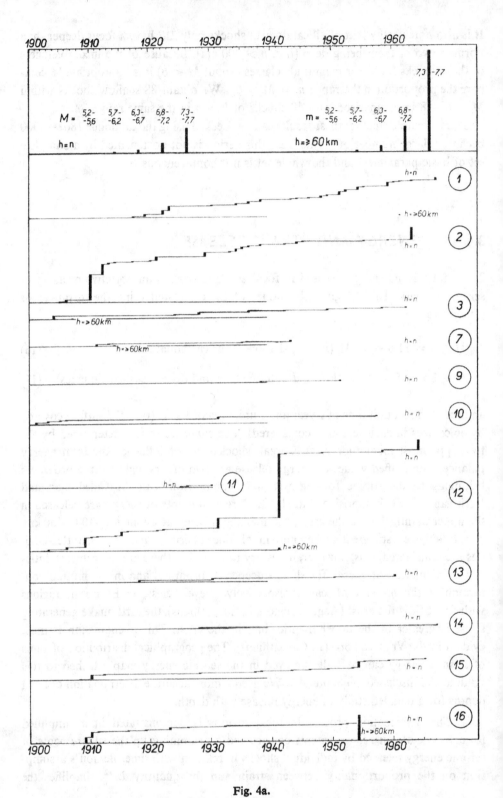

Fig. 4a.

Fig. 4. Cumulative curves of the value $\Sigma E_n^{1/2}$ for individual seismic regions (strain-release curves).

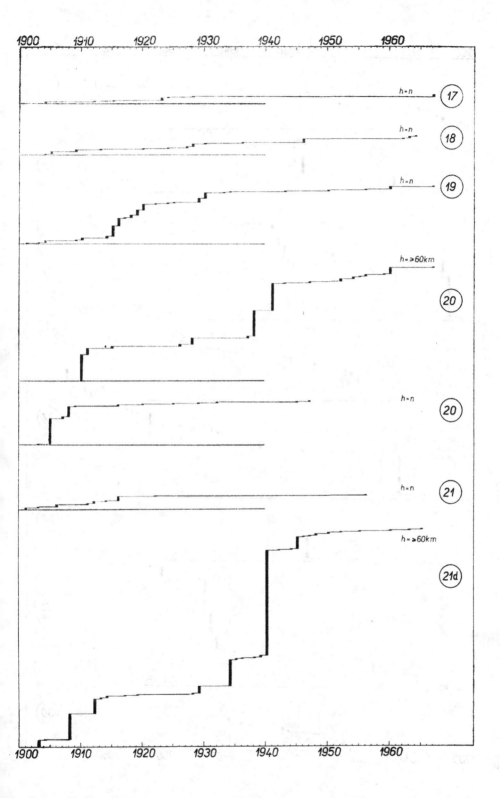

Fig. 4 b.

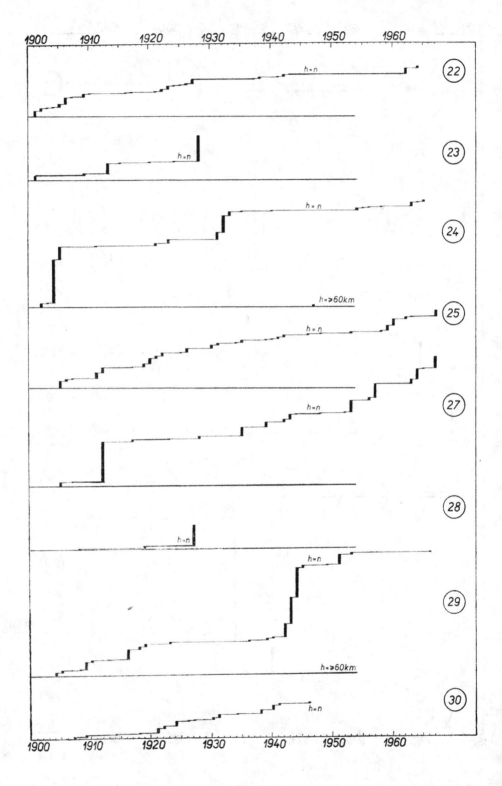

Fig. 4c.

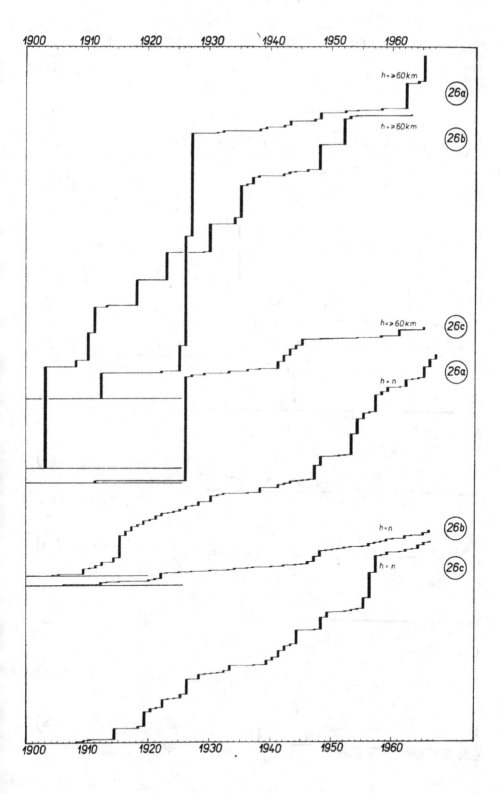

Fig. 4d.

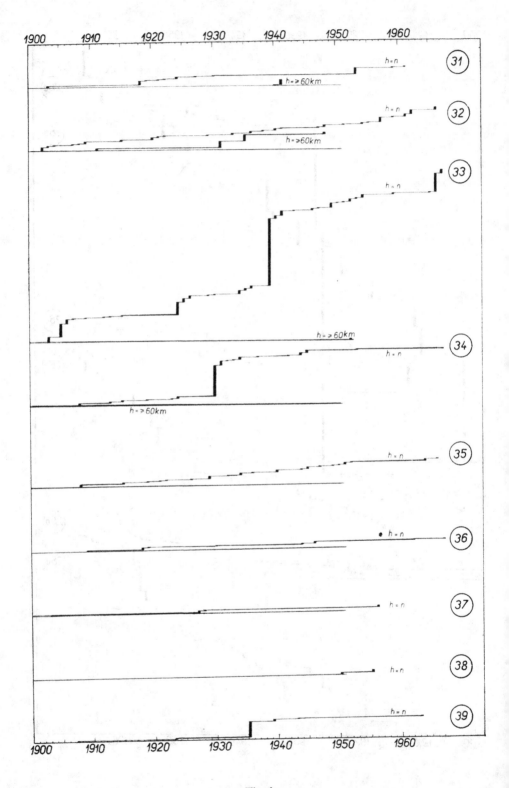

Fig. 4e.

real behaviour of stressed material in the crust and upper mantle, but no more appropriate procedure has been proposed as yet. The "Benioff curves" were constructed for all European seismic regions and reproduced in Fig. 4 or in a previous paper [21]. It should merely be added that the properties of the curves and the values of the standardized energy release can be correlated with the quantities b or M_{max} (see Table 3). In general, a high strain rebound rate and a broad band between the minimum and maximum values of a curve correspond to a low coefficient of b. A constant strain rate is well developed in regions Nos. 1, 15, 26b, 26c and 35 respectively. On the other hand there are regions with only one outburst of seismic activity during 1901–1955 and consequently an estimation of the trend is questionable.

4.   Seismological maps

The mapping of earthquake parameters or other derived characteristics provide us with a very convenient survey of the geographical distribution of seismic activity. Naturally, the information on European earthquakes has been also used in the construction of the corresponding seismic maps valid for the area under investigation. In fact, there exist, no maps of this area which would uniformly survey the European seismicity except two small maps in Gutenberg-Richter's Seismicity of the Earth. At the beginning of the work there were two possible ways of presenting the data, either in the form of an atlas, i. e. to divide Europe into many section maps (see e. g. the Atlas of Earthquakes of the USSR, Moscow 1962) or in the form of single maps of the whole investigated area. The second way was chosen because the project of the seismicity investigation of Europe should yield first of all an overall survey and provide seismologists with unified parameters. Moreover, a survey map gives a better possibility of studying the seismotectonic relations between units of different tectonic types. Uniform parameters can be simply applied later in regional and more detailed studies, i. e. also in compiling regional maps.

The first question in constructing the maps is the selection of an appropriate scale. After several attempts the scale 1 : 5 000 000 was used for the working draft of the maps and 1 : 7 500 000 for the final prints. The latter was accepted for practical reasons, since a compromise had to be made between the requirement of legibility of different signs and reasonable expenses for construction and print.

From the point of view of the scientific content we can divide the maps involving seismological information as follows:

A. *Seismic maps:*

1. *Epicentre map.*
2. *Maximum intensity map.*
3. *Energy release map.*
4. *Seismicity map.*

B. *Seismogeological maps:*

1. *Seismotectonic map.*
2. *Seismic zoning map.*

The individual kinds of seismic maps need not be explained in greater detail, except that of A 4, because the term *seismicity* is used with different meanings, e. g. for maps with isolines or numbers corresponding to the energy release per unit area

171

and time, to the sum of energy contributions from surrounding foci to a surface unit, to the product of the energy of a shock and the surface of the macroseismic area, to the product of the epicentral intensity I_0 and the macroseismic area, to the quotient of the radius of the macroseismic area to I_0, to the number of shocks corresponding to a given magnitude class, etc.

Our understanding of the term "seismicity" is close to the definitions based on the characteristics of seismic activity derived from the basic parameters such as the magnitude, intensity and number of events. These characteristics are calculated using conversion formulae and are subjected to some generalization and rounding off. It follows that a seismicity map does not represent original parameters compared with maps A1 and A2, respectively.

In this volume only maps of the type A1, A2 and A3 are included, the seismicity map will be published separately because its construction has not yet been standardized. All seismological maps have a common geographical basis which contains general geographic information, i. e. altitudes, rivers, towns, frontiers, volcanoes and additional information of seismological interest, i. e. seismological stations operating during 1901 – 1955.

Recently, a seismotectonic map of Europe has been compiled by V. Beloussov and A. Sorsky in accordance with the recommendations of the E.S.C. to the scale 1 : 5 000 000, and published in 1967 by the Academy of Sciences of the USSR. The seismic zoning map should furnish information on the probability of occurrence of different intensities and must be based on all previous types of maps mentioned sub A and B1. A special working group, under the chairmanship of Professor J. P. Rothé, was set up within E.S.C. Sub-Committee for the Seismicity of the European Area to develop prapropriate methods of construction of the map and to compile such a map for the European area. The main problems are how to extrapolate the seismic zones known at present along lines which might be active in the future and how to estimate the frequency and intensity of future shocks.

4.1.   EPICENTRE MAPS

The most convenient mapping of seismological parameters is in the form of epicentre maps because they represent the original basic information on earthquake activity. The scale chosen for the presentation of the whole European area on one map is not large enough to permit all epicentres from the interval 1901 – 1955 to be placed on a single map without reducing its legibility. The epicentres are distributed in irregular clusters, therefore the corresponding signs were plotted on two sheets according to the magnitude class: 1. $M = 4 \cdot 1 - 4 \cdot 6$ (one class), 2. $M = 4 \cdot 7 - 8 \cdot 3$ (seven M classes, four colours). Copies of the epicentre maps are given in the appendix. The key to each map explains the different symbols used. The symbols were so designed as to express the epicentre location, magnitude, depth, accuracy of position and accuracy of the classification. However, they must not cover each other up when overlapping and all of them must be large enough to be legible. Hence, only open symbols with different contours can be used. Circles of different diameter were suggested for shallow shocks, triangles for very shallow ($h < 5$ km) or intermediate and deep-focus earthquakes. Dotted contours mean low accuracy of the location. The vertial line crossing the symbol corresponds to an instrumental epicentre, the dot in the centre of the sign means class A in the location. In the most active zones there are also earthquakes with identical epicentre coordinates, in such a case the number of shocks corresponding to one point is written beside the symbol.

Naturally, the maps reflect the qualities of the original material. It can be seen at first glance that the data in the lowest magnitude class and possibly also in the class $M = 4 \cdot 7 - 5 \cdot 1$ are not homogeneous over the whole area. The condition of homogeneity for $M > 4\frac{1}{4}$ is fulfilled only for continental Europe plus Algeria. The exception is Greece where the lowest magnitude class corresponds to $I_0 = V$, i. e. to a value below the threshold given in the first recommendations of the ESC. For the same reason the data for the Eastern Mediterranean region are not complete. Many of the weak oceanic shocks might escape the recording capability of a few stations at the beginning of the 20th century and they are evidently not complete as far as the two lowest magnitude classes are concerned. Also the homogeneity of weak intermediate depth shocks is very questionable. We can observe a sudden drop in the number of such shocks, due to the lack of observations at $M = 5\frac{1}{2}\pm$ in the Vrancea zone (E. Carpathians) and at $M = 6\pm$ in other zones of upper mantle activity, i. e. in the regions of Crete and of the Tyrrhenian Sea, respectively. On the other hand, many epicentres with very shallow depth of focus ($h < 5$ km) are not plotted on the maps even when the macroseismic intensity was $I_0 > VI$, the magnitude, however, did not exceed $M = 4 \cdot 0$ and that was decisive. The advantage of an objective classification using magnitude instead of intensity can be demonstrated by comparing the final epicentre maps and the first drafts compiled in 1960 when I_0 was used as a classification quantity [29]. In the first drafts the activity of some zones was overestimated, e. g. in Albania, Algeria and Sicily where high intensities correspond to relatively low magnitudes; an opposite phenomenon can be observed in regions with incomplete macroseismic data.

4.2.   MAP OF MAXIMUM OBSERVED INTENSITY

This type of seismic map represents a survey of maximum macroseismic intensities which were reported and are reliable. This map must not be mistaken for a map of seismic zoning, although it precedes it as an essential stage of preparation. It should be compiled for a definite interval but as a matter of fact this is, impossible for

Europe as a whole. We depend on national maximum intensity maps because there are no national catalogues of shaken localities listing all intensities observed. Individual maps are not related to the same interval and therefore the resulting European map can only summarize all available observations without separating shorter intervals. The original maps are not uniform as far as the content is concerned and this fact introduces further inhomogeneities into the final draft. Some countries have published simple "zoning maps" (e. g. Turkey, Cyprus, Georgia, Rumania, etc.) which, in fact, correspond to ordinary maximum intensity maps.

Unfortunately, for some regions the intensity or "seismic zoning" maps were completely missing, e. g. for Great Britain, Belgium, the Netherlands, Fennoscandia, Libya, Egypt, Lebanon and Syria, inadequate, but applicable maps were available for Yugoslavia, Albania and Turkey. In these regions it was necessary to assume some intensity distribution on the basis of the epicentre maps and of some isoseismal maps of individual shocks. Consequently, the isolines of maximum intensities are not constructed for each intensity grade in some parts of the area one isoline includes two intensity grades. The contours of the lines are adjusted to the scale of the map so that the details are smoothed. Extrapolation was avoided as much as possible in order to obtain isolines reflecting the observations of maximum intensities. An attempt was made, however, to transform the map of maximum felt intensities into a simple seismic zoning map with the zones of intensities VII, VIII and $\geq$ IX which are of practical interest. It is impossible to separate the highest intensities X and XI within the third zone because of the relatively small areas affected in history by such intensities and because of the possible influences of local subsoil conditions. The large faults were taken into account in extending the zones, but no simple procedure can be elaborated. Each step depends to a large extent on personal judgement and opinion and has many weak points. The lines defining the three intensity zones are drawn separately in the map of maximum observed intensities and the degree of generalization can be easily compared.

For a practical application of the map with both systems of isolines the number of different intensity observations or at least the frequency of earthquake occurence should be known. Hence, it is recommended to combine this map with epicentre maps, magnitude-frequency relations and statistical estimations of the occurence of largest shocks, the last values being valid, however, for regions which are mostly too large for solving practical tasks.

The following maximum intensity or seismic zoning maps, available in 1966, were used:

Albania — (see also Yugoslavia); C. Morelli: Carta sismica dell'Albania, R. Acad. d'Italia, Com. It. di St. per i probl. del soccorso alle popul., Vol. X, 1942; C. Morelli: I terremoti in Albania, ibid., Fasc. II;

J. Mihailović: Mouvements séismiques épiro-albanais, UGGI, Com.

Nat. du Royaume des Serbes, Croates et Slovènes, Sec. Séism., S. B., Monogr. et Tr. Sc., F. 1, Beograd 1927.

Algeria – J. P. Rothé, A. Grandjean: Maximum intensity map of Algeria and Tunis, 1963.

Austria – M. Toperczer, E. Trapp: Karte der seismischen Zonen in Österreich, 1951.

Belgium – no general map available, some information in J. M. Van Gils: La séismicité de la Belgique, Atlas de Belgique, commentaire de la planche 10, Ac. Royale de Belgique, Bruxelles 1956.

Bulgaria – K. Kirov, E. Grigorova: Karta na maximalnite izoseisti za Blgariya, Seyzmichno rayonirovaniye na Blgariya, Izvestiya na BAN, ser. fiz., VI (1956),

I. Petkov, E. Grigorova: Carte des zones séismiques en Bulgarie, in Les épicentres en Bulgarie... etc., Proceed. of the 7th meeting of the ESC in Jena, Veröff. d. Inst. f. Bodendynamik u. Erdbebenforschung in Jena, p. 38, Heft 77, Berlin 1964.

Czechoslovakia – V. Kárník: Seismic maps of Czechoslovakia, National Atlas of ČSSR, Praha 1967.

V. Kárník: Neue seismische Karten der Tschechoslowakei. Travaux de l'Inst. Géophys. de l'Ac. Tchécosl. Sc. No. 88, 1958.

Cyprus – N. N. Ambraseys: Map of seismic regionalization of Cyprus, 1962, mimeographed.

Danemark – 0

Egypt – 0

Finland – Fennoscandia. N. I. Nikolaev: O svyazi seysmichnosti Baltiyskogo shchita i norvezhskikh Kaledonid s neotektonikoy, Vestnik Mosk. univ., ser. geol., IV, No. 3, 1966, 20–36.

France – J. P. Rothé: Seismicity maps of France, Proc. 3^d World Conf. on Earthq. Eng., 1965; also in [112].

Germany – W. Sponheuer: Untersuchung zur Seismizität von Deutschland, Veröff. d. Inst. f. Bodendyn. u. Erdbebenf., H. 73, 1962.

Great Britain – 0

Greece – A. Galanopoulos: Isolines of max. intensity felt in Greece over the period 1800–1960, Athens, 1965.

Holland – 0

Hungary – E. Bisztricsány, D. Csomor and Z. Kiss: Earthquake zones in Hungary, Ann. Univ. Sc. Budap. de R. Eötvös nomin., Sec. Geol., T. IV, Budapest 1961, 35–38.

Iceland — 0

Ireland — 0

Israel — E. Arieh: Draft map of seismic zones in Israel, 1963.

Italy — M. de Panfilis, F. Peronaci: Carta sismica d'Italia relativa at periode 1900
to 1960, 1 : 1 000 000, mimeographed, 1966, Roma.
M. Baratta: Abbozzo di carta sismica d'Italia, 1934.

Lebanon — 0

Libya — 0

Morocco — J. Debrach-: Maximum intensity map of Morocco, 1963.
J. P. Rothé: Le séisme d'Agadir et la séismicité du Maroc, 1962.
J. Debrach: Carte des intensités maximales au Maroc, Inst. Sc. Chérifien,
1962.

Norway — 0

Poland — 0

Portugal — H. Amorim Ferreira: Regioes sismicas de Portugal continental, Servico
Met. Nac., RT 682, Geo 32, 1962.

Rumania — G. Petrescu, C. Radu: Carte du rayonnement séismique, 1960.
G. Petrescu, P. Ionescu-Andrei, C. Radu: Asupra problemei standardului
de stat STAS 2923-52, Studii si cerc. de astr. si seism., No. 2, VII (1962).

Spain — J. M. Munuera: Estudio previo para efectuar el cálculo de construcciones
sismo — resistentes, Ist. Geogr. y Cat., Serv. di Sism., 1962.
A. Rey Pastor: Mapa sismotectónico de la Península Ibérica (1800 to
1951).
J. M. Munuera: Mapa seismico español de intensidad probable, Mem.
Ist. Geogr. y Cat. XXXIV/II, Madrid 1965.

Sweden — 0

Switzerland — F. Gassmann: Über die Gefährdung von Bauwerken durch Erdbeben
in der Schweiz, Jahresber. Schweiz. Erdbebendienstes, 1962, 29 — 36.

Syria — 0

Tunis — (see also Algeria); N. N. Ambraseys: The seismicity of Tunis, Annali di
Geof., XV (1962), No. 2 — 3.

Turkey — T. C. Imar Iskan Bakanligi Türkiye Deprem Bölgeleri, 1963;
I. Ketin: Türkiye nin sismotektonik haritasi, 1962.

N. Pinar: Carte montrant les rélations entre structures tectoniques et répartition des centres séismiques en Turquie, 1960.

S. Omote, M. Ipek: Türkiyenin sismisitesi, Technical Univ., Istanbul, No. 19, 1964.

USSR — Stroitelnye normy i pravila, SN i P II-A, 12−62, Moskva 1963.

E. I. Byus, A. D. Tskhakaya: Seysmologischeskie osnovy seysmorayoniro-vaniya Kavkaza, Byul. Sovieta po seysmologii, No 8, AN SSSR, 1960.

S. V. Medvedev: Opyt novogo rayonirovaniya Moldavskoy SSR po zonam seysmicheskoy aktivnosti, Trudy Geofiz. inst. AN SSSR, No. 5 (132), 1949.

Atlas zemletryaseniy v SSSR, Soviet po seysmologii AN SSSR, 1962.

S. V. Evseev: Deyaki zauvazhennya pro seysmichnist Skhidnikh Karpat, AN SSSR, Geol. zhurnal, XX (1960), vyp. 2, 40−50.

Yugoslavia — R. L. Nedeljković: Carte séismologique de Yugoslavie, Trav. de l'Inst. Séismol. de Beograd, 1950.

4.3.   ENERGY RELEASE MAP

The total seismic energy released by earthquakes is an objective measure of seismic activity or efficiency of a focal zone. Table 2 (or Table 10 of the first volume) contains the sums of energy calculated for all investigated regions [1]. However, the table cannot give information about the spatial distribution of the released seismic energy. Therefore, the values ΣE_n (1901−1955) were calculated also within each unit area of the size $0.5°$ long. $\times$ $0.5°$ lat. and reduced to the surface of $1000 \, km^2$ and to the period of one year. The size of the unit area roughly corresponds to the mean error in epicentre location. The resulting value $\log \Sigma E_n$ was attributed to the mean position of all foci involved within the unit area. The sums of energy were calculated separately for shallow foci ($h = 50 \, km$) and intermediate and deep foci ($h > 50 \, km$) respectively, according to the conversion formulae $\log E = 1.5M +$ $+ 11.4$, $\log E = 2.4 \, m + 5.8$.

It is possible to draw isolines of equal values of $\log \Sigma E_n$ in some parts of the map. The values in unit areas are discrete points, they represent no continuous distribution and thus the interpolation becomes difficult and somewhat questionable. The isolines are constructed only in order to facilitate a survey of the whole area. In addition to the values of $\log \Sigma E_n$ the largest magnitude corresponding to the period 1901−1955 is given in each unit area, too, because identical values of energy can originate as the cumulative effect of many weak shocks or as a result of one stronger shock.

The map gives clear evidence of the decisive role of the strongest shocks in the energy output of a region. Tens of shocks by two magnitude classes lower than the largest shock cannot in any way influence the total sum.

Separate section maps with the sums of energy corresponding to upper mantle shocks are added. Although the number of foci is very small the total energy release is larger than in the crust. In interpreting this result we must not neglect the unknown error in the mutual adherence of both magnitude energy formulae, valid for M and for m, respectively.

4.4.   SEISMIC BELTS, REGIONAL SURVEY

Earthquake epicentres are frequently arranged into narrow belts indicating active lines. These belts can also be traced by following the shape of the isolines of maximum intensities. Seismic maps in general provide the best opportunity for delineating the belts according to the geographical distribution of seismicity. The seismogenetic lines are usually associated with mountain chains, rift zones, boundaries marking differentiating movements, etc. Some belts or focal lines, however, intersect the known geologic structures. The correlation between seismic and tectonic phenomena raises numerous problems which will not be discussed here, they are left to experts in seismotectonics working on the seismotectonic map of the European area under the leadership of Professor V. V. Beloussov and to other specialists who will use the seismological data for such purposes.

The longest and most active focal line crosses Turkey and coincides with the North Anatolian fault system; it extends possibly towards Greece and the Ionian Islands as assumed by Galanopoulos [17, 24]; it seems also that its eastern termination divides into three branches. At 39°N, 40°E it is joined by another seismogenetic line, very well marked by epicentres but not so well by geological evidence; this line extends to the Gulf of Iskanderun where one branch continues to Cyprus and the other one southward to the Jordan rift zone. An active structural arc, the Cretan furrow, surrounds the Aegean area, beginning at the northern Ionian Island of Kérkira, following the west coast of Greece to Crete and then bending to Ródhos; in this belt volcanic activity was observed as well as shocks down to 200 km. The lines of epicentres delineate other major belts, e. g. along the Adriatic coast, or from NE Italy through the valleys of the Mur (E. Alps) and the Váh (W. Carpathians), along the chain of the Apennines to Calabria and Sicily, with secondary branches on both sides. The European area involves also a small fraction of the North Atlantic Ridge (regions Nos. 1 and 2) which belongs to the world-wide system of oceanic ridges. It is interconnected near the Azores by a very active belt in a parallel direction with the Strait of Gibraltar.

We can find several smaller belts throughout Europe, e. g. in the Caucasus (No. 32), in the Vardar or Struma valleys (regions Nos. 23 and 24), in Albania (No.

25), two arcuate belts in the W. Alps, a system of seismogenetic lines in Algeria, lines following the chains of the Atlas in NW Africa or the lines bordering the Iberian Peninsula, etc.

The European area as a whole has already been discussed by Gutenberg and Richter as part of a much larger unit, the major Alpide belt. This discussion will not be repeated here and the reader should refer to the famous book by the two authors [23]. Hence only some complementary results based on additional, more complete information will be reviewed in the following text. In the regional discussion we follow the numbering of regions, sometimes grouped into larger seismic provinces. The commentary to individual seismic regions is illustrated by summation graphs of $E^{1/2}$ which represent the time sequence of shocks with $M \geqq 5·2$ (Fig.4). Some of them show a monotonous release of the accumulated stress, the slope of the curves can be taken as constant with the possibility of a realistic forecasting of future activity. The second type is characterized by one outburst of seismic energy preceded and followed by a period of decreased activity. The width of the stripe limited by the maximum and minimum values of the curve varies and is proportional to the activity of a region. This conclusion can be simply proved by comparing the cumulative graphs and e. g. the magnitude-frequency relations. This slope of the strain release curves can be employed as a suitable characteristic of earthquake activity as demonstrated by Galanopoulos [27]. The deviations from a mean trend are very sensitive indications of the earthquake generating process as shown by Tsuboi [28].

The most recent publications on regional seismicity which provide more details on the specific problems of the corresponding region are given in "References". It must be pointed out that the list does not represent a complete bibliography; further references of older literature may be found in the listed publications. National or regional catalogues of basic seismological parameters are quoted in the lists attached to the catalogues 1901–1955 (Part I) and 1801–1900 (Part II), respectively.

The compilation of a complete seismological bibliography of a country represents a special task and one additional volume would be necessary for the European area; a good example of the extent of such a compilation is given by the "Bibliographie séismologique de Yugoslavie 1667–1947" by M. Uzelac (Publ. de l'Inst. Séismol. de Beograd, 1948) [147]. Finally, it must be noted that the following commentary on individual regions relates mainly to the 20th century (1901–1967).

Regions Nos 1 and 2 (Jan Mayen, Iceland)

The two regions form only a minor part of the Atlantic belt. The activity is associated with the Mid-Atlantic Ridge which crosses also Iceland and the Jan Mayen Isle. The close association of the ridge and the earthquake foci has been very well demonstrated during the last decade when a denser network of seismological stations and improved methods of epicentre determination decreased the inaccuracy of epicentre locations (see also the epicentre maps of USCGS or BCIS for every year or the map of Linden for the Arctic area [52]). The largest shocks during 1901 – 1955 occurred in region No 1 on April 8, 1922, $M = 6.4$, Oct. 10. 1923, $M = 6.5$, June 6, 1951, $M = 6.3$. In region No. 2 on Jan. 22, 1910, $M = 7.3$, May 6, 1912, $M = 7.0$, March 28, 1963, $M = 6.9$ [52, 54, 55]; the shocks are bound to southern and northern shores or to the Hekla region (volcanic?).

Region No. 3 (Fennoscandia)

This region includes the Baltic shield which is characterized by a low seismicity. Most epicentres are scattered along the Norwegian coast and in the Oslo Väner region. Some minor activity is observed from the Bothnian Gulf to Karelia (particularly near Kuusamo). Båth [56] considers the postglacial uplift as the cause of the Scandinavian earthquakes, but Kvale [57] stresses the tectonic origin of Scandinavian shocks and presents arguments for the relation of epicentres to fault lines of different ages. Recently, Nikolaev [60] has also given evidence of the primary importance of tectonic movements in the Baltic shield seismicity.

The strongest shocks of the 20th century: Oct. 23, 1904, Oslofjord, $M = 6.4$, $I_0 = \text{VIII}$; June 10, 1929, Norwegian Sea, $M = 6.0$ [53, 56 – 68].

Regions Nos. 4 and 5 (Russian platform)

These regions have the lowest seismicity within the whole area. Some weak shocks were reported on the Russian platform, their macroseismic intensities, however, did not exceed $I_0 = \text{VI}$ so that these regions appear as aseismic in our epicentre maps. There are only a few isolated shocks with $M < 5$ in the Ural Mts. [33].

Region No. 6 (Bohemian Massive)

The earthquakes occurring in this old Hercynian block are rare and have not exceeded $I_0 = \text{VII}$, $M \leq 5$ in the 20th century. The strongest shock ever reported was that on March 6, 1872, near Gera with $I_0 = \text{VIII}$. The epicentres follow the mountain chains in the northern (foci near Trutnov – Náchod and Opava) and western parts of the Bohemian Mass (Vogtland, Leipziger Bucht). Weak shocks of $I_{max} = \text{VI} - \text{VII}$ ($M_{max} = 4\frac{3}{4}$) occur in swarms between Kraslice and Aš in the western part of the Krušné Hory mountains (Erzgebirge). The duration of the swarms is between several days and several months [69 – 78].

Region No. 7 (Schwäbische Alb)

It comprises the region between the Main and the Danube. Earthquake activity is concentrated in the Hohenzollerngraben. The region is known for the strongest earthquake on Nov. 16, 1911, $M = 6.1$ which was studied in great detail by Gutenberg, Sieberg and Lais. The activity since then has not exceeded $M = 5\frac{1}{2} \pm$. Minor activity has been observed in the region of Reis [76 – 78].

Region No. 8 (Rhinegraben)

The epicentres are more associated with the axis of the graben than with its bordering faults. In the southern part there is some indication of another seismic line oriented in the SW–NE direction. The level of seismicity has been rather low, $M < 5\frac{1}{4}$ in the 20th century, but a few destructive shocks occurred during the historical era, e. g. that of Basel in 1356. This earthquake remains as an isolated event in the seismic history of the region and seismologists face an unsolved question of the possibility of recurrence of destructive shocks. In the northern part, in Odenwald, damaging swarms of shocks have taken place; the greatest period was that of 1871–1875 near Gross-Gerau [76–80].

Region No. 9 (Rhineland, the Netherlands, Belgium)

The seismicity of this region is relatively low, the strongest shock of $I_0 = VIII$ ($M = 5\cdot4$) originated on March 14, 1951, near Euskirchen. This zone of the Ardennes-Rheinland is the most active in the whole region. Ritsema (1966) concludes that the larger shocks are not clearly tied to the areas of maximum recent movements and that earthquakes originate when the continuous movement, which is the primary phenomenon in the region, becomes stagnant. Series of shocks are associated with the belt between Charleroi and Mons in Belgium. The foci are evidently very shallow with regard to very limited shaken areas, $I_{max} = VII$; it is a region of a heavy subsidence and of coal mines; this fact reveals the possibility of the origin of rockbursts in the abandoned mines. Charlier (1951) and Van Gils (1956) give three epicentral zones in Belgium: a) the Brabant anticlinorium, b) "Sillon houiller du Hainaut" and c) the province of Liège [76–78, 81–88].

Region No. 10 (Great Britain and Ireland)

Most shocks are known in Scotland where the association of epicentres with tectonic lines is evident. The strongest shock of the whole region, however, was located outside the British Isles in the North Sea, June 7, 1931 ($M = 6\cdot0$). Strong shocks occurred also in Wales and England. Isoseismal maps published by Davison show relatively large shaken areas giving evidence of foci at the bottom of the crust. Near Comrie in Scotland also swarms of minor shocks were observed in 1788, 1801, 1839–1846, 1924–1949; the swarm shocks are, however, of very shallow origin.

Ireland is practically without noticeable earthquakes and is only seldom shaken from foci in the Irish Sea.

According to Tillotson, the most prominent seismological line includes Inverness, Comrie, Carnarvon and Colchester epicentral zones. Another line may be formed by the centres at Ullapool, Lancashire, Herefordshire and the Midland counties, another one is represented by Glasgow, Lancashire, S. Wales, and SW England; there is the possibility of a line including the centres at Carlisle, Mansfield, Malvern, Pembroke and Camelford. There are tracts of land that are almost aseismic (e. g. N. Scotland, NE England); on the other hand there exist records of 10 earthquakes of intensity VIII to IX having occurred since 1750; probably the greatest of the British Isles shocks occurred on 22 April 1884 near Colchester ($I_0 =$ = IX R. F.) [89, 90].

Region No. 11 (Central and Western France)

The epicentres are probably associated with large fractures of the Hercynian structures in Bretagne, Vendée, Limousin and Limagne, some damage is reported during historical time, the largest shock in the 20th century was that on July 30, 1926, $M = 5.6, I_0 =$ VIII on the western coast of Normandy. Most epicentres are disposed along the coastal lines of Normandy and Bretagne and are associated with the continental slope [91, 92, 112, 114].

Region No. 12 (W. Spain, Portugal, Atlantic to 20°W)

This region is complex both in structure and in the pattern of seismic activity; therefore a subdivision was made with the meridian 10°W as the line of separation. The western part includes a belt of high seismicity stretching from the SW edge of the Iberian peninsula to the Azores. A linear distribution of epicentres farther to the east indicates a possibility of an extension of this active belt as far as Algeria. The continuation intersects known geologic structures of the Western Mediterranean. This belt deserves more attention, because the strongest shock in the North Atlantic occurred there on Nov. 25, 1941, $M = 8$. It was probably the seat of the destructive Lisbon earthquake on Nov. 1, 1755. Minor seismic activity is associated with the coastal area of Portugal. The total balance of the coastal zone is determined mainly by the shock on April 23, 1909 ($M = 6.6, I_0 = X$) and its aftershocks.

The E−W belt is also responsible for severe shocks in the south of Portugal (province of Algarve) which are known from historical records. The largest destruction occured in 1722, the following years were quieter, the strongest shock in the 20th century was that on April 1, 1918, $M = 5.8$ (coast of Algarve) or on March 15, 1964, $M = 6.8$ (Gulf of Cádiz). The central part of the Iberian peninsula, the Meseta Central, is aseismic [93].

Region No. 13 (Morocco, Canary Islands)

Earthquakes in the western part of this region (Canary Islands) are of volcanic origin with a very limited shaken area. As a matter of fact, also tectonic shocks on the continent have very shallow foci, e. g. that of Agadir on Feb. 29, 1960, though destructive it reached only a magnitude of $M = 5.8$. The Agadir earthquake has been thoroughly investigated by Rothé, Debrach and others who have published several papers discussing also the problem of the seismicity of Morocco. Medium seismicity

$(I_0 \leqq \text{VIII}, M \leqq 5\frac{1}{2})$ is associated with the chains of the Atlas mountains and some epicentres are also scattered in the sea, not only in the prolongation of the Atlas, and in the coastal area of Morocco [37, 94–96].

Regions Nos. 14 and 15 (*Algeria, Tunisia*)

The seismic activity is concentrated in the coastal areas and the epicentres are associated with the structural features of the Atlas mountains extending from Agadir to the Gulf of Gabès. The mountains are divided into two main chains separated by high plateaux. For the two extreme ends of the Atlas mountains a long "periodicity" of damaging shocks seems to be typical, i. e. in the regions of Agadir $(M_{max} = 6)$ and Tunis (Zaghouan fault, $M_{max} = 5\frac{1}{2}$), respectively.

There are about three zones of destructive shocks in Algeria. The first can be delineated by the towns Oran – Mascara – Rélizane, the second extends from the Massif de Dahra to the Mts. of Hodna and Aurès, and the third corresponds to the line Kerrata – Constantine – Guelma. All three zones were shaken by destructive earthquakes, the heaviest ones being on June 24, 1910, $M = 6\cdot6$, $I_0 = $ X (Bibans) and on Sept. 9, 1954, $M = 6\cdot5$, $I_0 = $ X (Orléansville).

The Saharian Atlas displays a relatively low seismicity, $M_{max} = 5\cdot2$ (Feb. 11, 1928). Region No. 15 includes also the Baleares Islands which probably belong to the continuation of the "Oválo Bético – Rifeňo" loop; the seismicity of the Baleares is very low and only the northern part of the largest island Mallorca has been shaken by local shocks of the intensity of about $I_0 = $ VII [37, 97–104].

Region No. 16 (*Oválo Bético – Rifeňo*)

Shallow shocks occur in a belt forming a loop which is supposed to be a continuation of a system of similar loops farther to the north (the Pyreness) and north-east (the Alps). This region aroused the interest of seismologists in 1954 when the deepest shock ever recorded outside the Pacific area occurred below Sierra Nevada at a depth of 655 km $(m = 7\cdot0)$. The activity is concentrated in several focal zones. In S. Spain such zones are in Granada (Sierra Nevada) and along the line Murcia – Alicante, with destructive shocks on Dec. 25, 1884, $I_0 = $ XI and on June 16, 1910, $M = 6\cdot3$, in the first zone; the strongest shock in the second zone was on March 21, 1911, $M = 5\cdot5$. A parallel line of minor activity appears in the valley of the Guadalquivir. The seismicity of this belt has been torroughly studied by J. Bonelli Rubio, J. Martin Romero, A. Rey Pastor, L. Chacon Alonso, and others [105–109].

Another active earthquake zone is associated with the Er Rif mountains on the African coast. The 20th century shocks were damaging along the Mediterranean coast S of Tetuán (Jan. 21, 1909, $I_0 = $ IX, Aug. 9, 1930, $M = 5\cdot2$, $I_0 = $ VIII) and at Melilla (e. g. Sept. 8, 1927, $M = 5\cdot0$, $I_0 = $ VII, May 12, 1952, $M = 5\cdot2$) and Alhucemas (Oct. 11, 1926, $M = 5\cdot6$, $I_0 = $ VIII). Historical records speak about destructions caused by earthquakes near Tangier and Melilla [94, 96, 110, 111].

Region No. 17 (the Pyrenees, Catalonia)

The epicentres of the largest shocks of the region are associated with the central part of the Pyrenees; according to Rothé they are located immediately north of the contact between the primary axis of the Pyrenees and the secondary structures of the foreland [112].

The recent damaging shock on Aug. 13, 1967, $I_0 = $ VIII—IX, $M = 5.7$, coincides with the most active zone of the Pyrenees. The strongest shocks of the 20th century were slightly stronger: July 13, 1904, $M = 5.8$, July 10, 1923, $M = 5.9$. A zone of minor activity can be traced along the chain of the Iberian mountains [93, 110—112].

Region No. 18 (the Alps)

The seismicity of the Western Alps was thoroughly studied by Rothé et al. He divides the epicentres into three distinct focal lines of an arcuate form. The eastern, called "arc piémontais", is situated in Italy and follows the inner chains of the Alps towards Lago Maggiore. The most active central line, "arc briançonnais", coincides with the crystalline core of the Central Alps from the Ligurian coast to Mont Blanc and the Simplon Pass. This zone is separated by an aseismic nucleus from the western zone which is not well developed and is determined by the positions of epicentres along the western side of the Alps from Marseille to Genève. Destructive earthquakes were located in the southern and northern parts of the central zone on April 29, 1905, $I_0 = $ IX, $M = 5.7$, in the valley of Chamonix; on June 11, 1909, $M = 6.2$, $I_0 = $ IX, in Provence; on Jan. 25, 1946, $M = 6.1$, $I_0 = $ VIII, Valais. The strongest shocks in the 19th century: Haut Valais, July 20, 1855, $I_0 = $ VIII—IX; Liguria, Feb. 23, 1887, $I_0 = $ X. Some minor shocks occur southward from the coast indicating a continuation of the seismic zone off the continent. These epicentres were mostly determined by the ISS and their position is uncertain; but the reality of the prolongation of the active zone off the coast was confirmed by a heavy shock on July 19, 1963, $M = 6.2$ (43.4°N, 8.2°E).

In the Eastern Alps there is a very sharply defined seismogenetic fault extending from the "Karnische Alpen" through the valleys of the Mur and the Leitha northeast to the West Carpathians. The level of activity decreases from the SW end to NE. Very active foci of Carnia-Friuli (NE Italy) are located in the SW prolongation of this line but this extension intersects superficial geological structures. The strongest shocks occurred on March 27, 1928, $M = 5.8$, Oct. 18, 1936, $M = 5.6$ in NE Italy. The famous Villach earthquake of 1348 ($I_0 = $ X?) also belongs to this zone. In the NE part $M_{max} = 5\frac{1}{2}$. A stable zone of minor but concentrated activity appears in Tirol near Innsbruck ($M_{max} = 5.3$, Oct. 7, 1930). The zones of increased activity correspond to the borders between the Central Alps and the "Kalkalpen" [92, 112—119].

Regions Nos. 19 and 20 (Italy and Sicily)

The main activity follows the Apennines which develop in the south into an arcuate structure resembling the Pacific type in the following features: An oceanic deep of the Tyrrhenian Sea with foci down to 450 km displayed on a steeply inclined conical surface (apex approx. at 40°N, 12°E, $h = 700$ km, dip 60°), shallow earthquakes forming the arc and active or extinct volcanoes along the Calabrian arc, however, some of them are also south of Sicily. The seismicity of the Tyrrhenian Sea has been investigated in detail by Peterschmitt [43]. There is a remarkable apparent coincidence in the periods of the upper mantle activity in the Tyrrhenian Sea and in the E. Carpathians though this phenomenon is difficult to explain.

The strongest shallow shocks in S. Italy occurred on Sept. 8, 1905, ($M = 7·3$), and Dec. 28, 1908, ($M = 7·0$) in Calabria and in the Strait of Messina. The strongest deep shock: April 13, 1938, $m = 7·1$, $h = 275$ km; the deepest shock: Feb. 17, 1955, $m = 5·6$, $h = 450$ km. We can observe a migration of epicentres after the two earthquakes in 1905 and 1908; they were succeeded by weaker shocks in region No. 19 with epicentres shifted successively to the north in the following sequence: Jan. 13, 1915, $M = 6·8$; May 17, 1916, $M = 5·8$; Aug. 16, 1916, $M = 5·8$; June 19, 1919, $M = 6·2$; Sept. 7, 1920, $M = 6·3$. A similar succession of five strong shocks in Calabria during 1783−1785 has been noted by Davison [171].

The strongest shocks in the Apennines occured on Jan. 13, 1915, $M = 6·8$ ($I_0 = XI$) near Avezzano, July 23, 1930, $M = 6·5$, Sept. 7, 1920, $M = 6·3$, June 29, 1919, $M = 6·2$. The main belt of the Apennines branches on both sides into many short zones with somewhat weaker but damaging earthquakes. A number of shocks appear in the NW part of the Adriatic Sea and at the adjacent coast between Ancona and Rimini. In this zone an exceptional earthquake occurred on Nov. 30, 1934, the first arrivals at all surrounding stations were of a compressional type as described by Caloi. Another active zone extends from the central belt eastwards to the Monte Gargano.

The region of Etna is characterized by swarms of volcanic tremors. Some of them were destructive, the largest one occurred on May 8, 1914, $M = 4·9$; their pleistoseismal area is always very small.

Earthquakes of very shallow origin are frequent near Velletri (SE Rome) and swarms originate also in the region of Alba [41−44, 120−122].

Region No. 21 (the Carpathians)

This region involves the Carpathian province, i. e. the East Carpathians, the West Carpathians, the Hungarian Basin and the Banat. In general most epicentres follow the Carpathian chains or are associated with young volcanic mountains on the inner side of the Carpathians. Damaging shocks originated also in the subsiding basins of Hungary and S. Slovakia (e. g. SE of Budapest on July 8, 1911, $M = 5·6$,

$I_0 = IX$; Jan. 12, 1956, $M = 5\cdot4$, $I_0 = VII$; E of Budapest in 1868, $I_0 = IX$; near Komárno 1763, $I_0 = IX$).

Region No. 21 bears evidence of the insufficient length of the period of 55 years for a full manifestation of the earthquake generating forces. Several zones, known for the damaging shocks in the 19th or 18th centuries, show very low activity $(M \leq 4)$ in the 20th century. A similar situation exists in the other marginal regions of the Alpide belt in the European area.

Zátopek found weak, very shallow foci earthquakes as characteristic for Ruthenia; they are caused by the movements of small blocks the boundaries of which influence the shape and size of the isoseismals [129].

The known focal zones are: 21a — the Little Carpathians ($M_{max} = 5\cdot7$, Jan. 9, 1906), West Beskydy, East Beskydy ($M_{max} = 5\cdot2$, May 26, 1914), Mukachevo—Rakhovo ($M_{max} = 4\cdot7$, Jan. 5, 1908), 21b — the line Bakony—Nógrad—Hegyalja ($M_{max} = 5\cdot6$ on July 8, 1911), Komárno ($I_{max} = IX$, June 28, 1763), 21c — region of Timişoara ($M_{max} = 5\cdot8$, April 2, 1901), 21d — a remarkably stable focus lies below the sharp bend of the E. Carpathians (Vrancea $45\frac{3}{4}°N$, $25\frac{3}{4}°E$) at a depth between 100 km and 150 km. This focal zone is being compared with a similar zone in the Hindukush ($h = 220$ km). It has been studied by many Rumanian seismologists who have published numerous papers on the mechanism, magnitude, frequency, deep structure and other phenomena relating to this remarkable focus. It seems likely that the foci are associated with a very steep surface inclined below the bend of the Carpathians. The energy of shallow foci is negligible compared with the intermediate ones. Shallow shocks originate a) along the outer boundary of the Carpathians (line Rimnicu Sarat—Bacau), b) near Campulung, $M_{max} = 6\cdot4$ on Jan. 26, 1916, c) in Transylvania (uncertain, dispersed).

The heaviest intermediate-depth shocks occurred on Nov. 10, 1940, $m = 7\cdot3$, $h = 130$ km, March 29, 1934, $m = 6\cdot9$, $h = 150$ km, and Oct. 6, 1908, $m = 6\cdot8$, $h = 150$ km. Petrescu and Radu (1962) mention the possibility of a systematic vertical migration of foci in the Vrancea zone [123—143].

Region No. 22 (Yugoslavia without Macedonia)

The main geological structures are of the NW-SE direction and the pattern of earthquake activity follows roughly the same direction. The region may be divided into two belts, one associated with the coast and the other involving northern and central Yugoslavia. The belts correspond rouhgly to the Outer Dinaridy and Central Dinaridy, respectively. The first zone 22a is characterized by frequent strong earthquakes along the coast between Dubrovnik and Zadar ($M_{max} = 6\cdot2$ on March 15, 1923, $M = 6\cdot0$, Jan. 7 and 11, 1962, Makarska). Further subdivision within zone 22a creates focal zones a) between Dubrovnik and Split (foci Makarska, Imotski, Mostar, Pagoda), b) Sarajevo—Zenica (shallow focus at Treskavica), c) Zvornik (upper valley of the Drina). In zone 22a approximately every 20 years a shock of

I_0 = IX occurs. In the interior of the country the earthquakes are not so frequent as at the coast but their maximum magnitudes are about the same. At least three focal zones can be found in the zone 22b, one between Zagreb and Ljubljana (upper valley of the Sava), one near Banja Luka – Brod and the other south of Belgrade (Shumadia). Zone 22b is known for the destructive Zagreb earthquake on Nov. 9, 1880 (I_0 = X?) [144–148].

Region No. 23 (Bulgaria)

In keeping with the subdivision introduced by the Bulgarian seismologists there are the following focal zones in region No. 23:

1. Sabla – Kaliakra – Kolarovgrad – Trnovo with two stable foci: Sabla (March 31, 1901, I_0 = X) and another near Trnovo (June 14, 1913, I_0 = X, M = 6·8); the region between these two centres has a minor seismicity.

2. Yambol – Topolovgrad (Feb. 15, 1909, M = 5·9).

3. Maritsa valley (Plovdiv – Dimitrovgrad), the most active zone with two destructive shocks in April 1928 (M = 6·8 and 7·0) accompanied by vertical movements up to 3 m in the pleistoseismal area. This subsidence zone corresponds to the system of grabens of Plovdiv and Marbas. The focus near Edirne (W. Turkey) may be aligned to this zone.

4. Rodopy Mountains (Chepelare – Kurdzhali), M_{max} = 5± [149–161].

Region No. 24 (Macedonia, Struma valley, Khalkidhiki)

This is a complex region and not well delineated as evidenced by discrepancies which were found later in empirical relations between M and I. The magnitude-intensity relation of earthquakes in the Vardar valley and westwards of it resembles more the earthquakes in region No. 25; on the other hand, the Struma valley may belong to No. 23 and the shocks in Khalkidike are more associated with the activity of the belt extending from the Marmara Sea to Central Greece. From the geological point of view, too, it seems likely that the central and western parts of Macedonia belong to the Dinaride zone and the eastern part to the Rodopy Mass. The Vardar zone became well known by the destructive Skopje earthquake in 1963 (M = 6·3), another strong shock occurred near Valandovo on March 8, 1931, M = 6·7. The Skopje earthquake stimulated an intensive study of this shock by international teams (reports by Zátopek, Ambraseys, Despeyroux, Muto, Okamoto, Hisada, Arsovski, Grujić, Gogjić, Shebalin, Hadžievski and others) as well as a study of the seismic history of the zone.

Much stronger, however, were two shocks on April 4, 1904, M = 7·2 and 7¾ in the frontier region between Macedonia and Bulgaria (Struma zone). The main shock was the strongest shallow shock in continental Europe in the 20th century.

The magnitudes of shocks in the central and western parts of region No. 24 did not surpass the threshold value of $M = 6\frac{1}{2}\pm$ which resembles the character of region No. 25.

Another group of strong earthquakes lies in the prolongation of the North Anatolian fault zone towards the eastern coast of Khálkidhiki, the largest shocks were on Nov. 8, 1905, $M = 7\cdot0$, and Sept. 26, 1932, $M = 6\cdot9$ [159 – 162, 169 – 174].

Region No. 25 (Albania)

According to Morelli (1942) we can distinguish three focal zones in Albania: a) Titograd – Shkodër, b) Durrës – Elbasan – Starovo, c) Vlorë – Gjirokastër – Ioánnina. We should add another zone, Ohrid – Debar – Peshkopi, which is also well marked by epicentres. All four zones are very active and in all of them the macroseismic intensity reached $I_0 = IX – X$. The earthquakes near Vlorë, in the third zone, are characterized by a small shaken area, i. e. by very shallow foci. The largest shock of the region occurred on Feb. 18, 1911, $M = 6\cdot7$, the recent shock near Debar (Nov. 30, 1967) attained $M = 6\cdot4$. In a recent paper Sulstarova and Koçiaj emphasize the existence of a seismogenetic zone Debar-Elbasan-Vlorë [163, 170, 222].

Region No. 26 (Aegean region)

This large and most active seismic region has been divided into three zones 26a (Central Greece – Pelopónnisos), 26b (Crete – Cyclades) and 26c (SW Turkey) according to the seismic energy release pattern. Zone 26a includes focal zones belonging to two belts, one in the northern part extending from the Tríkeri Canal and Thessalía to Levkás where it joins the second zone associated with the Ionian Islands, Messíni and Kíthira. The second belt, the Cretan furrow, has an arcuate form typical of the Pacific island arcs and has some of their characteristic features (volcanoes, gravity anomalies, intermediate depth earthquakes). It involves two large centres of activity in both wings (Kíthira-Ródhos).

The strongest shallow shocks in 26a were on Oct. 6, 1947, $M = 6\cdot9$ (Messíni), April 30, 1954, $M = 6\cdot8$ (Thessalia), Aug. 12, 1953, $M = 6\cdot7$ (Kefallínia). The strongest intermediate-depth shocks: Aug. 30, 1926, $m = 7\cdot4$, $h = 100$ km, July 1, 1927, $m = 7\cdot3$, $h = 120$ km (Kíthira). The largest subcrustal shock in region 26b originated very close to the above mentioned shocks, south of Kíthira, on Aug. 11, 1903, $m = 7\cdot7?$, other strong shocks ($m_{max} = 7\frac{1}{4}$) with foci deeper than normal ($h < 200$ km) are associated with the Cretan furrow. Shallow shocks of this zone are relatively weaker, $M_{max} = 6\frac{1}{2}$. Whereas there is no evident coincidence in the crustal activity within the whole of region No. 26, the upper mantle activity rises simultaneously in zones 26a and 26c respectively (see Fig. 4).

Zone 26c involves the seismic lines of Western Anatolia which form a relatively dense pattern mainly in the E-W direction. Pinar (1951) distinguishes ten epicentral

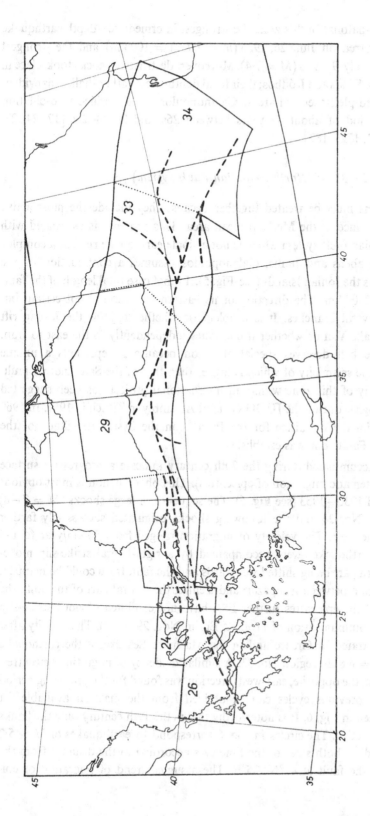

Fig. 5. The boundaries of the North Anatolian Fault system area.

lines in Western Anatolia. In this zone the strongest intermediate – depth earthquake of region 26 occurred on June 26, 1926 ($m = 7.7$, $h = 100$ km) and the strongest shallow shock on July 9, 1956 ($M = 7.4$). Moreover, destructive shocks took place in the islands Lésvos, Khíos and Ródhos; their focal depths were rather shallow as evidenced by very limited pleistoseismal areas. Galanopoulos (1966) verified an oscillation pattern with a period of about 52 years between 26a and 26b + 26c [17, 24, 27, 30, 35, 36, 40 – 42, 172 – 182].

Regions Nos. 27, 29, 33, 34 (North Anatolian Fault system)

These four regions must be treated together because they include the most active seismogenetic province of the Mediterranean area. This province is associated with the North Anatolian fault system which is not a single earthquake rift but a complex of depressions, grabens and faults. Galanopoulos assumes a continuation of the fault zone as far as the Ionian Islands (see Fig. 5). In that case total length of the fault system is about 2500 km. The direction of its eastern extension is not certain but it divides into several branches. It is problematic whether it joins the Jordan rift system west of Lake Van or whether it continues independently to the east to Iran. This seismoactive belt deserves special attention because it represents a unique continental unit the seismicity of which is higher than that of the San Andreas fault. A systematic study of this zone within an international project has been suggested by some seismologists during the IUGG General Assembly in Zürich in 1967. In 1969 the Turkish National Committee for the Project on the Systematic Study of the North Anatolian Fault zone was established.

The data accumulated during the 20th century indicate a progressive surface faulting (or a systematic migration of epicentres). This phenomenon is most obvious during the period 1930 – 1953 (see Fig. 6). The sequence of large shocks ($M = 7 – 8$) started in region No. 34 and the following shocks originated successively farther and farther to the west. The velocity of migration ranged from 50 km/year to 145 km/year. Most earthquakes were accompanied by a right-lateral strike-slip movement, the northern part being shifted westwards and the fault trace could be mapped. However, the region between Adapazari and Yenice, the central part of the fault, the Varto region and the epicentres further ESE have not exhibited obvious strike-slip motion (written communication of Prof. K. Ergin, Nov. 29, 1967). The activity after 1955 was concentrated in Greece and in the Marmara Sea except the earthquake of July 1966 ($M = 6.8$) in region No. 33. A similar tendency of migrating epicentres, but propagating in the opposite, east-west direction was found for the preceding period 1904 – 1930. No previous cycles can be derived from the material available for 1801 – 1900 as seen in Fig. 6. It is noteworthy that in the 19th century only the flanks of the fault were active. The circles in Fig. 6 correspond to earthquakes of $M > 5.2$ and are displaced on both sides of the time axis according to the distance from the central point of the fault at 41°N, 35°E. The assumed trend of migration of epi-

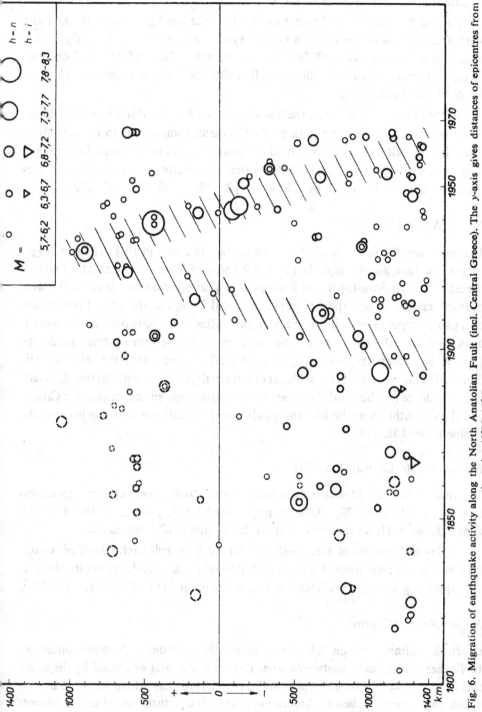

Fig. 6. Migration of earthquake activity along the North Anatolian Fault (incl. Central Greece). The *y*-axis gives distances of epicentres from the point 41°N, 35°E to the east (+) an to the west (−).

centres is marked by dashing. The graph also gives information about the largest magnitudes in different parts of the belt.

Several remarkable shocks were studied by Pinar and Lahn who published also a tectonic analysis of the North Anatolian system. A comprehensive, purely seismological study is, however, still lacking. The strongest shock of $M = 8.0$ occurred in Eastern Anatolia on Dec. 26, 1939. It is also the strongest shallow shock of the whole Mediterranean area.

One of the lines, in which the main fault branches at its eastern end continues in the north-eastern direction towards the Caucasus, transversally to its main structures. In this belt there is a condensed epicentral zone in the volcanic Dzhavakhetskoye nagorye which is remarkable for series of numerous weak shocks with very shallow foci studied in detail by Tskhakaya [30, 34, 35, 40, 183 – 193, 179].

Region No. 28 (Crimea)

The most remarkable series of strong earthquakes occurred in 1928. The two largest shocks, on June 26 and Sept. 11 $(M = 6.0$ and $6.8)$, were accompanied by serious damage between Sevastopol and Feodosiya. The aftershocks and later shocks were of lower magnitude and the foci were distributed along the SE coast of the Crimea on a plane steeply inclined below the peninsula. Minor seismicity outside this zone is associated with the continental slope, too; few epicentres were located inside the Crimea. It is supposed, according to geological evidence, that the seismogenetic fault marks the separation of two blocks of upheaval (peninsula with continental shelf) and subsidence (sea bottom). There are no earthquake epicentres between the Crimea and the Carpathians or the Balkans which would indicate a connection between the Alpine ridges [33, 194].

Region No. 30 (Central Anatolia)

The Inner Anatolian Masses include three distinct focal zones: a) near Afyonkarahisar $(M = 5.9$, Nov. 20, 1924), b) near Kirsehir $(M_{max} = 6.6$, April 19, 1938), c) near Kayseri. Historical sources report destruction in all three zones.

This distribution of foci indicates that the large belt Crete – Gulf of Argolis may extend through western and central Anatolia (see maps). However, there is no supporting geologic evidence for this assumption [184 – 186, 188, 191 – 192].

Region No. 31 (Cyprus)

Cyprus lies within a seismic belt which can be followed from E. Anatolia southwest. Its further continuation westwards from Cyprus is not well evidenced by dispersed foci $(h = n, h = i)$ between Cyprus and Crete. Cyprus has a long recorded history which was recently studied by Ambraseys [195]. The statistics based on old sources shows a frequency of about 180 years for destructive shocks with $I_0 = IX$, the maximum intensity reached exceptionally on the island is $I_0 = X$. The earthquake activity appears to be concentrated along the southern and eastern coast of the island. Historical reports lead to an apparent concentration of seismic activity in the largest centres of culture, i. e. Paphos, Curium, Limassol, Amathus, Arnaka, Famagusta, Salamis. The strongest shocks (Sept. 29, 1918, $M = 6.5$, Sept. 10, 1953, $M = 6.3$) have their epicentres at the southern coast of Cyprus. There is also a notable upper mantle activity W of Cyprus, the magnitudes of these shocks are $m \leq 6\frac{3}{4}$; these epicentres are scattered without any visible regularity, probable due to a low accuracy of location [31, 39, 41, 195].

Region No. 32 (Caucasus)

The main activity of this region is associated with the massif of the Great Caucasus. The seismic belt extends on both sides into the sea. We can observe three zones of elevated seismicity: a) The long belt between Dusheti and Shemakha with highest activity at both ends; it is a contact region of differential vertical movements between the Caucasus and the Alazan-Kura depression. The focal zone of Shemakha was very active in the 19th century, after 1902 only weak shocks were recorded. It seems that the foci are deeper towards the Caspian Sea, a few of them east of the Apsheron peninsula were located at a depth of 150 km, $m_{max} = 6.3$. b) Zone of Elbrus (June 29, 1921, $M = 5.6$). c) Zone of Dzhavakhetskoye nagorye which belongs to region No. 33.

The general distribution of epicentres in Armenia and Georgia forms a belt which intersects the main NW-SE direction of geological structures. This belt, which may be considered as a branch of the North Anatolian fault, stretches from Erzurum to Dzhavakhetskoye nagorye, Gori, Dusheti and Grozny.

There are several smaller zones of seismic activity along the northern slopes of the Great Caucasus or south of Makhachkala.

The magnitudes of shocks in the Caucasus are relatively small, $M < 6\frac{3}{4}$. A detailed study of the seismicity of the Caucasus is published in "Zemletryasenia v SSSR" (Earthquakes in the USSR) by Byus, Tskhakaya, Lebedeva, Levitskaya and others or in a monograph by Kirillova et al. [33, 196−200].

Region No. 35 (SE Anatolia)

This focal zone represents probably the northern continuation of the Jordan rift and the "Syrian Graben" structure. It may join the North Anatolian fault system in a nodal point ($39\frac{1}{2}°N$, $40\frac{1}{2}°E$) and continue towards the Caucasus or extend northeast to the seismic region of Mus−Lake Van. The existence of a focal line is well documented by epicentres. The activity is not equally distributed in time. Historical records located the centre of the activity near Antakya-Iskanderun but the 20th century observation shows a more regular distribution of epicentres [188, 191, 192].

Region Nos. 36, 37 (Levantine countries)

The regions comprise the Jordan Rift Valley, a tectonic complex of pleistocene grabens, with still continuing movements. The evidence of the 20th century data is not sufficient for a realistic investigation into the seismicity and we must deal with earthquakes prior to 1901, too. The primary result suggests that the rift zone itself is not an area of the highest seismicity, epicentres are often associated with its secondary branches. The activity increases from south to north, the maximum energy was released north of the Dead Sea if the shocks after $1900 (M_{max} = 6.0$, July 1927) are considered. Heavy shocks were observed prior to this date between Jerusalem and Jaffa or north of the Lake Tiberias and farther along the coast as far as Tripoli. Some epicentres are scattered in the Syrian desert but their location is very uncertain. The average activity of the region in comparison with that of other regions of the Eastern Mediterranean is rather low [38, 201–205].

Region No. 38 (Egypt)

No shocks with $I_0 \geq$ VI originated in Egypt during 1901–1955. The country was, however, shaken by shocks with epicentres north of the Nile delta, e. g. on Sept. 12, 1955 $(M = 6.1, h = n)$, south of Crete. The coast of the Red Sea formed by the rift zone is less active but destructive shocks were associated with the Nile valley and with the depression of the El Faiyûm according to historical reports. Instrumental records of the Helwan station give evidence of an epicentral zone north of Cairo. According to Madwar and Ismail (1953) 68 earthquakes were registred during 1922–1952 from distances smaller than 330 km; macroseismic intensities reported for eleven of those shocks did not exceed $I_0 =$ V [206, 207].

Region No. 39 (Libya)

Our knowledge of earthquakes in Libya is very limited. The country is without a seismological service and without any seismological station, thus all information must be derived from the instrumental data of more distant stations. The most prominent epicentral zone was located south of Misurata where the strongest shock of $M = 7$ originated on April 19, 1935 with two aftershocks of $M = 6$. Some epicentres are dispersed in the sea. The second focal zone of this region follows the coastal line of Cyrenaica. The effects of a recent shock on Feb. 21, 1963, $M = 5.3$, were described in a special report by K. Minami [208].

5. Earthquake mechanism

The motion observed in the earthquake foci belongs evidently to the basic parameters of seismicity. Ideally, the earthquake catalogues should contain a heading of nodal plane solution for each shock. We shall not discuss the problems of theoretical models, the technique of the construction of fault plane diagrams and difficulties in their interpretation or inconsistencies in the original data. In this field there are many questions still open, nevertheless the results of fault plane solutions may provide us with clues to the question how the earthquakes occur in the European focal zones.

Considering the importance of a critical study of the European earthquake mechanisms the European Seismological Commission established a special Working Group within the Sub-Commission for the Seismicity of the European Area in 1964. Dr A. R. Ritsema, the Reporter of the Working Group, compiled a review of all publications and started with a preliminary evaluation of the published information [209–211, 224, 225]. The Wickens-Hodgson uniform treatment of the world data, including 91 of the European events, contributed substantially to the testing of the consistency and reliability of the results [212].

The total number of European shocks with published nodal plane solutions amounted to 282 in 1966, the number of solutions was 344 [209,212] because for 52 events from one to four solutions were known. Ritsema reviewed the results published up to May 1966*) in the table given below:

EUROPEAN EARTHQUAKE MECHANISMS

	Number of earthquakes	Number of solutions[1]
Total	282	344
Basic data unknown	129	130 (23 ambiguous)
Known basic data	153	214
	10	0
	91	1
	39	2

*) In June 1968 there were already 557 solutions for 305 earthquakes available.

	Number of earthquakes	Number of solutions[1]
Investigated by the Wickens-Hodgson programme	7	3
	6	4
	91	138
	52	1
	31	2
	8	3
Conditions (arbitrary)		
Number of basic data $\geqq 24$	55	
Score $\geqq 80 \cdot 0$	72	
Variability of A, B and C axes $< 30°$	45	

		Dip-slip	Transcurrent
Three conditions fulfilled	25	12	13
Shallow shocks	22	9	13
Deep shocks	3	3	0

[1] A solution is called independent if the position of one of the axes A, B or C differs by more than 30°

According to this table only 25 solutions may be classified as reliable using the three criteria of Ritsema [209]. Three of these shocks are situated in northern Iran, nine in Turkey, eight in Greece, one in southern Spain, one in Central Europe and three in the Atlantic Ocean. The three deep shocks (Spain 660 km, Greece 120 km, Turkey 70 km) display dip-slip fault motions; of the remaining shallow earthquakes thirteen are transcurrent and nine dip-slip. The great majority of the transcurrent motion earthquakes indicate an east-west movement of a dextral kind except two shocks in SW Turkey with a reversed fault motion [209]. For most dip-slip motion earthquakes the stress direction is oriented perpendicular to geological structures and the motion is of the reverse fault motion type.

Vvedenskaya with a group of her coworkers has published many papers on earthquake mechanisms. In a recent summary, the stress field in the earthquake foci of the Earth is discussed [214]. The results are expressed in terms of the directions of principal pressure and principal tension. This would imply that a double couple is assumed as the model. In the majority of world seismic provinces there is a close correlation between the orientation of geologic structures and the direction of the principal axis of pressures or tensions. Mostly the direction of principal pressure is normal to structures and acts horizontally (Pacific, SE Asia, Carpathians), in the rift zones of the Atlantic and East Africa the tension perpendicular to the structures prevails. A confusing mixture of mechanism types has been found in the Mediterranean, explained by Vvedenskaya et al. [214, 215] by the complex tectonic conditions; mostly horizontal tension are observed, however, without any systematic relation to the direction of superficial structural features.

A. R. Ritsema found some characteristic features in the pattern of the Mediterranean fault plane solutions [223]. His conclusions can be summarized as follows:

In North Africa and in the Azores line, between the Straits of Gibraltar and the North Atlantic mid-ocean ridge, most earthquakes are of the transcurrent fault type. The fault motions are dextral more or less along West-East oriented fault planes.

Most of the earthquakes of the Calabrian arc and the Apennines display normal faults with a dip-slip fault motion in an approximately West-East direction.

Most of these Central European shocks are of the transcurrent fault type and the direction of maximal pressure is directed approximately NNW−SSE.

Most of the earthquakes of the Dalmatian coast and Western Greece are of the transcurrent fault type, and show sinistral fault motions roughly along NW−SE striking fault planes.

Most of the shocks of the Aegean zone are of the transcurrent fault type. The general picture is somewhat confused. Papazachos is of the opinion that the acting forces are tensional and approximately parallel to the regional tectonic lines.

In Turkey, the most conspicious feature is the very consistent fault planes in the north of the country. These motions are in agreement with those along the North Anatolian fault zone, as derived from field studies. In the western part of the country and in the southern part of the Aegean Sea the direction of movement seems to deflect towards NE−SW. There is an increase of dip-slip fault motions towards the south. Most of these are of the compressive type and the azimuths of the greatest pressure stress changes from NE−SW in the West to N−S in the East. Three intermediate depth shocks show either a thrusting of the mainland of Greece over the Ionian Sea basin in the SW, and/or an underthrusting of Greece by the Ionian Sea basin, or an upward block motion of the Ionian Sea side and/or an downward motion of the mainland side.

The mantle earthquakes in the Carpathian bend of Rumania seem to be of the dip-slip reversed fault type, with a horizontal component of the greatest compressive stress in roughly the WNW−ESE direction.

The mantle earthquakes of the Calabrian arc in the South Tyrrhenian Sea occur at two separate depth levels: 40−100 km and 210−320 km with one exceptional case at 450 km depth. Contrary to the Rumanian cluster, the eleven shocks that have been studied from this centre are of the horizontal tension type, i. e. normal faults with dip-slip character.

The only mantle earthquake occurred in Southern Spain at a depth of 600 km. The fault plane solution is well-defined: one of the nodal planes is practically horizontal, the other vertical, striking N−S. The motion direction is either hanging wall towards the East, footwall towards the West, or the East block up, the West block down.

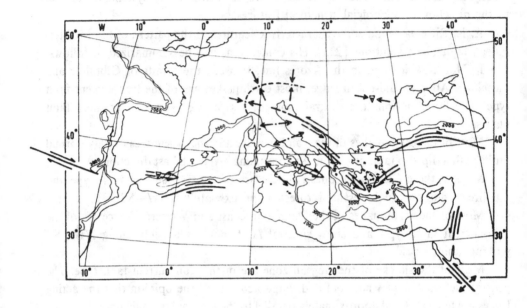

Fig. 7. The seismo-tectonic stress field in the Mediterranean area (after Ritsema, 1969).

Figure 7 shows the generalized picture of the motion and the stress directions given by Ritsema on the basis of the existing earthquake mechanism studies. According to Ritsema the present stress field seems to have the prevailing W−E to WNW−ESE direction [223].

All discussions during the recent symposium on earthquake mechanisms [211] as well as the recent reviews [212−216, 223−225] reveal several open questions (orthogonality of nodal planes, interpretation of nodal planes in terms of fault planes, reliability of original data and of interpretation, etc.). Although a proper solution of the whole problem cannot be presented immediately there is a real hope that, within the near future, the work of teams of specialists and the accessibility of adequate data from standard instruments in combination with computer programs will provide us with reliable fault plane solutions.

Table 8, compiled in consultation with Dr. A. R. Ritsema, gives a list of European shocks for which fault plane solutions were available in Jan. 1968. The corresponding parameters are published in several papers quoted above.

Table 8

EUROPEAN SHOCKS WITH FAULT PLANE SOLUTIONS

Date	Time	Date	Time
	h m		h m
1911, Nov. 16	21 26	1943, May 2	01 08
1915, Jan. 13	06 53	1943, June 20	15 33
1928, Mar. 27	08 32	1943, Oct. 16	13 08
1929, Nov 1	06 57	1943, Nov. 26	22 21
1930, Oct. 7	23 27	1944, Jan. 5	07 44
1931, May 20	02 22	1944, Feb. 1	03 23
1933, Feb. 8	07 07	1944, May 27	23 52
1934, Mar. 29	20 07	1945, Mar. 12	20 52
1934, June 8	03 17	1945, Sept. 7	15 47
1934, Nov. 30	02 58	1945, Dec. 9	06 09
1935, Jan. 31	13 39	1946, Apr. 5	20 53
1935, Feb. 25	02 51	1947, May 11	06 32
1935, Mar. 18	08 40	1947, Jun. 4	00 30
1935, Apr. 19	15 23	1947, July 7	22 35
1935, June 27	17 19	1947, Aug. 15	04 11
1935, June 28	09 10	1947, Aug. 15	04 58
1935, July 13	00 04	1947, Aug. 30	22 31
1936, Oct. 18	03 10	1947, Oct. 6	19 55
1936, Oct. 19	07 05	1948, Feb. 9	12 58
1937, Dec. 12	17 35	1948, May 29	04 49
1938, Apr. 13	02 43	1948, Jun. 29	16 06
1938, Apr. 13	02 46	1948, July 24	06 03
1938, Apr. 19	10 59	1948, Sep. 11	08 53
1939, Feb. 11	11 17	1949, July 23	15 03
1939, May 8	01 46	1950, June 20	01 19
1939, Sept. 15	23 16	1950, Sept. 05	04 09
1939, Oct. 15	14 05	1951, Jan. 30	23 07
1939, Nov. 21	08 49	1951, Apr. 8	21 38
1939, Dec. 26	23 57	1951, May 15	22 54
1940, June 24	09 57	1951, May 16	02 27
1940, Oct. 22	06 37	1951, May 19	15 54
1940, Nov. 8	12 01	1951, June 6	16 11
1940, Nov. 10	01 39	1951, Aug. 8	20 57
1940, Nov. 11	06 34	1951, Aug. 13	18 33
1940, Nov. 19	20 27	1951, Nov. 2	21 55
1940, Nov. 23	14 50	1951, Dec. 8	04 14
1941, Mar. 16	16 35	1952, June 3	05 53
1941, Nov. 25	18 04	1952, Oct. 5	10 54
1942, June 21	04 38	1952, Dec. 17	23 04
1942, Nov. 28	10 38	1952, Dec. 26	23 56
1942, Dec. 20	14 03	1953, Feb. 7	22 31

Table 8 (*continuous*)

Date	Time	Date	Time
	h m		h m
1953, Feb. 12	08 15	1956, July 10	03 01
1953, Mar. 18	19 06	1956, July 30	09 15
1953, Aug. 9	07 41	1956, Aug. 15	12 03
1953, Aug. 11	03 32	1956, Oct. 31	14 03
1953, Aug. 12	06 08	1956, Nov. 2	16 05
1953, Aug. 12	09 24	1956, Nov. 5	19 45
1953, Aug. 12	12 05	1957, Feb. 19	07 44
1953, Aug. 12	13 39	1957, Mar. 8	12 14
1953, Aug. 12	14 09	1957, Mar. 8	12 21
1953, Aug. 12	16 09	1957, Mar. 8	23 35
1953, Aug. 13	03 22	1957, Mar. 16	00 43
1953, Sept. 5	14 18	1957, Mar. 28	22 26
1953, Sept. 7	03 58	1957, Apr. 24	19 10
1953, Sept. 10	04 06	1957, Apr. 25	02 26
1953, Oct. 21	18 39	1957, Apr. 26	06 34
1954, Mar. 29	06 17	1957, May 20	19 58
1954, Apr. 30	13 03	1957, May 21	11 44
1954, Aug. 3	18 18	1957, May 26	06 33
1954, Sept. 9	01 04	1957, May 27	11 01
1954, Oct. 1	13 30	1957, July 2	00 42
1954, Oct. 6	14 31	1957, July 7	05 48
1954, Oct. 11	17 45	1957, Oct. 30	01 42
1954, Nov. 23	13 00	1957, Oct. 30	07 30
1954, Dec. 23	16 27	1957, Nov. 27	03 08
1955, Jan. 3	01 07	1957, Dec. 13	01 45
1955, Feb. 17	19 40	1957, Dec. 23	23 38
1955, Apr. 13	20 46	1958, Jan. 2	02 08
1955, Apr. 19	16 47	1958, Jan. 16	04 18
1955, Apr. 21	07 18	1958, Apr. 3	02 24
1955, May 1	21 23	1958, Apr. 3	07 19
1955, June 5	14 56	1958, May 3	20 18
1955, July 16	07 07	1958, May 9	02 40
1955, Sept. 12	06 09	1958, May 16	09 19
1955, Nov. 12	05 32	1958, May 27	18 28
1956, Jan. 6	12 15	1958, June 18	01 15
1956, Jan. 12	05 46	1958, June 24	06 07
1956, Feb. 1	15 11	1958, June 30	08 43
1956, Feb. 20	20 32	1958, July 8	05 02
1956, Apr. 12	22 34	1958, July 17	03 37
1956, Apr. 18	12 52	1958, Aug. 16	19 13
1956, May 15	22 56	1958, Aug. 27	15 16
1956, May 18	22 08	1958, Sept. 25	07 20
1956, July 9	03 12	1958, Nov. 15	05 43

Table 8 (*continuous*)

Date	Time	Date	Time
	h m		h m
1959, Jan. 29	23 24	1961, Dec. 3	18 31
1959, Apr. 25	00 27	1962, Jan. 7	10 03
1959, Apr. 25	01 06	1962, Jan. 11	05 05
1959, May 1	08 24	1962, Jan. 21	02 52
1959, May 14	06 36	1962, Jan. 26	08 17
1959, May 20	19 49	1962, Jan. 30	17 15
1959, May 31	12 16	1962, Feb. 27	21 34
1959, June 10	04 16	1962, Mar. 18	15 30
1959, June 26	13 45	1962, Mar. 25	21 38
1959, Aug. 17	01 33	1962, Apr. 4	20 59
1959, Oct. 5	20 34	1962, Apr. 10	21 37
1959, Oct. 7	08 31	1962, Apr. 11	10 47
1959, Nov. 15	17 09	1962, Apr. 25	04 45
1959, Dec. 1	12 39	1962, Apr. 28	11 18
1959, Dec. 23	03 29	1962, Apr. 28	12 43
1960, Jan. 3	20 20	1962, July 6	09 16
1960, Jan. 4	12 52	1962, Aug. 6	04 24
1960, Jan. 26	20 27	1962, Aug. 21	18 09
1960, Feb. 1	11 60	1962, Aug. 21	18 19
1960, Feb. 19	12 30	1962, Aug. 28	11 00
1960, Feb. 21	08 14	1962, Aug. 30	07 46
1960, Feb. 29	23 41	1962, Sept. 1	19 21
1960, Mar. 12	11 54	1962, Sept. 4	22 59
1960, Mar. 28	02 52	1962, Sept. 10	09 36
1960, Apr. 10	22 05	1962, Oct. 4	19 4o
1960, Apr. 24	12 14	1962, Nov. 9	02 15
1960, May 26	05 10	1963, Jan. 14	18 32
1960, June 25	14 29	1963, Jan. 27	19 35
1960, Aug. 17	15 28	1963, Mar. 11	07 27
1960, Oct. 13	02 21	1963, Mar. 22	23 57
1960, Oct. 28	04 18	1963, Mar. 24	12 44
1960, Nov. 5	20 21	1963, Mar. 28	00 16
1961, Jan. 17	01 52	1963, Apr. 25	13 36
1961, Jan. 28	07 18	1963, May 20	17 01
1961, Apr. 19	00 16	1963, July 16	18 27
1961, Apr. 28	20 48	1963, July 17	11 57
1961, Apr. 29	09 29	1963, July 19	05 45
1961, May 23	02 45	1963, July 26	04 17
1961, Aug. 25	12 22	1963, July 29	06 10
1961, Sept. 15	01 46	1963, Sept. 12	08 18
1961, Sep. 18	01 04	1963, Dec. 2	06 49
1961, Nov. 18	03 18	1963, Dec. 20	23 22
1961, Nov. 23	01 12	1964, Jan. 19	09 13

Table 8 (*Continous*)

Date	Time	Date	Time
	h m		h m
1964, Jan. 30	17 45	1964, Dec. 22	04 35
1964, Feb. 16	00 17	1964, Dec. 31	16 18
1964, Feb. 17	12 19	1965, Mar. 4	05 31
1964, Feb. 23	22 41	1965, Mar. 9	17 57
1964, Mar. 14	02 37	1965, Mar. 10	01 36
1964, Mar. 15	22 30	1965, Mar. 13	04 08
1964, Mar. 26	04 40	1965, Mar. 27	
1964, Apr. 11	16 00	1965, Mar. 30	17 34
1964, Apr. 29	04 21	1965, Mar. 31	09 47
1964, May 28	20 52	1965, Apr. 5	03 12
1964, June 14	12 15	1965, Apr. 9	23 57
1964, Jun. 30	12 30	1965, Apr. 27	14 09
1964, Jul. 26	20 22	1965, May 25	03 28
1964, July 17	02 34	1965, June 13	20 01
1964, Aug. 25	13 47	1965, July 6	03 18
1964, Aug. 27	22 27	1965, Aug. 23	14 08
1964, Oct. 6	14 31	1965, Sep. 19	
1964, Oct. 13	18 42	1965, Nov. 2	03 27
1964, Oct. 22	23 30	1965, Nov. 28	05 26
1964, Oct. 24	15 09	1965, Dec. 15	12 07·2
1964, Oct. 27	19 46	1965, Dec. 15	12 07·3
1964, Nov. 8	10 33	1965, Dec. 20	00 08
1964, Nov. 9	08 05	1965, Dec. 21	10 00

6.   Tsunamis

Certain coastal regions of the Mediterranean Sea and of the Atlantic belonging to our area under investigation are endangered by tsunamis. From the whole area only some rather self-contained regions are subject systematically to the effects of seismic sea-waves. The most prominent regions are the Aegean and Adriatic coasts, the Ionian sea coasts, and the Eastern Mediterranean coast of North Africa and Portugal. On the Eastern Mediterranean coast the following regions are subject to tsunamis: a) the Gulf of Corinth, b) the Euboean Gulf, c) the area between Himara an Durrës,

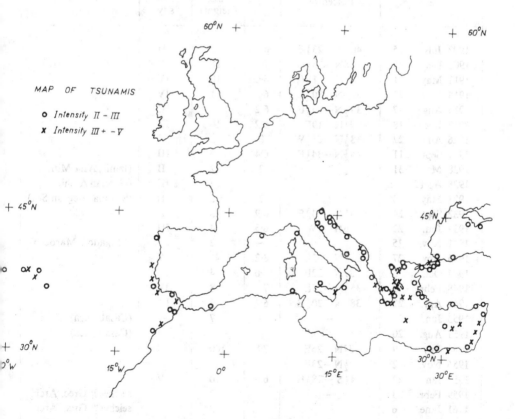

MAP OF TSUNAMIS

o Intensity II - III
x Intensity III+ - V

Fig. 8. Observations of tsunamis in the European area (according to N. N. Ambraseys).

d) the Sea of Marmara, e) the area between Cyprus, Jubeil and Acre, f) the area between Khíos and Izmir and g) the south of the Greek Archipelago. It is very probable that the generation of tsunami cannot be attributed to a specific and single mechanism. We can give several reasons for the origin, e. g. submarine tectonic movements of the ocean bottom, land-slides, volcanic eruptions, resonance in ocean trenches, etc. A certain mechanism may prevail in specific regions but a closer study has not yet been made.

The European Seismological Commission initiated a systematic study of the seismic sea-waves in the European area by establishing a Sub-Commission for Tsunamis in 1962 at the Jena meeting. The first chairman, Prof. A. Galanopoulos, was succeeded in 1963 by Dr N. N. Ambraseys who prepared the first Progress Report 1963 — 1964, proposing a mechanism of data collection, summarizing the information needed and listing the main parameters of some tsunamis that occurred during the period 1900 — 1960. Ambraseys concludes that the quality of data collected is far

Table 9

Date		Epicentre	M	h_{max} (height)	Intensity	
1902 July	5	$40\frac{3}{4}N-23\frac{1}{4}E$	6·6	2	II	
1908 Dec.	28	$38N-15\frac{1}{2}E$	7·5	18	V	
1911 May	11	$6N- 1W$	6·6	10	IV	
1914 Nov.	27	$38\frac{3}{4}N-20\frac{1}{2}E$	6·2	11	IV	
1915 Aug.	7	$38\frac{1}{2}N-20\frac{1}{4}E$	6·2	5	III	
1920 Dec.	18	$41N-19E$	6·2	10	IV	
1926 Aug.	22	$38\frac{1}{2}N-28\frac{1}{2}W$	?	2	II	
1927 Sept.	11	$44\frac{1}{2}N-34\frac{1}{2}E$	6·4	5	III	
1928 Mar.	31	?	?	2	II	(Izmir, Asia Minor)
1928 Apri 23—25		—		7	III	(Grecian Archipelago)
1928 May	3	?	?	3	II	(Struma, Aegean Sea)
1932 Sept.	26	$40\frac{1}{2}N-23\frac{3}{4}E$	6·9	3	III	
1939 Jun.	22	$6N-1W$	6·5	4	III	
1941 Nov.	25	—	—	2	—	(Mogador, Marocco)
1945 Nov.	27	$24\frac{1}{2}N-63E$	8·2	45	V	
1947 Oct.	6	$37N-22E$	7·0	4	III	
1948 Febr.	9	$35\frac{1}{2}N-27E$	7·1	8	IV	
1948 Apr.	22	$38\frac{1}{2}N-20\frac{1}{4}E$	6·4	2	III	
1953 Jun.	29	—	—	7	—	(Casablanca)
1953 Aug.	20	—	—	1	—	(Casablanca)
1956 July	9	$37N-26E$	7·8	100	V	
1956 Nov.	2	$39\frac{1}{2}N-23E$				
1957 Jan.	22	$4\frac{1}{2}S-28\frac{1}{2}E$	6·3	10	IV	
1959 Febr.	23	—		3		seiches?; Grec. Arch.
1961 June	6	—		3		seiches?; Grec. Arch.

from being adequate for the tsunamis regionalization. We are including his list of tsunamis as well as the corresponding map [217—220], Fig. 8, Tab. 9.

Seismic sea-wave intensity scale

I: *Very light*: Wave so weak as to be perceptible only on tide gauge records.

II: *Light*: Waves noticed by those living along the shore and familiar with the sea. On very flat shores generally noticed.

III: *Rather strong*: Generally noticed. Flooding of gently sloping coasts. Light sailing vessels carried away on shore. Slight damage to light structures situated near the coast. In estuaries reversal of the river flow for some distance upstream.

IV: *Strong*: Flooding of the shore to some depth. Light scouring on man-made ground. Embankments and dykes damaged. Light structures near the coast damaged. Solid structures on the coast slightly damaged. Big sailing vessels and small ships drifted inland or carried out to sea. Coasts littered with floating debris.

V: *Very strong*: General flooding of the shore to some depth. Quay-walls and solid structures near the sea damaged. Light structures destroyed. Severe scouring of cultivated land and littering of the coast with floating items and sea animals. With the exception of big ships all other types of vessels carried inland or out to sea. Big bores in estuary rivers. Harbour works damaged. People drowned, waves accompanied by strong roar.

VI: *Disastrous*: Partial or complete destruction of man-made structures for some distance from the shore. Flooding of coasts to great depths. Big ships severely damaged. Trees uprooted or broken by the waves. Many casualties.

It is evident that the foci responsible for tsunamis in the Mediterranean are located close to the shore-line. Because of the short travel times of seismic waves an early-warning system cannot be established in the Mediterranean although it may be effective along the Atlantic coast.

A critical analysis of data on the seismic-sea waves in the Eastern Mediterranean is published by Ambraseys [119, 220]. The paper contains a historical catalogue of tsunamis observed since the second millenium B. C. till 1961 A. D. and is accompanied by an extensive bibliography of 229 items.

Another summarizing, short paper accompanied by a catalogue of tsunamis comes from W. H. Berninghausen [218]. It relates to the area of the eastern Atlantic south of the Bay of Biscay. Most tsunamis were of a local nature and were reported mainly from the Azores and the Iberian peninsula. This fact is supposed to be due to the inhomogeneity of data rather than to the absence of tsunamis in other parts of the area. Long waves caused by tropical cyclones may sometimes be mistaken for tsunami as well as seiches originating in closed basins of water like bays, harbours, lakes; these free oscillations of the water body may be induced by a strong wind or by the arrival of earthquake waves from a distant source.

We give here only a brief note about tsunamis because a competent study of the tsunamis in the European area is at present under way within the E. S. C. and the results will be published later.

7.   Conclusions, recommendations

The object of the present study was to compile and unify the basic information on earthquakes in the European area and to present it in a form suitable for further research work on a regional scale. The difficulties involved need not be repeated; they left their negative traces on some sections of the work. The inhomogeneities were sometimes too great to be overcome. Nevertheless, a step forward has been made and the *earthquake catalogues* 1801—1900 and 1901—1955 and the *seismic maps* are ready for further applications. It must be emphasized once again that work of this kind is never completed to the full satisfaction of either the author or the reader. Some uncertainties, doubts or simply errors and gaps always remain and can only be removed by continuous investigation. Our activities are directed towards a more and more accurate determination of the earthquake risk in any arbitrary point of an earthquake zone. At present, we can answer the questions put before us by everyday practice in a relatively modest way. It is evident that a much greater understanding of the physical processes generating earthquakes is needed. This basic research must, however, be combined with a continuous observation of natural phenomena including the realistic information which can be gained from historical records. Earthquakes are caused by long lasting forces within a geological time scale and in some regions not even a 155 year period is sufficient for the estimation of the level of earthquake activity. The recording of very weak shocks with ultra-sensitive seismographs has proved to be not so simple and effective as was assumed and we face the whole problem in only a slightly improved situation compared with 20—30 years ago. In general, we can say that promising lines of investigation may be e. g. *the regularities in time, space and magnitude distributions of earthquakes, theory and experiments with the generation of shocks, properties of seismic waves and of rocks under extreme conditions and the investigation of other geophysical phenomena connected with the origin of earthquakes.*

But what is needed first, particularly in the European area? It may sound trivial, but a sophisticated network of seismic stations, equipped with uniform instruments, is absolutely necessary. A satisfactory state has not yet been reached even in countries with the highest seismicity, so that this idea must be stressed again and again. *Detailed studies of seismicity* should be commenced in regions where seismogenetic forces act with the greatest violence, thus allowing a relatively rapid accumulation of data. In this connection the North Anatolian fault system has been mentioned several times as a promising experimental area. It is a large region but special investigations may start also within more limited focal zones, e. g. in Macedonia. The European Seismological Commission should encourage such projects and assist in their implementation. Evidently, other organizations, e. g. UNESCO or the European Association of Earthquake Engineering, would be interested in the practical consequences of the projects.

The immediate task of the *Sub-Commission for the Seismicity of the European Area* is to prepare the first version of the seismic zoning map of Europe aimed at practical applications in city planning and the protection of the population. Naturally, this version will be further improved according to the new achievements of basic research. The earthquake catalogue 1901–1955 [1] will be continued for the years after 1955 by the BCIS in Strasbourg.

It is further hoped that the study of the mechanism of earthquakes will throw more light on the problem of earthquake origin. The possibilities of the E. S. C. are limited and activity is mainly concentrated on organizational work; the results, however, will be determined by the research work of individual laboratories and researchers.

Concluding this stage of the work carried out under the auspices of the European Seismological Commission I wish to pay tribute to the efforts of the collaborating scientific community of European seismologists and to the spirit of mutual cooperation without which the given task could never have been implemented.

8.   References

[1] V. Kárník: Seismicity of the European area, Part I, Academia, Praha 1968, 364 pp.

[2] S. V. Medvedev, W. Sponheuer, V. Kárník: Seismic intensity scale, version 1964, Soviet Geophys. Com., Moscow 1965, mimeographed.

[3] Ch. Davison: On scales of seismic intensity and on the construction and use of isoseismal lines, Bull. Seism. Soc. Am., *11* (1921), No. 2, 95—166.

[4] Z. Suzuki: A statistical study on the occurrence of small earthquakes, I, II, III, IV. The Science Rep. of the Tohoku Univ., Series 5, Geophysics *5* (1953), No. 3, 177, *6* (1955), No. 2, 105, *10* (1958), No. 1, 15, *11* (1959), No. 1, 10.

[5] T. Asada, Z. Suzuki, Y. Tomoda: Notes on the energy and frequency of earthquakes, Bull. Earthq. Res. Inst., XXIX (1951), P. 2, 289.

[6] S. Vinogradov: Akusticheskie nablyudeniya processov razrusheniya gornykh porod v rudnike Anna (Chekhoslovakiya), Izvestiya AN SSSR, ser. geofiz., 1963, No. 4, 501—512.

[7] C. Tsuboi: Earthquake province-domain of sympathetic seismic activities, Journ. Phys. Earth., *6* (1958), 35—49.

[8] C. Tsuboi: On seismic activities, First Paper, Geophys. Notes, *3* (1949), No. 4, 1—22.

[9] C. Lomnitz: Transition probabilities between seismic regions, Geophys. Journ., *13* (1967), No. 4, 387—391.

[10] J. V. Riznichenko: Metod summirovaniya zemletryaseniy dlya izucheniya seysmicheskoy aktivnosti, Izvestiya AN SSSR, ser. geofiz., No. 7, 1964.

[11] S. Suyehiro: Deep earthquakes in the Fiji region, Papers in Met. and Geoph., *13* (1962), 216.

[12] V. Rudajev: Seismizität der Gebirgsschläge bei Kladno, Freib. Forschungshefte C 225, 1967, 7—19.

[13] K. Mogi: Study of elastic shocks caused by the fracture of heterogeneous materials and its relation to earthquake phenomena, Bull. Earthq. Res. Inst., *40* (1962), 125, 831; *41* (1963), 615.

[14] S. Vinogradov: Experimentalnoye izuchenie raspredeleniya chisla razryvov po energii pri razrushenii gornykh porod, Izvestiya AN SSSR, ser. geofiz., No. 2, 1962, 171.

[15] M. Båth: Earthquake energy and magnitude, Physics and Chemistry of the Earth, Vol. 7 (1966), 117—165.

[16] J. V. Riznichenko: On quantitative determination and mapping of seismic activity, Ann. di Geof., *12* (1959), 227.

[17] A. G. Galanopoulos, B. C. Papazachos et al.: Aftershock sequences and crustal structure in the region of Greece, Scientific Progress Report No. 8, Nat. Obs. of Athens, Seism. Inst., Athens, June 1967.

[18] S. Suyehiro: A search for small, deep earthquakes using quadripartite stations in the Andes, Bull. Seism., Soc., Am., 57 (1967), 447.

[19] M. Båth, S. Duda: Earthquake volume, fault plane area, seismic energy, strain, deformation and related quantities, Annali di Geof. XVII (1964), No. 3, 353—368.

[20] C. Radu, V. Tobyáš: Contribution to the investigation of small intermediate-depth earthquakes in the Vrancea region, Studia geoph. et geod. *12* (1968), 402—406.

[21] V. Kárník: Magnitude, frequency and energy of earthquakes in the European area, Trav. Inst. Géophys. Ac. Tchécosl. Sc., No. 222, 1965, 247—273.

[22] V. Kárník: Crustal and upper mantle seismic activity in the European area, Studia geoph. et geod., *11* (1967), 324—334.

[23] B. Gutenberg, C. F. Richter: Seismicity of the Earth and associated phenomena, Princeton Univ. Press 1949.

[24] A. G. Galanopoulos, B. C. Papazachos: Progress Report 1964—65, Nat. Obs. Athens, Seism. Inst., Univ. of Athens, Seism. Lab., 1966, 18 pp.

[25] K. Mogi: Regional variations in magnitude-frequency relation of earthquakes, Bull. Earthq. Res. Inst., *45* (1967), 313—325.

[26] R. Ikegami: On the secular variation of magnitude-frequency relation of earthquakes, Bull. Earthq. Res. Inst., *45* (1967), 327—338.

[27] A. G. Galanopoulos: The seismic efficiency of Greece, Prakt. Ac. Athens, T. 31 (1956), 368—375.

[28] C. Tsuboi: Time rate of earthquake energy release in and near Japan, Proc. Jap. Ac. *41* (1965), 392—397.

[29] V. Kárník: Epicentre maps for Europe, $I_0 \geq VI$, 1901—1955, Studia geoph. et geod., *5*, (1961), 133—137.

[30] V. Kárník: Seismicity of Europe, Progress Reports II, III, IV, IUGG Monographs No. 9 (1961), No. 23 (1963), No. 29 (1965).

[31] A. Sieberg: Erdbebengeographie, Handbuch der Geophysik, Bd. IV, Abschn. VI, Berlin 1932, 708—744.

[32] F. Montandon: Les tremblements de terre destructeurs en Europe, Genève 1953, 195 pp.

[33] E. F. Savarensky, I. E. Gubin, D. A. Kharin (editors): Zemletryaseniya v SSSR, AN SSSR, Soviet po seysmologii, Moskva 1961, 410 pp.

[34] C. F. Richter: Elementary seismology, San Francisco 1958, 768 pp.

[35] V. Kárník: Comparison of seismic activity of European earthquake zones, Izv. AN SSSR, Fizika Zemli, No. 7, 1969, 70—77.

[36] P. Caloi: Struttura geologico-sismica dell'Europa centromeridionale, dell'Italia e del Mediterraneo centro-occidentale, Ann. di Geof., V (1952), No. 4.

[37] G. P. Gorshkov: O seysmichnosti Afriki, Byull. Sov. po Seysm. AN SSSR, No. 13, 1963, 40 pp., a report publ. also by Unesco, 1961, 56 pp.

[38] N. N. Ambraseys: On the seismicity of South-West Asia, Revue pour l'étude des calamités, No. 37 (1961), 18—30.

[39] A. Sieberg: Untersuchungen über Erdbeben und Bruchschollenbau im östlichen Mittelmeergebiet, Denkschriften d. med.-naturw. Ges. zu Jena, *18* (1932), 2. Lief., 161, 273.

[40] Unesco Seismological Survey Missions, Part III, Report of the Mission to the Mediterranean and Middle East, IUGG Monograph, No. 18 (1962).

[41] B. Gutenberg, C. F. Richter: Deep focus earthquakes in the Mediterranean region, Geof. pura e applic., XII, 1948, F. 3—4, 1—4.

[42] A. Cavasino: Note sul catalogo de terremoti discruttivi dal 1501 al 1929, nel bacino Mediterraneo, Publ. comm. ital. per lo studio delle grandi calamita, V. II, Rome 1931, 29—60.

[43] E. Peterschmitt: Quelques données nouvelles sur le séismes profonds de la mer Tyrrhénienne, Ann. di Geof., IX, No. 3 (1956), 305—334.

[44] P. E. Valle: Contributo allo studio delle carratteristiche sismiche del Mediterraneo centro-orientale, Ann. di Geof., 1, No. 2 (1948), 266—277.

[45] E. J. Gumbel: Simplified plotting of statistical observations, Trans. Am. Geophys. Union, *26* (1945), 69.

[46] J. M. Nordquist: Theory of largest values applied to earthquake magnitudes, Trans. Am. Geophys. Union, 26 (1945), 29.

[47] B. Epstein, C. Lomnitz: A model for the occurrence of large earthquakes, Nature, 211 (1966), No. 5052, 954.

C. Lomnitz: Statistical prediction of earthquakes, Rev. in geoph. 4 (1966), No. 3, 377.

[48] N. V. Smirnov, I. V. Dunin-Barkovskiy: Kratkiy kurs matematicheskoy statistiki dlya tekhnicheskikh prilozheniy, Fizmatgiz, Moskva, 1959.

[49] V. Kárník: Magnitude-intensity relations for European and Mediterranean seismic regions, Studia geophys. et geod., 9 (1965), No. 3, 236.

[50] V. N. Gayskiy, A. P. Katok: Primeneniye teorii extremal'nykh znacheniy dlya ocenki povtoryayemosti sil'nykh zemletryaseniy, Dinamika zemnoy kory, AN SSSR, Nauka, Moskva 1965, 9—14.

[51] V. Kárník, Z. Hübnerová: The probability of occurrence of largest earthquakes in the European area, Pure and appl. geoph., 70 (1968), 61—73.

[52] N. A. Linden: O karte seysmichnosti Arktiki, Seysm. i glats. issled. v period MGG, No. 2, Moskva 1959, 7—17.

[53] G. D. Panasenko: Seysmichnost Kolskogo poluostrova i severnoy Karelii, Izv. AN SSSR, ser. geofiz., No. 8, 1957.

[54] E. Tryggvason: Seismicity of Iceland and the surrounding ocean, Natturufraedingnum, 25, 1955, 194—197.

[55] E. Tryggvason, S. Thoroddsen, S. Thorarinsson: Report on earthquake risk in Iceland, Timariti Verkfraedingafélas Islands, H. 6, V. 43, 1958, 19 pp.

[56] M. Båth: Seismicity of Fennoscandia and related problems, Gerl. Beitr. z. Geoph., B. 63 (1953), H. 3, 173—208.

[57] A. Kvale: Norvegian earthquakes in relation to tectonics, Årbok for Universitetet i Bergen, mat.-naturv. serie, 1960, No. 10.

[58] E. Sahlström: A seismological map of northern Europe, Sver. Geol. Unders., Årbok, Ser. C, No. 365 (1930), 1—8.

[59] S. Miyamura: A note on Fennoscandian seismicity, Geophysica 7 (1962), No. 4, 1—11.

[60] N. I. Nikolaev: O svyazi seysmichnosti Baltiyskogo shchita i norvezhkikh Kaledonid s neotektonikoy, Vestnik Moskov. univ., ser. IV, geol., No. 3, 1966, 20—36.

[61] E. Pentillä: On the local earthquakes in Finland, Univ. of Helsinki, Publ. in Seismology No. 40, 1960, 91—96.

[62] E. Pentillä: Some remarks on earthquakes in Finland, manuscript, 1963, 3 pp.

[63] S. Miyamura, E. Pentillä: Seismic events located in and near Finland, Geophysica 9 (1964), No. 1, 1—48.

[64] E. Pentillä, P. Saastamoinen: On the depths of earthquakes in the Baltic Shield, Publ. of the Inst. of Seism., Univ. of Helsinki, No 64, 1963.

[65] E. Vesanen: On outline of the development of the seismograph station network and the study of the seismicity of the Baltic Shield, ibid., No. 67, 1964.

[66] H. Renquist: Finlands Jordskalv, Fennia, 54, No. 1, Helsinki 1930, 1—113.

[67] I. Lehmann: Danske jordskaelv, Medd. Dansk Geol. For., B 13, H. 2, Copenhagen 1956, 88—103.

[68] M. Nurmia, E. Vesanen: Finland's contribution to the seismotectonic map of Europe, Paper pres. at the Alicante meeting, 1959, 2 pp.

[69] T. Olczak: Seismic phenomena on the territory of Poland during the period 1895—1960, manuscript, 1960, 8 pp.

[70] A. W. Janczewski: Earthquakes in Upper Silesia, P. II., Arch. Gornictwa, Vol. 1, Warsaw 1956, 321—344.

[71] T. Olczak: Seismic phenomena on the territory of Poland during the period 1901—1950, Acta geoph. Pol., X (1962), No. 1, 3—11.

[72] A. Zátopek: Sur la propagation des séismes se produisant dans les Alpes orientales à travers le massif de Bohème, BCIS, S. A., Trav. Sci., F. 17, 1950, 123—132.

[73] V. Kárník: Earthquake swarm in the region of Kraslice in 1962, Studia geoph. et geod., 7 (1963), 288—296.

[74] V. Kárník: XII. Seismology, IGY and IGC in Czechoslovakia, Praha, 1960, 219—238.

[75] J. Špaček: Les tremblements de terre dans la région frontière Silésie-Moravie, U. G. G. I., Sect. de Séism., B, Fasc. 4., Strasbourg 1933, 74—90.

[76] W. Sponheuer: Die Tiefen der Erdbebenherde in Deutschland, Ann. di Geof., XI (1958), Nos. 3—4, 157.

[77] W. Sponheuer: Erdbeben und Tektonik in Deutschland, Freiberger Forschungshefte, C. 7, 1953, 5—15.

[78] W. Sponheuer: Untersuchung der Seismizität von Deutschland, Veröff. Inst. Bodendyn. u. Erdbebenforsch., Jena, H. 72 (1962), 23—54.

[79] W. Hiller, J. P. Rothé, G. Schneider: La séismicité du Fossé Rhénan, Ann. de l'Inst. de Phys. du Globe, N. S. Géophys., T. VIII (1957), 11—17.[1])

[80] G. Schneider: Die Erdbeben in Baden-Württenberg 1955—1962, Veröff. d. Landes-erdbebendienstes Baden-Württenberg, 1964, mimeographed, 46 pp.

[81] F. Robel, L. Ahorner: Rheinische Erdbeben I., Geol. Inst. der Univ. Köln, 1958, 15 pp.

[82] M. Bider: Die Erdbebentätigkeit in Basel und Umgebung seit dem grossen Erdbeben, Basler Jahrbuch 1956, 17—44.

[83] J. M. van Gils: La séismicité de la Belgique, Acad. Royale Belgique, Com. Nat. Géogr., Atlas de Belgique, Planche 10, 1956, 11 pp.

[84] A. R. Ritsema: Materials for the construction of a seismotectonic map of Europe, region of the Netherlands, mimeographed, 1959, 6 pp.

[85] F. H. van Rummelen: Oerzicht v. d. tussen 600 en 1940 in Z. Limburg en omgeving ge-voelde aardbevingen, Med. Jaarversl. Geol. Bur., 1942—43, No. 15 (1945).

[86] Ch. Charlier: Etude systématique des tremblements de terre belges récents, (1900—1950), IVe Partie, La Séismicité de la Belgique, Obs. Roy. de Belgique, Publ. du Serv. Séism. Grav., Série S, No. 10, 151, 60 pp.

[87] A. R. Ritsema: Note on the seismicity of the Netherlands, Proc. Koninkl. Nederl. Ak. Wetensch., S. B., 69 (1966), No. 2, 235—239.

[88] M. Schwarzbach: Die Erdbeben des Rheinlandes, Kölner Geol. Hefte, H. 1, Köln 1951, 28 pp.

[89] E. Tillotson: British earthquakes and the structure of the British Isles, manuscript, 1952, 58 pp.

[90] C. Davison: A history of British earthquakes, Cambridge, 1924, 416 pp.

[91] J. P. Rothé: Note sur la séismicité de la France métropolitaine, manuscript, 1962, 5 pp.

[92] J. P. Rothé: La séismicité des Alpes occidentales, Ann. Inst. Phys. Globe, Strasbourg, T. IV, 1948, 89—105 pp.

[93] J. M. Munuera: Datos basicos para un estudio de sismicidad en el area de la Peninsula Iberica, Memorias del Instituto Geografico y Catastral, V. XXXII, 1963, 97 pp.

[94] J. Debrach: Tremblements de terre marocains (1932—1951), Notes et Docum. Sc., Casablanca 1952, 3 pp.

[95] F. Duffaud, J. P. Rothé, J. Debrach, P. Erimesco, G. Choubert, A. Faure-Muret:

---

[1]) Several regional studies on seismicity of SW Germany were prepared as doctoral thesis at the Technische Hochschule, Stuttgart and are deposited there.

Le séisme d'Agadir du 29 février 1960, Notes et Mém. du Serv. Géol. du Maroc, No. 154, 1962, 68 pp.

[96] J. DEBRACH: Sur quelques alignements séismiques remarquable du Maroc, Publ. BCIS, S. A. Tr. Sc., F. 19, 1956, 307—313.

[97] A. GRANDJEAN, R. GUIRAUD, J. POLVECHE: Le séisme de M'Sila, Publ. Serv. Géol. de l'Algérie, N. S., Bull. No. 33 (1965), Alger 1966.

[98] J. P. ROTHÉ: Le tremblement de terre d'Orléansville et la séismicité de l'Algérie, La Nature, No. 3237, Janvier 1955.

[99] J. P. ROTHÉ: Les séismes de Kerrata et la séismicité de l'Algérie, Bull. Serv. de la Carte géologique d'Algérie, 4e série, Géophysique, No. 3, Mende, 1950, 1—40.

[100] A. GRANDJEAN, J. LAGRULA: Sur la séismicité de l'Algérie, C. R. Acad. Sc. Paris, t. 259 (7. sept. 1964), groupe 10, 1749—1751.

[101] A. HÉE: La fréquence des tremblements de terre en Algérie, Publ. BCIS, S. B. Monographies, F. 5, 59—110.

[102] N. N. AMBRASEYS: The seismicity of Tunis, Ann. di Geof. XV (1962), Nos. 2—3, 233—244.

[103] A. GRANDJEAN: Epicentres des séismes Algériens, Alicante meeting, 1959, manuscript, 11 pp.

[104] J. P. ROTHÉ: La séismicité de l'Algérie, Congr. Géol. Int. Alger 1952, S. IX, F. IX, 1954, 267—274.

[105] J. M. BONELLI RUBIO, L. CHACON ALONSO: Estudio de la sismicidad de la zona Murciano-Alicantina, Inst. Geogr. y Cat., Madrid 1961, 19 pp.

[106] A. REY PASTOR: Sismicidad de la comarca costera Alicantina, Inst. Geogr. y Cat., Madrid 1948, 24 pp.

[107] J. M. BONELLI RUBIO, L. ESTEBAN CARRASCO: Resultados provissionales del estudio del caracter sismico de la falla del Giuadalquivir, Inst. Geogr. y Cat., Madrid 1953, 19 pp.

[108] J. MARTIN ROMERO: Comportamiento del terreno ante la vibracion sismíca en la demarcacion Alicantina, Inst. Geogr. y Cat., Madrid 1959, mimeographed, 19 pp.

[109] A. REY PASTOR: Estudio sismotectonico de la region sureste de España, Inst. Geogr. y Cat., Madrid 1951, 52 pp.

[110] A. REY PASTOR: Sintesis de sismicidad de la Peninsula Ibérica, Inst. Geogr. y Catastr., Madrid 1954.

[111] J. M. MUNUERA: Datos basicos para un estudio de sismicidad en el area de la Peninsula Iberica, Memorias del Inst. Geogr. y Cat., V. XXXII, 1963, 97 pp.

[112] J. P. ROTHÉ: Cartes de séismicité de la France, Ann. de l'Inst. de Phys. du Globe, N. S., Géophys., T. VIII, 1967, 3—10.

[113] J. P. ROTHÉ: Les séismes des Alpes Françaises en 1938 et la séismicité des Alpes occidentales, Ann. Inst. Phys. Gl., Strasbourg, T. III, 1941, 1—105.

[114] J. P. ROTHÉ, N. DECHEVOY: La séismicité de la France de 1951 à 1960, ibid. 19—84.

[115] M. TOPERCZER: Ein Beitrag zur Seismotektonik der Ostalpen, Skizzen zum Antlitz der Erde, Wien 1953, 72—80.

[116] M. TOPERCZER, E. TRAPP: Ein Beitrag zur Erdbebengeographie Österreichs, Österr. Ak. Wiss., Mitt. d. Erdb.-Kom., N. F., Nr. 65, Wien 1950, 59 pp.

[117] F. MONTANDON: Les séismes de forte intensité en Suisse, Revue pour l'étude des calamités, Genève 1943, 5, Nos. 18—19, 107—169, 6, No. 20, 3—49.

[118] D. DI FILIPPO, F. PERONACI: Terremoti di frattura e relazioni an la tettonica nella Alpi Orientali, Ann. di Geof., XV (1962), 195—224.

[119] E. WANNER: Die Erdbebenherde in der Umgebung von Zürich, Ec. Geol. Helv., 38 (1945), No. 1, 151—161.

[120] D. D. FILIPPO, F. PERONACI: Ingadine preliminare della natura fisica del fenomeno che ha originato il periodo sismico Irpino dell Agosto 1962, Ann. di Geof., V. XVI, No. 4, 1963, 625—644.

[121] R. Malaroda, C. Raimondi: Linee di dislocazione e sismicità in Italia, Bull. Geod. Sci. Aff., V. 16, 1957, 273—323.

[122] M. DE PANFILIS: Attività sismica in Italia dal 1953 al 1957, Ann. di Geof., XII (1959), No. 1, 21—148.

[123] E. BISTRICSÁNY, D. CSOMOR, Z. KISS: Earthquake zones in Hungary, Ann. Univ. Sc. Bud. de R. Eötvös Nom., S. Geol., T. IV, 1961, 35—38.

[124] V. KÁRNÍK, L. RUPRECHTOVÁ: Seismicity of the Carpathian region, Trav. de l'Inst. Géophys. de l'Ac. Tchécosl. Sc., No. 182, 1963, 143—187.

[125] I. ATANASIU: Cutremurele de pamînt din Romînia, Bucarest 1961, 194 pp.[1])

[126] C. RADU: Regimul seismic al regiunii Vrancea, Studii si cerc. geol. geof. geogr., ser. geof., 3 (1965), No. 2, 231—279.

[127] E. A. SAGALOVA: O seysmicheskom rezhime rayona Vrancha, Geof. sb. IG AN USSR, vyp. 1 (3), 1962, 68—74.

[128] S. V. MEDVEDEV: O posledstviyakh Karpatskikh zemletryaseniy 1940 g., Trudy Geof. Inst. AN SSSR, No. 1 (128), 1948.

[129] A. ZÁTOPEK: Les tremblements de terre en Slovaquie et ancienne Russie subcarpatique 1923—1938, BCIS, S. A., F. 17, 1950, 115—121.

[130] D. CSOMOR, Z. KISS: Die Seismizität von Ungarn, Studia geoph. et geod., 3, No. 1 (1959), 33—42.

[131] V. V. EZ: O tektonicheskikh usloviyakh vozniknoveniya silnykh zemletryaseniy v Chekho-slovatskihk Karpatakh, Izv. AN SSSR, Ser. geofiz., No. 2, 1964, 161—173.

[132] D. CSOMOR, Z. KISS: Magyarorszag szeizmicitasa, Geof. Közl., XI, No. 1—4, 1962, 51—75.

[133] T. IOSIF, C. RADU: Kharakteristika obuslovlennykh napryazheniy glubokogo ochaga vo Vranche, Izv. AN Mold. SSR, No. 4, 1962, 91—106.

[134] G. PETRESCU, C. RADU: Seysmichnost territorii Rumunskoy Narodnoy Respubliki za period 1901—1960 g., Izv. AN Mold. SSR, No. 4, 1962, 68—90.

[135] G. PETRESCU, C. RADU: Seismicitatea teritoriului R. P. R. in perioda anterioera anului 1900, Probl. de Geof., II, 1963, 79—85.

[136] L. CONSTANTINESCU, D. ENESCU: The characteristic features of the mechanism of the Carpathian earthquakes and their seismotectonic implications (in Rumanian), Studii si cerc. de geof., 1, No. 1, 1963, 51—98.

[137] V. KÁRNÍK: Die Seismizität der Westkarpaten, Trav. de l'Inst. Géoph. de l'Ac. Tchécosl. Sc., No. 134, 1960, 265—283.

[138] A. ZÁTOPEK: Les relations séismotectoniques dans les Carpathes occidentales, Trav. de l'Inst. Géophys. de l'Ac. Tchécosl. Sc., 1960, No. 135, 285—297.

[139] M. V. GZOVSKY, G. P. GORSHKOV, G. A. SHENKAREVA: Variant sopostavleniya seysmich-nosti s tektonikoy Vengrii, Ann. Univ. Sc. Budap., de R. Eötvös Nom., Ser. geol., T. V., 1961, 55—63.

[140] G. PETRESCU, C. RADU: Seismicitatea si raionarea seismica a teritoriului RPR in intervalul 1900—1958, Studii si cerc. de astr. si seism., VI (1961), No. 2, 225—245.

[141] L. CONSTANTINESCU, D. ENESCU: Fault-plane solution for some Rumanian earthquakes and their seismotectonic implications, Journ. of Geoph. Res., 69, (1964), 667—674.

[142] C. RADU, G. PURCARU: Considerations upon intermediate earthquakes generating stress systems in Vrancea, Bull. Seism. Soc. Am. 54 (1964), 79—85.

---

[1]) The publications on the seismicity of the East. Carpathians incl. the Vrancea region are very numerous and can be found mainly in the Rumanian journal Studii si cerc. de geol. geogr., ser. geof. or Studii si cerc. de astr. si seism. or Comun. Ac. Rep. Pop. Rum. The authors are G. Petrescu, C. Radu, T. Iosif, L. Constantinescu, D. Enescu, etc.

[143] S. V. Evseev: Naris seismichnosti Zakarpattya, Geol. zhurnal, T. XXIII, vyp. 3, 1963, 63—70.

[144] V. Ribarić: Študija seizmičnosti ozemlja SR Slovenije, manuscript, Astr.-geofiz. observ. Univ. v Ljubljani, 1963, 95 pp.

[145] M. Janković: Seismic activity of central Yugoslavia, manuscript, 1964, 6 pp.

[146] J. Michailović: Caractère séismique de la côte sud de l'Adriatique yugoslave, Publ. de l'Inst. Séism. de Beograd, 1947, 22 pp.

[147] M. Uzelac: Bibliographie séismologique de Yougoslavie, 1667—1947, Publ. Inst. Séism. de Beograd, 1948, 93 pp., compl. list for 1947—1966, mimeographed, Beograd 1966, 17 pp.

[148] C. Morelli: La sismicità a Trieste, Tecnica Italiana, N. S., Vol. IV, No. 5, 1949, 8 pp.

[149] E. Grigorova, B. Grigorov: Epicentrite i seyzmichnite linii v NR Blgariya, Geofiz. inst. BAN, Sofia, 1964, 83 pp.

[150] E. Grigorova, K. Palieva: Seysmichnost na Gornooryakhovskata seyzmichna zona, Izv. na Geof. Inst. BAN (Bull. Inst. Géoph.), T. V/2, 1964, 139—151.

[151] E. Grigorova, K. Palieva: Seyzmichnost na chernomorskoto kraybrezhie Balchik-Blatnitsa, ibid. T. IV, 1963, 253—268.

[152] E. Grigorova, K. Palieva: Seyzmichnost na oblastta po srednoto techenie na r. Tundzha, ibid., T. III, 1962, 203—219.

[153] E. Grigorova: Seysmichnost na Manastirskite vzvisheniya, ibid., T. IV, 1963, 277—287.

[154] K. Kirov, E. Grigorova: Seizmichnostta na Marishkata dolina, ibid., T. II, 1961, 5—55.

[155] Ch. Rizhikova, K. Palieva: Pltnost i chestota na zemetreseniyata v Blgariya, ibid., T. IV, 1963, 241—251.

[156] K. T. Kirov, E. I. Grigorova, N. P. Ilev: Prinos k'm seismichnostta na Blgariya, ibid., T. 1, 1960, 137—183.

[157] I. N. Petkov: Korelatsiya na geofizichnite poleta s osobenostite na seismichniya rezhim v severoiztochna Blgariya i chast Dobrudja s ogled na seysmichnoto rayonirane, ibid., 221—230.

[158] L. V. Christoskov: Vrkhu povtoryaemostta na zemetreseniyata v Blgariya v zavisimost ot energetichnata im klasifikatsiya, ibid., T. III, 231—242.

[159] K. Kirov, K. Palieva: Seizmichnost na Strumskata dolina, ibid., T. II, 1961, 57—93.

[160] K. T. Kirov, E. Grigorova: Seizmichno rayonirane na Blgariya, Izv. na BAN, ser. geof., (Bull. de l'Ac. Sci. de Bulg.), T. 6, 1957, 389—405.

[161] I. N. Petkov: Seismicheskaya aktivnost teritorii NR Bolgarii po dannym statisticheskikh materialov, Stud. si cerc. de astr. si seism., VI (1961), 201—210.

[162] M. N. Critikos: Sur la séismicité de la Macédoine, Gerl. B. z. Geoph., 43 (1933), 371—379.

[163] C. Morelli: I terremoti in Albania, R. Acad. d'It., Boll. Comm. ital. di studio per i probl. del soc. alle popul., T. II, 1942, 1—25.

[164] J. Mihailović: Die Erdbebenkatastrophen in Albanien, Gerl. B. z. Geoph., 47 (1936), Heft 3, 252—266.

[165] J. Mihailović: La séismicité de la région du lac de Scutari, Geof. pura e appl., XIV (1949), F. 3—4, 3—11.

[166] M. Magnani: Tettonica e sismicita nella regione Albanese, Geol. pura e appl., VIII (1946), F. 1—2, 1—80.

[167] C. Morelli: Carta sismica del l'Albania, Reale Academia d'Italia, Vol. 10, Firenze, 1942, 120 pp.

[168] C. Morelli: La sismicita del l'Albania, Ist. Geofis. Trieste, No. 147, Roma 1943.

[169] J. Mihailović: Catalogue des tremblements de terre Epiro-Albanais, Trav. Inst. Séism. Beograd, Zagreb 1951, 73 pp.

[170] J. Mihailović: Mouvements séismiques Epiro-Albanais, Com. Nat. du Roy. des Serbes, Croates et Slovènes, S. B., F. 1, Beograd 1927, 78 pp.

[171] C. Davison: Great earthquakes, London 1936, 286 pp.

[172] A. G. Galanopoulos: Seismic geography of Greece, Ann. Géol. des Pays Hellén., *6* (1955), 83—121.

[173] N. D. Delibasis, A. G. Galanopoulos: Variation of the annual strain release pattern in the region of Greece, Prakt. of the Athens Ac., *40* (1965), 68—78.

[174] A. G. Galanopoulos: On mapping of seismic activity in Greece, Ann. di Geof., XVI (1963), No. 1, 36—100.

[175] A. G. Galanopoulos: Die Seismizität der Insel Leukas, Gerl. B. z. Geoph., *60* (1950), 1—15, *62* (1952), 256—263.

[176] A. G. Galanopoulos: Die Seismizität von Messenien, Gerl. B. z. Geoph., *61* (1950), 144—162.

[177] A. G. Galanopoulos: Die Seismizität der Insel Chios, Gerl. B. z. Geoph., *63* (1954), 253—264.

[178] B. Papazachos, N. Delibasis, N. Liapis, G. Moumoulidis, G. Purcaru: Aftershock sequences of some large earthquakes in the region of Greece, Ann. di Geof., XX (1967), No. 1, 1—93.

[179] A. G. Galanopoulos, B. C. Papazachos et al.: Aftershock sequences and crustal structure in the region of Greece, Seism. Inst. Nat. Obs. of Athens, Sc. Progress Rep. No. 8, June 1967, 59 pp.

[180] B. Papazachos: A contribution to the investigation of the earthquake mechanism in Greece, Seism. Inst. of the Athens Observ., 1961, 75 pp.

[181] A. G. Galanopoulos: The earthquake activity in the Greek area from 1950—1953, Prakt. of the Athens Acad., Vol. 30, Athens 1955, 38—49.

[182] N. A. Critikos: Sur la séismicité des Cyclades et de Crète, Ann. Obs. Nat. d'Athènes, T. IX, 1926, 37 pp.

[183] A. D. Tskhakaya: Seysmichnost Dzhavakhetskogo (Akhalkalakskogo) nagorya i prilegayushchikh rayonov, Trudy Inst. Geol. AN GrSSR, T. XVI, 1957, 177—219.

[184] E. Lahn: Seismological investigations in Turkey, Bull. Seism. Soc. Am., *39* (1949), No. 2, 67—71.

[185] E. Lahn: Relations entre tectonique et séismicité en Turquie, Bull. Soc. Géol. de France, T. XVII (1947), 5ᵉ série, 493—502.

[186] N. Pinar: Relations entre la tectonique et la séismicité de la Turquie, Bull. Inf. UGGI, 2 (1953), No. 1, 261—264.

[187] N. Pinar: Les régions sismiques de l'Anatolie occidentale, Publ. BCIS, S. A., Tr. Sc., F. 18 (1951), 5—14.

[188] N. Öcal: Die Seismizität der Türkei, Pure and Appl. Geoph., *57* (1964), 103—116.

[189] N. N. Ambraseys, A. Zátopek: Earthquake reconnaissance mission, the Varto-Üstükran (E. Anatolia) earthquake of 19 August 1966, mimeographed, Unesco, Paris, 1967, 45 pp.

[190] E. Lahn: Les zones séismiques de l'Anatolie orientale, Publ. BCIS, S. A., Tr. Sc., F. 18 (1952), 15—22.

[191] S. Omote, M. Ipek: Seismicity in Turkey, mimeographed, 1959, 9 pp.

[192] M. Ipek, Z. Uz, U. Güçlü: Sismolojik donelere göre Türkiyede deprem bölgeleri, mimeographed, 1965, 13 pp.

[193] F. Peronaci: Sismicita dell'Iran, Ann. di Geof., XI (1958), No. 1.

[194] J. V. Riznichenko: Problemy fiziki zemletryaseniy, Fizika Zemli No. 2, 1966, 3—24.

[195] N. N. Ambraseys: The seismic history of Cyprus, Revue pour l'étude des calamités, No. 40, 1965, 3—26.

[196] A. D. Tskhakaya: O glubinakh kavkazskikh zemletryaseniy, Izv. AN SSSR, ser. geofiz., No. 5, 1962, 577—584.

[197] E. I. Byus: Seysmicheskiye usloviya Zakavkazya, Inst. Geofiz. AN GrSSR, T. I., 1948, 304 pp, T. II, 1952, 174, pp, T. III, 1955, 130 pp.

[198] I. V. Kirillova, E. N. Lyustikh, V. A. Rastvorova, A. A. Sorskiy, V. E. Khain: Analiz geotektonicheskogo razvitiya i seysmichnosti Kavkaza, Izd. AN SSSR, Moskva 1960, 340 pp.

[199] E. I. Shirokova: O napryazheniakh deystvuyushikh v ochagakh zemletryaseniy Kavkaza, Izv. AN SSSR, ser. geof., No. 10, 1962, 1227—1306.

[200] A. D. Tskhakaya: Study of Caucasian earthquakes, Studia geoph. et geod., 5 1961 364, 370.

[201] N. Shalem: La seismicité au Levant, Bull. of the Res. Council of Israel, Vol. II, June 1952, No. 1, 5—16.

[202] N. Shalem: Seismicity and Erythrean disturbances in the Levant, Publ. BCIS, S. A., Tr. Sc., F. 19, 1956, 267—275.

[203] D. H. Kellner-Amiran: A revised earthquake catalogue of Palestine, Israel Expl. Journ., Vol. I, No. 4, 1950—1951, 223—246.

[204] J. Plassard, B. Kogoj: Catalogue des séismes ressentis au Liban, Annales-Mémoires de l'Observ. de Ksara, T. IV, Cah. 1, 1962, 12 pp.

[205] N. N. Ambraseys: A note on the chronology of Willis's list of earthquakes in Palestine and Syria, Bull. Seism. Soc. Am., 52 (1962), 77—82.

[206] M. R. Madwar, A. Ismail: Sur une zone séismique autour du Caire, Publ. BCIS, S.A., Tr. Sc., F. 18, 1952, 23—28.

[207] A. Ismail: Near and local earthquakes at Helwan 1903—1950, Fac. of Sc., Helwan Obs., Bull. No 49, 1960, 33 pp.

[208] K. Minami: Relocation and reconstruction of the town Barce, Cyrenaica, Libya, damaged by the earthquake of 21 february 1963, mimeographed, Unesco, Paris 1965, 16 pp.

[209] A. R. Ritsema: Mechanismus of European earthquakes, Tectonophysics 4 (1967), 247—259.

[210] A. R. Ritsema: Earthquake mechanismus: a European survey, Tectonophysics 4 (1967), 211—212.

[211] European Earthquake Mechanisms, edited by A. R. Ritsema, special issue, Tectonphysics, 4, (1967), No 3.

[212] A. J. Wickens, J. H. Hodgson: Computer re-evaluation of mechanism solutions 1922 to 1962, Publ. Domin. Obs., V. XXXIII, No. 1, 1967.

[213] H. J. Schäffner: Tabelle kinematischer Erdbebenherde, Freib. Forschungsh., 63 (1962), Suppl.

[214] L. M. Balakina, A. V. Vvedenskaya, L. A. Misharina, E. I. Shirokova: Napryazhe-noye sostoyaniye v ochagakh zemletryaseniy i pole uprugikh napryazheniy Zemli, Fiz. Zemli, No. 6, 1967, 3—15.

[215] E. I. Shirokova: Obshchiye zakonomernosti v orientatsii glavnykh napryazheniy v ocha-gakh Sredizemnomorsko-Aziatskogo seysmicheskogo poyasa, Fiz. Zemli, No. 1, 1967, 22—36.

[216] J. H. Hodgson, A. E. Stevens: Seismicity and earthquake mechanism, Research in Geo-physics, Vol. 2, ch. 2, MIT Press 1964.

[217] A. G. Galanopoulos: Tsunamis observed on the coasts of Greece from antiquity to present time, Ann. di Geof., XII (1960), Nos. 3—4, 369—386.

[218] W. H. Berninghausen: Tsunamis and seismic seiches reported from the eastern Atlantic south of the Bay of Biscay, Bull. Seism. Soc. Am. 54 (1964). No. 1, 439—442.

[219] N. N. Ambraseys: Data for the investigation of the seismic sea-waves in Europe, Progress Report I, 1963—64, ESC meeting in Budapest 1964.

[220] N. N. Ambraseys: Data for the investigation of the seismic sea-waves in the Eastern Medi-terranean, Bull. Seism. Soc. Am., 52 (1962), No. 4, 895—913.

217

[221] J. P. Rothé: The Seismicity of the Earth 1953—1965, La séismicité du globe 1953—1965, Unesco, Paris 1969.

[222] E. Sulstarova, S. Koçiaj: Le tremblement de terre du 30 novembre 1967 et la ligne sismogene Vlorë-Dibër, Bul. i Univ. Shtetëror të Tiranës, Ser. Shk. Nat., No 2, 1969, 65—94.

[223] A. R. Ritsema: Seismo-tectonic implications of a review of European Earthquake Mechanisms, Geol. Rundschau, 59 (1969), 36—56.

[224] N. Canitez, B. Üçer: A catalogue of focal mechanism diagrams for Turkey and adjoining areas, Istanbul Teknik Üniv., Maden Fak., No 25, 1967, 111 pp.

[225] L. Ahorner: Herdmechanismen rheinischer Erdbeben..., Sonderveröff. Geol. Inst. Univ. Köln, 13, (1967), 109—130.

ERRATA

V. Kárník: Seismicity of the European Area, Part 1, Praha 1968

Page 29, 12^{ts} *line,* "the first of these two formulae" should be replaced by "formula (1)".

Page 41, 3rd *line* from below, $\sqrt{(A_N + A_N)}$ should be changed to $\sqrt{(A_N + A_E)}$.

Page 49, caption for Fig. 4, (lower part) should read (right side); the same correction applies to *p.* 58, 3rd and 2nd *lines* from below.

Page 68, *formula* (9c), 4·48 should be replaced by 0·48.

Corrections and misprints in the catalogue (chapter 5·0):

Page	Date			Correct value
108	April?		21·9°E	29·1°E
112	March 29		$M = (5·6)$	$M = (5·4)$
116	April 4,	10^{h} 13^{m}	$M = (5·6)$	$M = (5.4)$
	April 11,	00^{h} 50^{m}	$M = (4·6)$	$M = (4·1)$
117	April 19		$I = \text{VII} - \text{VII}$	$I_0 = \text{VII} - \text{VIII}$
118	Oct. 12		48·2°N	48·6°N
121	May 23		$r = 35$ km	$r = 45$ km
123	Aug. 25		16·9°E	13·9°E
124	Sept. 8		$r = (700)$ km	$r = 260$ km
	Oct. 24		$M = (4·6)$	$M = (4·1)$
126	March 1,	09^{h} 26^{m}	$M = (5·5)$	$M = (5·2)$
	March 1,	17^{h} 45^{m}	$M = (5·2)$	$M = (5·5)$
127	June 2,	00^{h} 06^{m}	20·4°E	15·0°E
132	Aug. 13		$r = 40$ km	$r = 70$ km
136	Oct. 5		2·4°E	22·4°E
137	Nov. 16		$I_0 = (\text{IV} - \text{V})$	$I_0 = (\text{V} - \text{VI})$
	Dec. 4		10 55 37·7°N 15·0°E	10 05 38·0°N 14·8°E
142	May 17		8·8°E	8·8°W
	May 22		38·1°N	37·1°N
144	Dec. 24		1·9°E	15·9°E
157	July 16		34·3°E	45·3°N
160	May 7		22^{h}	21^{h}
161	Oct. 4,	02^{h} 07^{m}	28°N	38°N
162	Oct. 26		$r_S = 28$ km	$r_S = 70$ km
164	March 2		$r = 45$ km	$r = 54$ km
172	Aug. 16,	08^{h} 15^{m}	34·2°N	44·2°N
175	June 20		9·0°E	9·6°E
181	June 29		$I_0 = \text{X}$	$I_0 = \text{IX} - \text{X}$
185	June 12		30 3/4°E	20 3/4°E
190	Sept. 11		15 47	15 17
196	Oct. 20		$h = \text{sup}$	$h = n$

Page	Date			Correct value
196	Oct. 24		$h = n$	$h = \text{sup}$
198	June 17		$h = n$	$h = \text{sup}$
205	Dec. 13		$63.6°\text{N}$	$63.8°\text{N}$
208	Aug. 28	r_5 km and $h' = 9$ km belongs to Aug. 21		
210	Jan. 8		$r_5 = 22$ km	$r_5 = 16$ km
214	Oct. 1		$3.8°\text{E}$	$3.8°\text{W}$
215	Dec. 15		$r = (7)$ km	$r = (75)$ km
224	April 18		22 37	22 27
	April 19		$h = 4.6 (6)$	$M = 4.8 (6)$
231	Aug. 7		$44.0°\text{W}$	$44.0°\text{E}$
238	Nov. 21		$I_0 = \text{XI}$	$I_0 = \text{X}$
	Jan. 5		$h = n$	$h = \text{sup}$
242	July 5		$29.9°\text{N}$	$39.9°\text{N}$
257	June 28		$48.9°\text{N}$	$48.2°\text{N}$
			$r_5 = 200$ km	$r = 200$ km
267	June 11		$M = 5.8 (8)$	$M = 5.2 (8)$
269	Oct. 16		$3.6°\text{E}$	$3.6°\text{W}$
277	Nov. 8,	$12^h\ 02^m\ 42^s$	$26.8°\text{E}$	$26.2°\text{E}$
		$12^h\ 02^m\ 44^s$	$26.2°\text{E}$	$26.8°\text{E}$
279	March 17		09 09 33	08 09 33
285	Jan. 28		00 55	03 55
	Jan. 29		$r_5 = 36$ km	$r_5 = 25$ km
292	April 19		$M = 4.5 (1)$	$M = 4.2 (1)$
	July 1		$M = (4.7)$	$M = (4.9)$
293	Oct. 22			$I_0 = \text{VI}$
			Nov. 28	Nov. 29
296	Dec. 25		$47.0°\text{N}$	$57.0°\text{N}$
302			April 27	April 29
305			Sept. 17	Sept. 27
312	April 13		$36°\text{N}$	$35°\text{N}$
313	June 21		19 22 02	19 21 02
	July 31		$r_5 = 20$ km	$r_5 = 30$ km
320	May 13		22 59 57	22 29 57
324	Dec. 31		$I_0 = \text{V}$	$I_0 = \text{VI}$
325	Febr. 10		$56.7°\text{N}$	$66.7°\text{N}$
326	March 18,	$19^h\ 06^m$	$M = 7.5 (16)$	$M = 7.2 (16)$
	March 24			$M = 4.9,\ I_0 = \text{VI}$
336	March 21		$42.8°\text{N}$	$42.3°\text{N}$
	March 23			$r = 7$ km
338	May 4,	$16^h\ 45^m$	$M = 4.6 (3)$	$M = 5.6 (3)$
	May 9,	$09^h\ 25^m$	$M = 5.6$	$M = 4.6$
348	July 4		24 55 33	23 55 33

Additional material from *SEISMICITY of the European Area /2*,
ISBN 978-94-010-3080-9 (978-94-010-3080-9_OSFO4),
is available at http://extras.springer.com

Printed in the United States
by Baker & Taylor Publisher Services